THE BOOK OF THE YEAR

The Book of the Year

MIDDLE AMERICAN CALENDRICAL SYSTEMS

Munro S. Edmonson

University of Utah Press
Salt Lake City
1988

Library of Congress Cataloging-in-Publication Data

Edmonson, Munro S.
  The book of the year.

  Bibliography: p.
  Includes index.
  1. Indians of Mexico — Calendar.   2. Indians of
Central America — Calendar.   I. Title.
F1219.3.C2E3   1988   529'.3'09728   88-5642
ISBN 0-87480-288-1

CONTENTS

PREFACE

I am indebted to many people for help in the completion of this work. My friends and colleagues in the Department of Anthropology at Tulane have been more than generous with their time and expertise, specifically, E. Wyllys Andrews V, Dan M. Healan, Judith M. Maxwell, and especially Harvey M. and Victoria R. Bricker, whose technical advice, searching questions, and critical reading of the entire manuscript have been an organic necessity. Without them, indeed, the work as it stands would have been impossible and they have saved me from errors both literally and figuratively astronomical. A number of anthropology graduate students have contributed to the calculations, artwork, bibliography and data: Roland Baumann, George Bey, Jeffrey Colville, Laurie Greene, Craig Hanson, Weldon Lamb, Chris Pool, William Ringle, and Stewart Speaker. The late Donald Robertson of the Department of Art and William Bertrand of the School of Public Health and Tropical Medicine gave me access to important data and to the computer, respectively. I am also grateful to Thomas Niehaus, Martha Robertson, and Ruth Olivera of the Latin American Library for help with a massive bibliographic problem. The late Roger T. Stone and Doris Z. Stone, of Madisonville, Louisiana, have provided a number of long fruitful weekends in their mini-think tank. A number of colleagues elsewhere have been similarly generous and helpful: Miguel León-Portilla and Marta Foncerrada de Molina, of the Universidad Nacional Autónomo de México, Costanza Vega, of the Museo Nacional de Antropología, and Javier Noguez, of the Colegio de México, Gordon Brotherston, of the University of Essex, Lawrence Feldman, of the University of Missouri, Georges Baudot, of the University of Toulouse, George Stewart, of the National Geographic Society, Maarten Jansen, of the University of Leiden, Frank Lipp, of the New School for Social Research, Arthur J. O. Anderson, of the University of San Diego, and Louis K. Bell, of Atlanta, Georgia.

This work would have been impossible without Harvey M. Bricker's MAYA-CAL, a Mayan-Gregorian calendrical conversion program for the computer.

To all of these and to my wife, Barbara Edmonson, who has shared not only time and energy but the computer as well, I am deeply grateful.

INTRODUCTION

This book began as a planned monograph to make available to general scholarship the basic ethnohistorical data relevant to reconstructing the calendars of Middle America and relating them to the European calendar. But the project seemed to develop an impetus of its own, expanding to include the epigraphic data and generating questions that seemed answerable and answers that generated further questions. The final result is a comprehensive survey of the history of the unitary calendrical system of Middle America and a theory of its origin and development.

The work is divided into five parts. Chapter 1 deals with the nature of the Middle American calendar, its numerology, and structure. The chapter focuses, as does the rest of the work, on what has been called the *calendar round*, the cycle of 52 years of 365 days each that was the common heritage of all the peoples of Middle America, or Anahuac. An understanding of this cycle requires understanding the day count and the year count and the interaction between them. A synthesis of these cycles and the relationships among the sixty calendars documented are given in the summary Figure 3 of chapter 1.

Chapter 2 provides the epigraphic, ethnohistorical, and ethnographic texts necessary for correlating the Middle American calendars with one another and with the European calendar. The texts are presented in chronological order from the seventh century B.C. to the twentieth century A.D. Each text is dated in the Mayan calendars of Tikal or Mayapan and in the Julian or Gregorian Christian calenders. Although all of these dates are completely accurate within the calendar round, many of them are only approximate within European chronology or the Mayan Long Count. The calendar round placement of the dates is, however, sufficient to document the calendrical placement given in the summary Figure 3 of Chapter 1.

Chapter 3 presents the historical genealogy of the Middle American calendar system in a preliminary form. It is preliminary because it is restricted to the structural conclusions that can be drawn only from internal evidence: the numerological relationships among the New Years' dates and other organizational features of the calenders documented in Chapter 2 and summarized in Figure 3 of Chapter 1. Figure 8, of Chapter 3, which sums up the chronology of calendrical development, is refined by the argument of Chapter 4, but this refinement depends on external evidence: the solar astronomy of the period.

Chapter 4 relates the Middle American calendar to solar astronomy. It presents and documents a theory refining the genealogy and chronology of the development of the calendar that confirms and extends the preliminary conclusions presented in Chapter 3. The conclusions of Chapter 4 imply a number of corrections to the chronological assertions of the preceding chapters, but no attempt has been made to eliminate the inconsistencies. Letting them stand may help to clarify which conclusions derive from which sections of the analysis.

In general terms, this book is a comprehensive survey of the year calendar of Middle America as it is represented in pre-Conquest and post-Conquest written sources. It documents that the correlation constant of 584,283 is the only acceptable solution to the "correlation question" on the basis of this range of facts: J. E. S. Thompson's (1950) considered judgment on the point was correct. It also indentifies the original Middle American calendar round as an Olmec calendar that is first documented at Cuicuilco and inaugurated on the summer solstice of 739 B.C. Its sole intact descendant is the Quiche calendar of today. The Middle American calendar has been concerned with the prediction of solar astronomy over a period of 2,726 years, and it has calculated the length of the tropical year as accurately as we do (365.2422 days) since 433 B.C. Throughout its history, it has been the focal feature, greatest achievement, and almost the defining atribute of native America's most advanced civilization.

CHAPTER 1

THE CALENDAR

To the more sophisticated Aztec elite and their peers in other ethnic groups, the known world was the land between the seas — Anahuac — bounded by deserts to the north and west and jungles to the south and east. This was civilization, and its most important defining characteristic was its sense of time. Later scholarship has delineated much the same area on the basis of a number of objective cultural characteristics and renamed it Middle America, but certainly its distinctive calendar is one of the most salient of these features.

The peoples of Anahuac were and are numerous and diverse. In ancient times, most of the region to the west of the Isthmus of Tehuantepec was Otomanguean speaking. The area to the east of Veracruz and Chiapas was Mayan. In between was the heartland of the Olmecs, speakers of Zoquean languages. To the southeast of Guatemala (and outside of Anahuac) was the territory of Chibchan speech. The northern frontier expanded and contracted through time because of periodic pressures from the barbarian Chichimecs. These included Hokan-Siouan speakers from Texas and California and Uto-Aztecans from Arizona and New Mexico, and the ancestors of some peoples whose languages remain isolates: Tarascan and Cuitlatec.

The invading foreigners from the north tended to be culturally assimilated in time as they found their way south — some of them, like the Subtiaba, as far south as Nicaragua. Even the last of the Chichimec, the Nahua, were thoroughly naturalized although the Cora and Huichol, their western relatives, were not and remained outside the orbit of Anahuac and its calendar. The fifty-seven language groups that have been part of Anahuac are located on the map (Figure 1) and are listed and identified in Figure 2, the table of languages and calendars (Mixtec A, B, and C are counted as one). For clarity, Nahuatl is omitted from the map because it came to be spoken to various degrees throughout Anahuac. It is given separate treatment in the table for similar reasons: its calendrical relations with other groups are highly ramified and complex.

The relationship between the linguistic groups and their calendars is obviously significant but falls far short of identity. The reason is that language similarity is a function of longtime participation in the same ethnic system of interaction and mating exchange, whereas the calendar is a matter of religion. Some of the calendars appear to represent imperial (and thus multiethnic) religions. The Zapotec calendar was probably used by ten languages and the Quiche calendar by seven;

1

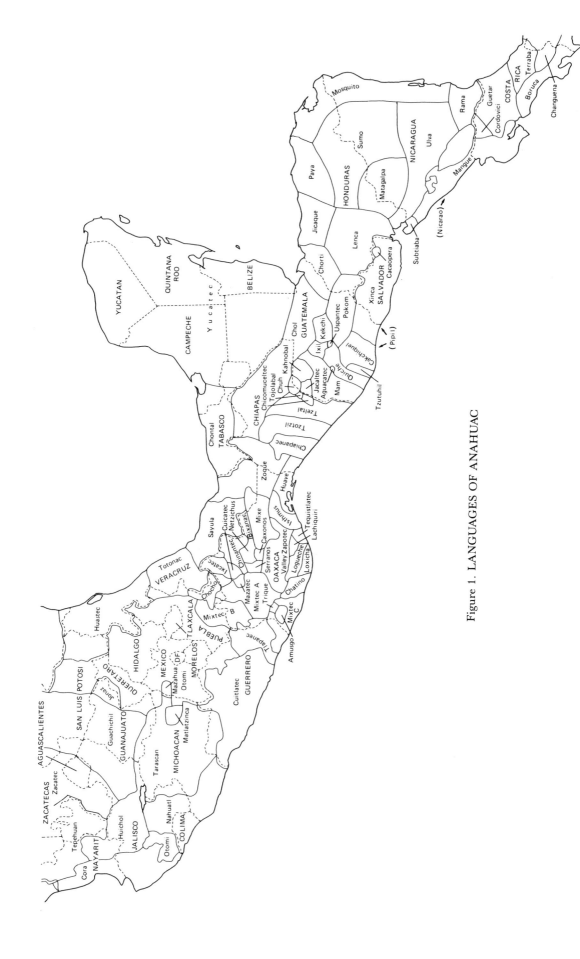

Figure 1. LANGUAGES OF ANAHUAC

Figure 2. LANGUAGES AND CALENDARS

FAMILY	LANGUAGE	CALENDAR	(NAHUATL)
Mayance	Yucatec	Valladolid	
		Mayapan	
		Campeche	
		Palenque	
		Tikal	
	Chol		
	Chontal		
	Chorti		
	Kekchi		
	Cakchiquel	Iximche	
		Cakchiquel	
	Pokom	Quiche	
	Quiche		
	Uspantec		
	Tzutuhil		
	Ixil		
	Aguacatec		
	Mam		
	Kanhobal	Kanhobal	
	Jacaltec		
	Chuh		
	Tojolabal		
	Chicomuceltec		
Otomanguean	Chiapanec	Chiapanec	
		Nimiyua	
Mayance	Tzeltal	Tzeltal	
		Cancuc	
	Tzotzil	Tzotzil	
		Guitiupa°	
		Istacostoc	
		Mitontic	
Zoquean	Mixe	Mixe	
		Mazatlan	
	Huave		
		Olmec	
	Zoque	Izapa	
	Sayula	Cuicuilco	
	Jicaque	Jicaque	
	Lenca		
	Xinca	Kaminaljuyu	
Hokan-Siouan	Tequistlatec		Nahuatl:
Unclassified	Subtiaba	Subtiaba?	(Uto-Aztecan)
Otomanguean	Cuitlatec	Tenango	
	Mixtec C	Yucuñudahui	Chalca
	Mixtec B		
		Cholula	Tepexic
		Toltec	Colhuacan
Unclassified	Tarascan	Tarascan	Tepanec
Otomanguean	Matlatzinca		
	Mazahua		
Mayance	Huastec	Huastec	Cuitlahuac
Zoquean	Totonac	Totonac	Teotitlan
Otomanguean	Otomi	Otomi	Tepepulco
	Mazatec	Mazatec	Aztec
			Texcoco
	Mixtec A	Tilantongo	Tlaxcalan
	Ixcatec		
	Triqui		
	Amusgo		
	Chocho		
	Tlapanec	Tlapanec	Metztitlan
		Teotihuacan	Colhua
	Isthmus	Zapotec	
	Logueche		
	Lachiguiri		
	Valley		
	Caxonos		
	Bixonos		
	Netzichus		
	Cuicatec		
	Chatino		
	Loxicha		
		Loxicha	
	Chinantec	Chinantec	
		Lalana	
	Mangue	Mangue?	
Chibchan	Cacaopera	Cacaopera?	

the Mixe calendar may have been used by as many as seven, but it is impossible to tell from the data. The Tilantongo calendar was used by six languages, the Kanhobal calendar by five, and the Yucuñudahui and Tarascan calendars by at least three each. In other cases, the calendars seem to reflect the religious schism of formerly unified ethnic groups. The Tzotzil have used at least five calendars, and the Tzeltal, Chinantec, Chiapanec, and Cakchiquel have had at least two apiece. In some cases, the two processes overlap: Yucatec has historically had five calendars, but one of them (Tikal) was shared with four other groups. Nahuatl used no less than twelve calendars, three of which may have been unique to Nahuatl (Texcoco, Metztitlan, and Colhua). The other nine were shared with at least one and (in the case of Tlaxcalan) as many as five other peoples.

The central importance of the calendar to the religions of Anahuac and the central importance of religion to everything else conspire to make the history of the calendar something of a key to the general cultural history of the region.

Most of the essential features of the native calendar of Anahuac are summarized in Figure 3 and its attendant caption. Perhaps the most remarkable of these features is its unity. Not only did it have a single origin, but despite its employment by nearly 100 ethnic groups speaking almost as many different languages, it has retained this unity over a period of more than 2600 years. This is not just a matter of pattern similarity but of precise mathematical accuracy in the measurement of time, and it is an astonishing civilizational achievement on a world scale.

All of the native peoples of central and southern Mexico and western Central America counted time in a special and sacred vigesimal number system illustrated in Figure 3 by its Yucatecan Mayan version with the number names of the days of the day count on the right-hand side of the figure. For reasons of convenience, they are listed in the order 2 to 20; the day Ik is day 2, and the day Ahau is day 20. The day Imix (day 1) appears at the bottom. They are listed in this order so that the figure can begin with the first day of the classical Mayan year (Ik), even though all the peoples of the region agreed that Ik or its equivalent is the second of the 20 days. Imix was everywhere the first day of the day count but not necessarily of the year.

Other names were used to number the 20 sacred days in other languages, but whatever other meanings and associations they may have had, they are precise numerical equivalents to the Yucatecan Mayan names, and the equivalent to Ik is always 2 and is always set apart as a vigesimal number. Using this count, then, the Middle Americans counted the days by 20s.

They also counted the days by 13s, normally using the secular cardinal numbers. In Middle America, this number system was generally decimal, though there are indications in a number of the languages of other numerical bases: 4, 5, 8, or even (in Yucatec itself) 11 (see Colville 1985). In at least two cases, the 13-day count used a special tridecimal number system sacred to this purpose (see Mixe,

Mixtec). And in two others (Teotihuacan and Tlapanec), the count was from 2 to 14 instead of from 1 to 13. In all cases, however, this count was numerically equivalent to the Yucatecan count by 13s. Across the top of Figure 3 are the numbers of the initial days of the nineteen columns in the figure, beginning with day 6 at the upper left.

The days of the 20-day count are conveniently designated by the Spanish term *veintena*. Those of the 13-day count constitute the *trecena*. There was also a 9-day count or *novena*. The trecena and veintena were counted concurrently, so the first day on the upper left of the figure is 6 Ik. Reading down the column from that date, the days are 7 Akbal, 8 Kan, 9 Chicchan, 10 Cimi, 11 Manik, 12 Lamat, 13 Muluc, 1 Oc, and so forth, until all the permutations of the veintena and trecena have been named and the date 6 Ik returns on the 261st day (at the top of the fourteenth column).

The 260-day cycle of the veintena and trecena is the sacred and universal day count of Anahuac. It was called *tonalpohualli* in Nahuatl and *tzol kin* in Yucatec and had a name in all the other languages of the area, too. Any particular day has always had and still has the same position in the day count everywhere. When it was 6 Ik in Tikal, it was 6 Ik in Nayarit and Costa Rica as well. Mistakes in the naming of the days are amazingly rare. Figure 3 covers a 380-day segment of the day count from 6 Ik to 8 Imix. Equivalent names in languages other than Yucatec are given in the calendar index in Chapter 5, at least where they are known or can be guessed. The English names of the Aztec days are given at the extreme right of Figure 3.

The second major feature of the calendar is the year of 365 days. It is composed of a separate count of veintenas using a different, or at least separate, set of numbers, illustrated in Figure 3 by the Tikal calendar numeration down the left-hand side of the figure, numbering the 20 days of the native month. In the Tikal calendar (and a number of others), they are enumerated from 0 to 19. In most of the modern calendars, they are counted from 1 to 20. In all cases, they are counted vigesimally, and the correspondence of the counts is calculable and thus reducible to a single base.

The months are numbered separately from 1 to 19, sometimes using secular cardinal or ordinal numbers but more often using names. In either case, the number names of the months constitute an additional novedecimal count designating 18 months of 20 days and a final one of 5 days. The Yucatecan names of the months are given across the bottom of Figure 3 from Pop (1) to Uayeb (19). The last 15 days of the nineteenth column are the first 15 of the next year from 0 to 14 Pop, because Uayeb has only 5 days.

Unlike the day count, which is permutative, the *year count* is combinatory. The first day in Figure 3, in the upper left corner, is 0 Pop. Reading down the first column, it is followed by 1 Pop, 2 Pop, 3 Pop, down to 19 Pop. The second column

Figure 3. NEW YEARS' DAYS (1549–50 JULIAN) (CORRELATION CONSTANT 584,283)
YEAR 6 IK TIKAL

COEFFICIENTS OF THE DAY IK

DAYS OF THE MONTH	6	13	7	1	8	2	9	3	10	4	11	5	12	6	13	7	1	8	2	Day	Meaning
0	TK KE			TH				XII	10	30					20	10	30			Ik	Wind
1	PA TA	TU	CL	YU								20		IV				20	10	Akbal	House
2	CP MY									II							VI			Kan	Iguana
3																				Chicchan	Serpent
4							OL	A												Cimi	Death
5						D	C	B				CK	TL	CN	QU NM	E			TK KE	Manik	Deer
6	20					MX	JI	ZA LX	ZO CE	ZE	LA	GU	IS	MI	CC	KA TO	HU		PA TA	Lamat	Rabbit
7		10	30			CO	ME	TM	TX	AZ	OT	TT	CT	TP	CU	IC	LC TG		MY CP	Muluc	Water
8				20	10	30														Oc	Dog
9			IX				20	10	30			20								Chuen	Monkey
10							XI			20		III	10	30	20					Eb	Grass
11								I		10			V				10	30	20	Ben	Cane
12																	VII			Ix	Jaguar
13																				Men	Eagle
14																				Cib	Buzzard
15																				Caban	Quake
16	30																			Etz'nab	Flint
17		20																		Cauac	Rain
18	VIII		10	30	20															Ahau	Flower
19			X		10	30	20					10	30							Imix	Alligator
	N.	O.	P.	Q.	R.	A.	B.	C.	D.	E.	F.	G.	H.	I.	J.	K.	L.	M.	X.		

MONTHS OF THE YEAR

Maya months: Pop, Uo, Zip, Zotz', Tzec, Xul, Yaxkin, Mol, Ch'en, Yax, Zac, Ceh, Mac, Kankin, Muan, Pax, Kayab, Cumku, Uayeb

Aztec months: Tlaxochimaco, Xocolhuetzi, Ochpaniztli, Teotleco, Tepeilhuitl, Quecholli, Panquetzaliztli, Atemoztli, Tititl, Izcalli, Cuauhuitleua, Tlacaxipehualiztli, Tozoztli, Hueitozoztli, Toxcatl, Etzcualiztli, Tecuilhuitontli, Hueitecuilhuitl, Nemontemi

CALENDARS IN FIGURES 3 AND 8			
Aztec	IV.E.20 = Texcoco	+ 1m (f1548)	AZ
Cakchiquel	II.G.1 = Quiche	– 5m	CK
Campeche	IV.N.20 = Palenque	+ 1d	CP
Cancuc	III.J.1 = Tzeltal	+ 5m (f1584)	CC
Chiapanec	II.K.1 = Quiche	– 1m (f1548)	CN
Chinantec	III.D.19 = Zapotec	+ 1m (f1548)	CE
Chalca	IV.L.20 = Huexotzinco	+ 1m	LC
Cholula	III.P.0 = Yucuñudahui	– 1m	CL
Colhua	IV.A.20 = Metztitlan	– 1m	CO
Colhuacan	IV.J.20 = Tepanec	+ 1m	CU
CUICUILCO	I.C.19 = Original		A
Cuitlahuac	IV.H.20 = Teotitlan	+ 1m	CT
Guitiupa	III.G.1 = Tzotzil	+ 3m (f1584)	GU
HUASTEC	III.M.0 = Tarascan	– 1m	HU
Istacostoc	III.H.1 = Guitiupa	+ 1m (f1584)	IS
Iximche	= Cakchiquel	+ (400)*	IM
Izapa	II.C.19 = Cuicuilco	+ 1d	B
Jicaque	III.B.0 = Kaminaljuyu	+ 1d	JI
KAMINALJUYU	II.B.0 = Olmec	+ 1d – t – (360)*	C
Kanhobal	III.L.1 = Quiche	+ 1d	KA
Kekchi	II.N.0 = Tikal	(f1548)	KE
Lachixola	III.F.19 = Chinantec	+ 2m (f1548)	LA
Loxicha	= Zapotec?	+ (9)*	LX
Mayapan	IV.N.1 = Campeche	+ i (f1548)	MY
Mazatec	= Aztec	(f1617)	MZ
Mazatlan	= Mixe	+ (13)* (f1901?)	ML
Metztitlan	IV.B.20 = Tilantongo	– 1m	ME
Mitontic	III.I.1 = Istacostoc	+ 1m (f1584)	MI
Mixe	III.A.0 = Jicaque	– 1m (1532)	MX
Nimiyua	II.L.1 = Chiapanec	+ 1m (f1548)	NM
OLMEC	I.B.19	+ (360)*	OL
Otomi	IV.F.20 = Aztec	+ 1m	OT
Palenque	III.N.20 = Tikal	+ t – 0	PA
Pokom	= Quiche	(f1548)	PK
Quiche	II.L.1 = Tikal	– 2m – 0 – (360)*	QU
TARASCAN	III.N.0 = Tula	– 1m	TA
Tenango	IV.L.0 = Totonac	+ 1d	TG
TEOTIHUACAN	II.Q.0 = Olmec	+ 1d – t – 105d + (14)*	TH
Teotitlan	IV.G.20 = Otomi	+ 1m	TT
Tepanec	IV.I.20 = Cuitlahuac	+ 1m	TP
Tepexic	IV.K.20 = Colhuacan	+ 1m	IC
Texcoco	IV.D.20 = Tilantongo	+ 1m (f1548)	TX
TIKAL	II.N.0 = Olmec	– 1m – 105d – t	TK
TILANTONGO	IV.C.20 = Zapotec	+ 1d – 0	TM
Tlapanec	II.J.0 = Teotihuacan	– 7m	TL
TOLTEC	III.O.0 = Cholula	– 1m	TU
TOTONAC	III.L.0 = Huastec	– 1m	TO
Tzeltal	III.E.1 = Kanhobal	– 7m (f1548)(f1584)	ZE
Tzotzil	III.D.1 = Tzeltal	– 1m (f1548)(f1584)	ZO
Valladolid	= Mayapan	+ (24)*	VA
YUCUÑUDAHUI	III.Q.0 = Teotihuacan	+ 1d	YU
ZAPOTEC	III.C.19 = Olmec	+ 2d + 1m	ZA

OTHER CALENDARS

A		= Cuicuilco
B		= Izapa
C		= Kaminaljuyu
D		= Unknown
E		= Dubious
CL	Aguacatec	= Quiche
CO	Amusgo	= Tilantongo?
CU	Bixanas	= Zapotec?
A	Cacaopera	= ?
CT	Caxonos	= Zapotec?
GU	Chatino	= Zapotec
HU	Chicomuceltec	= Kanhobal?
IS	Chocho	= Tilantongo
IM	Chol	= Tikal
B	Chontal	= Tikal?
JI	Chorti	= Tikal?
C	Chuh	= Kanhobal
KA	Cuicatec	= Zapotec?
KE	Cuitlatec	= Tenango?
LA	Huave	= Mixe?
LX	Huexotzinco	= Tepexic
MY	Ixcatec	= Tilantongo?
MZ	Ixil	= Quiche
ML	Isthmus	= Zapotec?
ME	Jacaltec	= Kanhobal
MI	Lachiguiri	= Zapotec?
MX	Lenca	= Jicaque?
NM	Logueche	= Zapotec?
OL	Mam	= Quiche
OT	Mangue	= ?
PA	Matlatzinca	= Tepanec
PK	Mazahua	= Otomi?
QU	Netzichus	= Zapotec?
TA	Nicarao	= Teotitlan?
TG	Pipil	= Teotitlan
TH	Sayula	= Mixe?
TT	Serrano	= Zapotec?
TP	Subtiaba	= ?
IC	Tepepulco	= Otomi
TX	Tequistlatec	= Mixe?
TK	Tlaxcalan	= Tilantongo
TM	Tojolabal	= Kanhobal?
TL	Trique	= Tilantongo?
TU	Tzutuhil	= Quiche?
TO	Uspantec	= Quiche?
ZE	Xinca	= Kaminaljuyu?
ZO	Zoque	= Mixe?

Abbreviations: d = day; f = frozen; i = initial naming; m = month; t = terminal naming; 0 = zero counting;
* = special features:

    (9)　= novena calendar
    (13)　= trecena calendar
    (14)　= 2–14 day count
    (24)　= 24-year katun
    (360) = tun and Long Count
    (400) = 400 day cycle

Calendars that also initiated writing systems are Capitalized.
Omitted from Figure 3: Iximche, Pokom, Mazatec, Mazatlan, and Valladolid.
Omitted from Figure 8: Chalca, Colhuacan, Cuitlahuac, Teotitlan, Tepanec, and Tepexic.

begins with 0 Uo. The year ends with 4 Uayeb, the fifth day in the nineteenth column. The figure continues to 14 Pop in the lower right-hand corner. The list of Nahuatl equivalents for the Yucatecan months is also given across the bottom of the figure, but their correspondence is only partial because the Nahua did not begin the year with 6 Ik. I shall return to this point.

Figure 3 also provides the equivalents for the dates of the day count and the year count in the European calendar, specifically the Julian calendar. The correspondence indicated is for 1549–1550. The figure marks the first day of each European month with a corresponding Roman numeral (from I to XII). As a further aid to locating European dates, the tenth, twentieth, and thirtieth days of the months are indicated in Arabic numerals (10, 20, 30). If one derives from the figure that January 1, 1550 Julian was 8 Ben 11 Ch'en Tikal, one is reading the figure correctly.

To get this result, find January 1 (marked I, in column 9, row 12, Figure 3). To get the day-count coefficient, count down by 13s from the starting position at the top of column 9, which is 10 (thus 10, 11, 12, 13, 1, 2, 3, 4, 5, 6, 7, 8). The day name will be found on the right side of the twelfth row—the row in which January 1 falls (hence Ben). The coefficient of the year count can be read from the left margin of row 12 (thus 11) and the month name from the bottom of column 9 (thus Ch'en).

Simple as they are, the day count and the year count of Anahuac interact in numerological patterns that are not immediately obvious, though they were thoroughly familiar to the day keepers and priests who used them. A slightly different display of the veintena of the day count will make the point clear. This is offered in Figure 4.

Figure 4. THE DAY COUNT

Type

I.	a. Alligator	f. Death	k. Monkey	p. Buzzard
II.	b. Wind	g. Deer	l. Grass	q. Quake
III.	c. House	h. Rabbit	m. Cane	r. Flint
IV.	d. Iguana	i. Water	n. Jaguar	s. Rain
V.	e. Serpent	j. Dog	o. Eagle	t. Flower

The day count can be read in the order of the accompanying letters. The English names are those of the Aztec day count, which is selected arbitrarily for this purpose because it is easier to "translate" than most of the other day-name systems, many of which are archaic and obscure. They remain, however, essentially a vigesimal number system—a graphic and evocative one, to be sure, but an enumeration of days common to all the Anahuac calendars. The Yucatecan day names of Figure 3 can thus be matched to Figure 4: Alligator is Imix, Wind Ik, House Akbal, and so forth, down to Flower, or Ahau. Some of the Yucatecan names actually have different meanings as names but not as numbers.

The Tikal year that begins Figure 3 starts with 6 Ik. That means, as the figure indicates, that all the other months of that year will also begin with Ik. Because of counting by trecenas, the coefficients of these initial days will always follow the numerical pattern that is indicated across the top of the figure (6, 13, 7, 1, 8, etc.), thus advancing by 7 every month or by 1 every 2 months. The 5-day month at the end of the year dictates that the next year will begin on 7 Manik, the year after that on 8 Eb and the next on 9 Caban. Then we return to Ik again, with 10 Ik. Referring to Figure 4, we see that this sequence (Wind, Deer, Grass, Quake) occupies the row designated in the figure as Type II. These are the possible New Years' days of the Tikal calendar: the days that begin the year. Other calendars may begin the year on other days and, hence, use different New Years' sets. The Mayapan calendar for example (MY in Figure 3) begins the year on Muluc (Water), and thus uses the New Years' days designated as Type IV in Figure 4 (Iguana, Water, Jaguar, Rain).

Figure 3 indicates the position of New Year's Day in forty-four native calendars. Fifteen of these use a Type IV New Year's, like those of Mayapan. They are lined up in the eighth row from the top, beginning with CO (Colhua) and ending with MY (Mayapan). Twenty more use Type III New Years' days. They are listed in the seventh row from MX (Mixe) to PA (Palenque), and beyond that (in the second row) to YU (Yucuñudahui). An additional eight use Type II New Years' days. They are listed in the sixth row from CK (Cakchiquel) to TK (Tikal), and (in the first row) TH (Teotihuacan). There is one calendar using Type I New Years' days which is located in the fifth row of Figure 3—OL (Olmec). The New Years' days of an additional thirteen calendars are known and correspond to those of one or another of the calendars already registered. These are listed under "other calendars" in the caption of Figure 3 and differ from the equivalents listed, only in using different languages. The inclusion of these and of Calendars A, B, and C brings the total to sixty. An additional thirty "possible" or inadequately documented calendars are also listed.

It will be apparent from the varying placement of the beginning of the year why the correspondence of the months from one calendar to another is only partial. In each calendar, the 5-day nineteenth month falls immediately before New Years' Day. For example, the first day of the eighteenth Aztec month falls on 2 Ch'en Tikal, but the first day of the first Aztec month falls on 7 Yax Tikal; the 5 days of Nemontemi fall on 2 to 6 Yax. On the other hand, all such relations among the native calendars are permanent and invariant, because none of them ever had intercalary leap-year days, at least before the Conquest. This is discussed further below.

A final complication of Figure 3 involves the naming of the years. The

Tikal calendar, indeed virtually all the Mayan calendars, named the years for their initial days. Thus the year 1549–50 was 6 Wind (Ik) Tikal and 8 Iguana (Kan) Mayapan. Many of the calendars, however, named the years for what were considered to be their last day: the final day of the eighteenth month. The nineteenth month is ignored for this purpose because it is generally viewed as a kind of calendrical hiatus between years — a time of special danger or even of horror — and its days were said to be useless, lost, or even nameless.

Thus the name day of the year may be its 360th day rather than its first. For example, Figure 3 indicates that the Aztec year began on January 17, 1550, Julian on 11 Water 7 Yax Tikal. This would be 11 Water 1 Izcalli Aztec. However, the Aztec called that year 6 Tochtli (Rabbit), naming it for its last day, 359 days later. Almost all of the calendars to the west and north of the Isthmus of Tehuantepec named the year terminally, like the Aztec, rather than initially, like the Maya. In either case, the day that named the year was said to be its year bearer. Tikal's year bearer and New Year were Type II. The Aztec New Year was Type IV, but its name day, and hence year bearer, was Type III.

Whether it is initial or terminal, the name day of the year always occurs twice during the year. The first day of the first month was repeated in the day count as the first day of the fourteenth month. The last day of the eighteenth month was anticipated by the same day of the day count as the last day of the fifth month. Thus the second name day of the Maya and the first name day of the Aztec became the occasion for a special celebration, which is easily confused with the beginning of the year (see chap. 2, 1555), a kind of "little New Year."

This feature is, however, a convenience for calculating terminal name days from the New Year and vice versa. Referring to Figure 3 again, if one begins from the Aztec New Year (AZ) on 11 Water 1 Izcalli and counts to the end of the fifth month (move up one row and right five columns), we land on the day marked CC. This is the Aztec first name day for 1550 and thus gives the name of the Aztec year, 6 Rabbit. (The day in question is 6 Muan Tikal, but it is 20 Hueitozoztli Aztec.)

What Figure 3 summarizes is the fact that the fundamental features of the calendars of Anahuac are unitary. Calendars differed in when they began the year and how they named it. They shared a common day count and a common organization of the 18 veintenas and one 5-day month to form the 365-day year count. Furthermore, the relation between one calendar and another was completely invariant, so that Figure 3 makes it easy to convert the name day of the year in one calendar to that of another. Thus a year 3 House Aztec is 3 Wind Quiche. Because Aztec is terminally named, find its first name day on the square marked (CC). Call that 3 House and count forward (right) to the QU (Quiche) column, using the number sequence across the top of the Table (3, 10, 4) then back (up) to Quiche New Year (4 House, 3 Wind). Because Quiche is initially named, 3 Wind is the Quiche year bearer. Between two terminally named or two initially named systems, it is even easier. For example, call the square (AZ) 3 House and count back (left) 2 (3, 9, 2) to (TM) and up 1 (from 2 House) to (ZA) to get Zapotec 1 Wind.

A single table of year bearer correspondences is thus a convenient guide to the dating of all the calendars in European time. Such a table, based on that of Caso (1967), is given in Figure 5. It gives the European year in which the Aztec year of the given name *begins*: the actual name day of the Aztec year will usually fall in the following European one. Note that two different Aztec years began in A.D. 156 and A.D. 1616, one on January 1 and one on December 31 Julian. The entire table is in the Julian calendar. In Gregorian time, it would be 1576 that contained the starting points of two Aztec years—an irrelevancy because the Gregorian calendar had not yet been introduced.

What is known about each of the Middle American calendars is summarized in the calendar index in Chapter 5. In order to place them in Figure 3 and to use that figure in calculating dates in them, it is necessary to know five things about them:

1. Which Type of New Year Day do they employ?
2. Which month of the year is the first (following the 5-day month)?
3. Are they named terminally or initially?
4. Do they begin the month with 0 or with 1?
5. Do they begin the day count with 1 or with 2?

Only two calendars, Teotihuacan and Tlapanec, counted the day count from 2 to 14, which disposes of question 5. The other four points can be reduced to a convenient code of three elements. The type of New Year's Day can be indicated by roman numerals, as in Figure 4. The months actually used to begin the native calendars are coded across the bottom of Figure 3 from A. (the first month Colhua, Xul in Yucatec) through N. (the first month Mayapan, Pop in Yucatec), and on to R. (Tzec in Yucatec). The name day of the year is given in Arabic numerals: 0 or 1 for initial naming, 19 or 20 for terminal naming, thus also specifying whether or not the calendar uses 0.

The Tikal calendar is II.N.0 (Type II New Year, beginning on N. [Pop], and named for its first day, which is numbered zero). The Zapotec calendar is III.C.19 (Type III New Year, beginning on month C. [Mol], using zero and terminal naming.) The Olmec calendar is I.B.19 (Type I New Year, beginning on month B. [Yaxkin], using zero and named terminally). The coded placement of each calendar is indicated in the headings of the alphabetized calendar index (chap. 5) and in the note to Figure 3.

For some kinds of calendrical problems, it is convenient to be able to calculate which day falls on, say, the 267th day of a native or Christian year. The vigesimal base of Figure 3 facilitates this calculation. One may readily find, for example, by counting by 20s in Figure 3, that the 267th day of the Tikal calendar is the Tepanec New Year (TP), whereas the 267th day of the Tepanec calendar is

Figure 5a. AZTEC AND CHRISTIAN YEARS (1041 B.C. TO A.D. 2025 JULIAN)

Aztec Year																														
2 Cane	-1041	-989	-937	-885	-833	-781	-729	-677	-625	-573	-521	-469	-417	-365	-313	-261	-209	-157	-105	-53	-1	52	104	155	207	259	311	363	415	467
3 Flint	-1040	-988	-936	-884	-832	-780	-728	-676	-624	-572	-520	-468	-416	-364	-312	-260	-208	-156	-104	-52	1	53	105	156	208	260	312	364	416	468
4 House	-1039	-987	-935	-883	-831	-779	-727	-675	-623	-571	-519	-467	-415	-363	-311	-259	-207	-155	-103	-51	2	54	106	157	209	261	313	365	417	469
5 Rabbit	-1038	-986	-934	-882	-830	-778	-726	-674	-622	-570	-518	-466	-414	-362	-310	-258	-206	-154	-102	-50	3	55	107	158	210	262	314	366	418	470
6 Cane	-1037	-985	-933	-881	-829	-777	-725	-673	-621	-569	-517	-465	-413	-361	-309	-257	-205	-153	-101	-49	4	56	108	159	211	263	315	367	419	471
7 Flint	-1036	-984	-932	-880	-828	-776	-724	-672	-620	-568	-516	-464	-412	-360	-308	-256	-204	-152	-100	-48	5	57	109	160	212	264	316	368	420	472
8 House	-1035	-983	-931	-879	-827	-775	-723	-671	-619	-567	-515	-463	-411	-359	-307	-255	-203	-151	-99	-47	6	58	110	161	213	265	317	369	421	473
9 Rabbit	-1034	-982	-930	-878	-826	-774	-722	-670	-618	-566	-514	-462	-410	-358	-306	-254	-202	-150	-98	-46	7	59	111	162	214	266	318	370	422	474
10 Cane	-1033	-981	-929	-877	-825	-773	-721	-669	-617	-565	-513	-461	-409	-357	-305	-253	-201	-149	-97	-45	8	60	112	163	215	267	319	371	423	475
11 Flint	-1032	-980	-928	-876	-824	-772	-720	-668	-616	-564	-512	-460	-408	-356	-304	-252	-200	-148	-96	-44	9	61	113	164	216	268	320	372	424	476
12 House	-1031	-979	-927	-875	-823	-771	-719	-667	-615	-563	-511	-459	-407	-355	-303	-251	-199	-147	-95	-43	10	62	114	165	217	269	321	373	425	477
13 Rabbit	-1030	-978	-926	-874	-822	-770	-718	-666	-614	-562	-510	-458	-406	-354	-302	-250	-198	-146	-94	-42	11	63	115	166	218	270	322	374	426	478
1 Cane	-1029	-977	-925	-873	-821	-769	-717	-665	-613	-561	-509	-457	-405	-353	-301	-249	-197	-145	-93	-41	12	64	116	167	219	271	323	375	427	479
2 Flint	-1028	-976	-924	-872	-820	-768	-716	-664	-612	-560	-508	-456	-404	-352	-300	-248	-196	-144	-92	-40	13	65	117	168	220	272	324	376	428	480
3 House	-1027	-975	-923	-871	-819	-767	-715	-663	-611	-559	-507	-455	-403	-351	-299	-247	-195	-143	-91	-39	14	66	118	169	221	273	325	377	429	481
4 Rabbit	-1026	-974	-922	-870	-818	-766	-714	-662	-610	-558	-506	-454	-402	-350	-298	-246	-194	-142	-90	-38	15	67	119	170	222	274	326	378	430	482
5 Cane	-1025	-973	-921	-869	-817	-765	-713	-661	-609	-557	-505	-453	-401	-349	-297	-245	-193	-141	-89	-37	16	68	120	171	223	275	327	379	431	483
6 Flint	-1024	-972	-920	-868	-816	-764	-712	-660	-608	-556	-504	-452	-400	-348	-296	-244	-192	-140	-88	-36	17	69	121	172	224	276	328	380	432	484
7 House	-1023	-971	-919	-867	-815	-763	-711	-659	-607	-555	-503	-451	-399	-347	-295	-243	-191	-139	-87	-35	18	70	122	173	225	277	329	381	433	485
8 Rabbit	-1022	-970	-918	-866	-814	-762	-710	-658	-606	-554	-502	-450	-398	-346	-294	-242	-190	-138	-86	-34	19	71	123	174	226	278	330	382	434	486
9 Cane	-1021	-969	-917	-865	-813	-761	-709	-657	-605	-553	-501	-449	-397	-345	-293	-241	-189	-137	-85	-33	20	72	124	175	227	279	331	383	435	487
10 Flint	-1020	-968	-916	-864	-812	-760	-708	-656	-604	-552	-500	-448	-396	-344	-292	-240	-188	-136	-84	-32	21	73	125	176	228	280	332	384	436	488
11 House	-1019	-967	-915	-863	-811	-759	-707	-655	-603	-551	-499	-447	-395	-343	-291	-239	-187	-135	-83	-31	22	74	126	177	229	281	333	385	437	489
12 Rabbit	-1018	-966	-914	-862	-810	-758	-706	-654	-602	-550	-498	-446	-394	-342	-290	-238	-186	-134	-82	-30	23	75	127	178	230	282	334	386	438	490
13 Cane	-1017	-965	-913	-861	-809	-757	-705	-653	-601	-549	-497	-445	-393	-341	-289	-237	-185	-133	-81	-29	24	76	128	179	231	283	335	387	439	491
1 Flint	-1016	-964	-912	-860	-808	-756	-704	-652	-600	-548	-496	-444	-392	-340	-288	-236	-184	-132	-80	-28	25	77	129	180	232	284	336	388	440	492
2 House	-1015	-963	-911	-859	-807	-755	-703	-651	-599	-547	-495	-443	-391	-339	-287	-235	-183	-131	-79	-27	26	78	130	181	233	285	337	389	441	493
3 Rabbit	-1014	-962	-910	-858	-806	-754	-702	-650	-598	-546	-494	-442	-390	-338	-286	-234	-182	-130	-78	-26	27	79	131	182	234	286	338	390	442	494
4 Cane	-1013	-961	-909	-857	-805	-753	-701	-649	-597	-545	-493	-441	-389	-337	-285	-233	-181	-129	-77	-25	28	80	132	183	235	287	339	391	443	495
5 Flint	-1012	-960	-908	-856	-804	-752	-700	-648	-596	-544	-492	-440	-388	-336	-284	-232	-180	-128	-76	-24	29	81	133	184	236	288	340	392	444	496
6 House	-1011	-959	-907	-855	-803	-751	-699	-647	-595	-543	-491	-439	-387	-335	-283	-231	-179	-127	-75	-23	30	82	134	185	237	289	341	393	445	497
7 Rabbit	-1010	-958	-906	-854	-802	-750	-698	-646	-594	-542	-490	-438	-386	-334	-282	-230	-178	-126	-74	-22	31	83	135	186	238	290	342	394	446	498
8 Cane	-1009	-957	-905	-853	-801	-749	-697	-645	-593	-541	-489	-437	-385	-333	-281	-229	-177	-125	-73	-21	32	84	136	187	239	291	343	395	447	499
9 Flint	-1008	-956	-904	-852	-800	-748	-696	-644	-592	-540	-488	-436	-384	-332	-280	-228	-176	-124	-72	-20	33	85	137	188	240	292	344	396	448	500
10 House	-1007	-955	-903	-851	-799	-747	-695	-643	-591	-539	-487	-435	-383	-331	-279	-227	-175	-123	-71	-19	34	86	138	189	241	293	345	397	449	501
11 Rabbit	-1006	-954	-902	-850	-798	-746	-694	-642	-590	-538	-486	-434	-382	-330	-278	-226	-174	-122	-70	-18	35	87	139	190	242	294	346	398	450	502
12 Cane	-1005	-953	-901	-849	-797	-745	-693	-641	-589	-537	-485	-433	-381	-329	-277	-225	-173	-121	-69	-17	36	88	140	191	243	295	347	399	451	503
13 Flint	-1004	-952	-900	-848	-796	-744	-692	-640	-588	-536	-484	-432	-380	-328	-276	-224	-172	-120	-68	-16	37	89	141	192	244	296	348	400	452	504
1 House	-1003	-951	-899	-847	-795	-743	-691	-639	-587	-535	-483	-431	-379	-327	-275	-223	-171	-119	-67	-15	38	90	142	193	245	297	349	401	453	505
2 Rabbit	-1002	-950	-898	-846	-794	-742	-690	-638	-586	-534	-482	-430	-378	-326	-274	-222	-170	-118	-66	-14	39	91	143	194	246	298	350	402	454	506
3 Cane	-1001	-949	-897	-845	-793	-741	-689	-637	-585	-533	-481	-429	-377	-325	-273	-221	-169	-117	-65	-13	40	92	144	195	247	299	351	403	455	507
4 Flint	-1000	-948	-896	-844	-792	-740	-688	-636	-584	-532	-480	-428	-376	-324	-272	-220	-168	-116	-64	-12	41	93	145	196	248	300	352	404	456	508
5 House	-999	-947	-895	-843	-791	-739	-687	-635	-583	-531	-479	-427	-375	-323	-271	-219	-167	-115	-63	-11	42	94	146	197	249	301	353	405	457	509
6 Rabbit	-998	-946	-894	-842	-790	-738	-686	-634	-582	-530	-478	-426	-374	-322	-270	-218	-166	-114	-62	-10	43	95	147	198	250	302	354	406	458	510
7 Cane	-997	-945	-893	-841	-789	-737	-685	-633	-581	-529	-477	-425	-373	-321	-269	-217	-165	-113	-61	-9	44	96	148	199	251	303	355	407	459	511
8 Flint	-996	-944	-892	-840	-788	-736	-684	-632	-580	-528	-476	-424	-372	-320	-268	-216	-164	-112	-60	-8	45	97	149	200	252	304	356	408	460	512
9 House	-995	-943	-891	-839	-787	-735	-683	-631	-579	-527	-475	-423	-371	-319	-267	-215	-163	-111	-59	-7	46	98	150	201	253	305	357	409	461	513
10 Rabbit	-994	-942	-890	-838	-786	-734	-682	-630	-578	-526	-474	-422	-370	-318	-266	-214	-162	-110	-58	-6	47	99	151	202	254	306	358	410	462	514
11 Cane	-993	-941	-889	-837	-785	-733	-681	-629	-577	-525	-473	-421	-369	-317	-265	-213	-161	-109	-57	-5	48	100	152	203	255	307	359	411	463	515
12 Flint	-992	-940	-888	-836	-784	-732	-680	-628	-576	-524	-472	-420	-368	-316	-264	-212	-160	-108	-56	-4	49	101	153	204	256	308	360	412	464	516
13 House	-991	-939	-887	-835	-783	-731	-679	-627	-575	-523	-471	-419	-367	-315	-263	-211	-159	-107	-55	-3	50	102	154	205	257	309	361	413	465	517
1 Rabbit	-990	-938	-886	-834	-782	-730	-678	-626	-574	-522	-470	-418	-366	-314	-262	-210	-158	-106	-54	-2	51	103	155	206	258	310	362	414	466	518

Figure 5b. AZTEC AND CHRISTIAN YEARS (1041 B.C. TO A.D. 2025 JULIAN)

Aztec Year																													
2 Cane	519	571	623	675	727	779	831	883	935	987	1039	1091	1143	1195	1247	1299	1351	1403	1455	1507	1559	1611	1662	1714	1766	1818	1870	1922	1974
3 Flint	520	572	624	676	728	780	832	884	936	988	1040	1092	1144	1196	1248	1300	1352	1404	1456	1508	1560	1612	1663	1715	1767	1819	1871	1923	1975
4 House	521	573	625	677	729	781	833	885	937	989	1041	1093	1145	1197	1249	1301	1353	1405	1457	1509	1561	1613	1664	1716	1768	1820	1872	1924	1976
5 Rabbit	522	574	626	678	730	782	834	886	938	990	1042	1094	1146	1198	1250	1302	1354	1406	1458	1510	1562	1614	1665	1717	1769	1821	1873	1925	1977
6 Cane	523	575	627	679	731	783	835	887	939	991	1043	1095	1147	1199	1251	1303	1355	1407	1459	1511	1563	1615	1666	1718	1770	1822	1874	1926	1978
7 Flint	524	576	628	680	732	784	836	888	940	992	1044	1096	1148	1200	1252	1304	1356	1408	1460	1512	1564	1616	1667	1719	1771	1823	1875	1927	1979
8 House	525	577	629	681	733	785	837	889	941	993	1045	1097	1149	1201	1253	1305	1357	1409	1461	1513	1565	1617	1668	1720	1772	1824	1876	1928	1980
9 Rabbit	526	578	630	682	734	786	838	890	942	994	1046	1098	1150	1202	1254	1306	1358	1410	1462	1514	1566	1618	1669	1721	1773	1825	1877	1929	1981
10 Cane	527	579	631	683	735	787	839	891	943	995	1047	1099	1151	1203	1255	1307	1359	1411	1463	1515	1567	1619	1670	1722	1774	1826	1878	1930	1982
11 Flint	528	580	632	684	736	788	840	892	944	996	1048	1100	1152	1204	1256	1308	1360	1412	1464	1516	1568	1620	1671	1723	1775	1827	1879	1931	1983
12 House	529	581	633	685	737	789	841	893	945	997	1049	1101	1153	1205	1257	1309	1361	1413	1465	1517	1569	1621	1672	1724	1776	1828	1880	1932	1984
13 Rabbit	530	582	634	686	738	790	842	894	946	998	1050	1102	1154	1206	1258	1310	1362	1414	1466	1518	1570	1622	1673	1725	1777	1829	1881	1933	1985
1 Cane	531	583	635	687	739	791	843	895	947	999	1051	1103	1155	1207	1259	1311	1363	1415	1467	1519	1571	1623	1674	1726	1778	1830	1882	1934	1986
2 Flint	532	584	636	688	740	792	844	896	948	1000	1052	1104	1156	1208	1260	1312	1364	1416	1468	1520	1572	1624	1675	1727	1779	1831	1883	1935	1987
3 House	533	585	637	689	741	793	845	897	949	1001	1053	1105	1157	1209	1261	1313	1365	1417	1469	1521	1573	1625	1676	1728	1780	1832	1884	1936	1988
4 Rabbit	534	586	638	690	742	794	846	898	950	1002	1054	1106	1158	1210	1262	1314	1366	1418	1470	1522	1574	1626	1677	1729	1781	1833	1885	1937	1989
5 Cane	535	587	639	691	743	795	847	899	951	1003	1055	1107	1159	1211	1263	1315	1367	1419	1471	1523	1575	1627	1678	1730	1782	1834	1886	1938	1990
6 Flint	536	588	640	692	744	796	848	900	952	1004	1056	1108	1160	1212	1264	1316	1368	1420	1472	1524	1576	1628	1679	1731	1783	1835	1887	1939	1991
7 House	537	589	641	693	745	797	849	901	953	1005	1057	1109	1161	1213	1265	1317	1369	1421	1473	1525	1577	1629	1680	1732	1784	1836	1888	1940	1992
8 Rabbit	538	590	642	694	746	798	850	902	954	1006	1058	1110	1162	1214	1266	1318	1370	1422	1474	1526	1578	1630	1681	1733	1785	1837	1889	1941	1993
9 Cane	539	591	643	695	747	799	851	903	955	1007	1059	1111	1163	1215	1267	1319	1371	1423	1475	1527	1579	1631	1682	1734	1786	1838	1890	1942	1994
10 Flint	540	592	644	696	748	800	852	904	956	1008	1060	1112	1164	1216	1268	1320	1372	1424	1476	1528	1580	1632	1683	1735	1787	1839	1891	1943	1995
11 House	541	593	645	697	749	801	853	905	957	1009	1061	1113	1165	1217	1269	1321	1373	1425	1477	1529	1581	1633	1684	1736	1788	1840	1892	1944	1996
12 Rabbit	542	594	646	698	750	802	854	906	958	1010	1062	1114	1166	1218	1270	1322	1374	1426	1478	1530	1582	1634	1685	1737	1789	1841	1893	1945	1997
13 Cane	543	595	647	699	751	803	855	907	959	1011	1063	1115	1167	1219	1271	1323	1375	1427	1479	1531	1583	1635	1686	1738	1790	1842	1894	1946	1998
1 Flint	544	596	648	700	752	804	856	908	960	1012	1064	1116	1168	1220	1272	1324	1376	1428	1480	1532	1584	1636	1687	1739	1791	1843	1895	1947	1999
2 House	545	597	649	701	753	805	857	909	961	1013	1065	1117	1169	1221	1273	1325	1377	1429	1481	1533	1585	1637	1688	1740	1792	1844	1896	1948	2000
3 Rabbit	546	598	650	702	754	806	858	910	962	1014	1066	1118	1170	1222	1274	1326	1378	1430	1482	1534	1586	1638	1689	1741	1793	1845	1897	1949	2001
4 Cane	547	599	651	703	755	807	859	911	963	1015	1067	1119	1171	1223	1275	1327	1379	1431	1483	1535	1587	1639	1690	1742	1794	1846	1898	1950	2002
5 Flint	548	600	652	704	756	808	860	912	964	1016	1068	1120	1172	1224	1276	1328	1380	1432	1484	1536	1588	1640	1691	1743	1795	1847	1899	1951	2003
6 House	549	601	653	705	757	809	861	913	965	1017	1069	1121	1173	1225	1277	1329	1381	1433	1485	1537	1589	1641	1692	1744	1796	1848	1900	1952	2004
7 Rabbit	550	602	654	706	758	810	862	914	966	1018	1070	1122	1174	1226	1278	1330	1382	1434	1486	1538	1590	1642	1693	1745	1797	1849	1901	1953	2005
8 Cane	551	603	655	707	759	811	863	915	967	1019	1071	1123	1175	1227	1279	1331	1383	1435	1487	1539	1591	1643	1694	1746	1798	1850	1902	1954	2006
9 Flint	552	604	656	708	760	812	864	916	968	1020	1072	1124	1176	1228	1280	1332	1384	1436	1488	1540	1592	1644	1695	1747	1799	1851	1903	1955	2007
10 House	553	605	657	709	761	813	865	917	969	1021	1073	1125	1177	1229	1281	1333	1385	1437	1489	1541	1593	1645	1696	1748	1800	1852	1904	1956	2008
11 Rabbit	554	606	658	710	762	814	866	918	970	1022	1074	1126	1178	1230	1282	1334	1386	1438	1490	1542	1594	1646	1697	1749	1801	1853	1905	1957	2009
12 Cane	555	607	659	711	763	815	867	919	971	1023	1075	1127	1179	1231	1283	1335	1387	1439	1491	1543	1595	1647	1698	1750	1802	1854	1906	1958	2010
13 Flint	556	608	660	712	764	816	868	920	972	1024	1076	1128	1180	1232	1284	1336	1388	1440	1492	1544	1596	1648	1699	1751	1803	1855	1907	1959	2011
1 House	557	609	661	713	765	817	869	921	973	1025	1077	1129	1181	1233	1285	1337	1389	1441	1493	1545	1597	1649	1700	1752	1804	1856	1908	1960	2012
2 Rabbit	558	610	662	714	766	818	870	922	974	1026	1078	1130	1182	1234	1286	1338	1390	1442	1494	1546	1598	1650	1701	1753	1805	1857	1909	1961	2013
3 Cane	559	611	663	715	767	819	871	923	975	1027	1079	1131	1183	1235	1287	1339	1391	1443	1495	1547	1599	1651	1702	1754	1806	1858	1910	1962	2014
4 Flint	560	612	664	716	768	820	872	924	976	1028	1080	1132	1184	1236	1288	1340	1392	1444	1496	1548	1600	1652	1703	1755	1807	1859	1911	1963	2015
5 House	561	613	665	717	769	821	873	925	977	1029	1081	1133	1185	1237	1289	1341	1393	1445	1497	1549	1601	1653	1704	1756	1808	1860	1912	1964	2016
6 Rabbit	562	614	666	718	770	822	874	926	978	1030	1082	1134	1186	1238	1290	1342	1394	1446	1498	1550	1602	1654	1705	1757	1809	1861	1913	1965	2017
7 Cane	563	615	667	719	771	823	875	927	979	1031	1083	1135	1187	1239	1291	1343	1395	1447	1499	1551	1603	1655	1706	1758	1810	1862	1914	1966	2018
8 Flint	564	616	668	720	772	824	876	928	980	1032	1084	1136	1188	1240	1292	1344	1396	1448	1500	1552	1604	1656	1707	1759	1811	1863	1915	1967	2019
9 House	565	617	669	721	773	825	877	929	981	1033	1085	1137	1189	1241	1293	1345	1397	1449	1501	1553	1605	1657	1708	1760	1812	1864	1916	1968	2020
10 Rabbit	566	618	670	722	774	826	878	930	982	1034	1086	1138	1190	1242	1294	1346	1398	1450	1502	1554	1606	1658	1709	1761	1813	1865	1917	1969	2021
11 Cane	567	619	671	723	775	827	879	931	983	1035	1087	1139	1191	1243	1295	1347	1399	1451	1503	1555	1607	1659	1710	1762	1814	1866	1918	1970	2022
12 Flint	568	620	672	724	776	828	880	932	984	1036	1088	1140	1192	1244	1296	1348	1400	1452	1504	1556	1608	1660	1711	1763	1815	1867	1919	1971	2023
13 House	569	621	673	725	777	829	881	933	985	1037	1089	1141	1193	1245	1297	1349	1401	1453	1505	1557	1609	1661	1712	1764	1816	1868	1920	1972	2024
1 Rabbit	570	622	674	726	778	830	882	934	986	1038	1090	1142	1194	1246	1298	1350	1402	1454	1506	1558	1610	1662	1713	1765	1817	1869	1921	1973	2025

December 29, 1549 Julian. It is sometimes convenient to be able to make the same calculation in the day count. Figure 6 is provided for this purpose.

What is the distance between 13 Monkey and 9 Cane? From Figure 6 we find the day Monkey (the eleventh day), then read across row 11 to the coefficient 13. Because the columns are arranged by 20s, it is easy to see that 13 Monkey is the 91st day of the day count. Similarly, locating 9 Cane (the 113th day), one finds that the two dates are 22 days apart. Note that the rows of coefficients in Figure 6 follow the same order as the top row of Figure 3: 1, 8, 2, 9, 3, 10, 4, 11, 5, 12, 6, 13, 7. Thus, to find 13 Monkey in Figure 3, we must begin with 6 Ik (Wind) in the upper-left corner and count down the first column by 13s until the day Monkey is reached in the tenth row (thus 6, 7, 8, 9, 10, 11, 12, 13, 1, 2). Then one counts across the tenth row in uinal order, beginning with 2, until one reaches 13 in the tenth column (thus 2, 9, 3, 10, 4, 11, 5, 12, 6, 13). Reading from the left and bottom margins, one can identify this as 9 Yax Tikal, and counting from AZ in column 10 (the Aztec New Year), and reading the Nahuatl month from the very bottom of the figure, one gets the corresponding date of 3 Izcalli Aztec. Because it falls between the 100th and 265th days of the Tikal year, 13 Monkey occurs only once in that year, but because it falls in the first Aztec month, it will occur again in the fourteenth month, on 3 Tepeilhuitl Aztec.

The designation of a day in the day count automatically refers to a date within the cycle of 260 days. It may, however, refer to a date in some other cycle as well. We have seen how such a designation may be used to name years. The day that initiates Figure 3, for example, also initiates the year 6 Ik in the calendar of Tikal. As the years pass, the numeral coefficient of this designation will advance by 1 each year, starting over when it reaches 13. The day name will advance one year bearer at the same time, following the sequence Ik, Manik, Eb, Caban (Wind, Deer, Grass, and Quake, in Figure 4). Because the year has four possible year bearers and thirteen possible numbers, its name will not repeat until 52 years (4 x 13) have gone by. This cycle of 18,980 days has come to be called a *calendar round*. It was called *xiuhmolpilli* in Nahuatl and *hunab* in Maya, and, of course, it had other names in other languages. Europeans have often called it the native "century."

All of the peoples of Anahuac used the calendar-round system of dating. Just as they differed about when the year began, so they also differed about when the calendar round began. Each calendar assumed that one of the four year bearers was the senior one, and in most instances, the occurrence of that year bearer with the coefficient 1 signaled the beginning of the calendar round.

This was a date that rarely occurred more than once in a lifetime, and it was treated with corresponding seriousness and ceremony. All fires were extinguished, and new fire was ritually ignited. Often new temples were dedicated, or other special events were scheduled to coincide with the New Fire ceremonies of the calendar round. In the Tikal calendar, the calendar round began on 1 Caban. The

Figure 6. THE TRECENAS OF THE DAY COUNT

		1	2	3	4	5	6	7	8	9	10	11	12	13
Alligator	1	1	8	2	9	3	10	4	11	5	12	6	13	7
Wind	2	2	9	3	10	4	11	5	12	6	13	7	1	8
House	3	3	10	4	11	5	12	6	13	7	1	8	2	9
Iguana	4	4	11	5	12	6	13	7	1	8	2	9	3	10
Serpent	5	5	12	6	13	7	1	8	2	9	3	10	4	11
Death	6	6	13	7	1	8	2	9	3	10	4	11	5	12
Deer	7	7	1	8	2	9	3	10	4	11	5	12	6	13
Rabbit	8	8	2	9	3	10	4	11	5	12	6	13	7	1
Water	9	9	3	10	4	11	5	12	6	13	7	1	8	2
Dog	10	10	4	11	5	12	6	13	7	1	8	2	9	3
Monkey	11	11	5	12	6	13	7	1	8	2	9	3	10	4
Grass	12	12	6	13	7	1	8	2	9	3	10	4	11	5
Cane	13	13	7	1	8	2	9	3	10	4	11	5	12	6
Jaguar	14	1	8	2	9	3	10	4	11	5	12	6	13	7
Eagle	15	2	9	3	10	4	11	5	12	6	13	7	1	8
Buzzard	16	3	10	4	11	5	12	6	13	7	1	8	2	9
Quake	17	4	11	5	12	6	13	7	1	8	2	9	3	10
Flint	18	5	12	6	13	7	1	8	2	9	3	10	4	11
Rain	19	6	13	7	1	8	2	9	3	10	4	11	5	12
Flower	20	7	1	8	2	9	3	10	4	11	5	12	6	13

year 6 Ik of Figure 3 can therefore be calculated to be the sixth year of the Tikal calendar round, which began in 1544.

Naturally, the Indians were thoroughly aware of when the calendar round began, and perhaps equally naturally, they usually failed to mention it. It is consequently very difficult to pin down the beginning point in most of the native systems. Those that can be specified are listed under *calendar round* in Chapter 5. We can nevertheless reconstruct complete calendar-round dates in all sixty of the calendars listed in Figure 3, even though we may not know when the calendar round began in a particular system. (The sixty calendars include thirteen of the "other" calendars whose new years are definitely known; an additional thirty calendars are inadequately documented.)

A complete calendar-round date requires specification of the year and of the day within it. The names of the years were commonly written by designating a day in the day count and marking it with a special sign indicating that it was the name of a year. I have indicated the presence of these year signs by the notation *anno*.

Figure 7. THE SIGNS FOR THE YEAR (*Anno*)

| Olmec | Zapotec | Teotihuacan | Xochicalco | Tilantongo |

Such a year name may then be followed with the name and number of the particular day in the 260-day day count. Because the year is 105 days longer than the day count, such a date remains ambiguous until we specify whether it is the first or second occurrence of these 105 days that is intended. That is normally done by adding the date in the year count, specifying the name or number of the month, and (redundantly) the number of the day within it. For the middle 155 days of the year, this year-count date is itself redundant, because those days occur only once in the given year, and it was often omitted, perhaps as a mark of sophistication and elegance. (It should be noted that year dates of this type were rarely used by the classic Maya, who preferred to date in terms of *tuns* and *katuns*.)

The astonishing thing about the variations displayed by the different calendars is their restriction to one quarter of the year. Not only are their New Year days confined to a single limited series of months, but they are limited to a single set of sequential year bearers. There would appear to be no inherent reason that some calendar could not have begun elsewhere in the year 1550, say on Men, or Cib, or Caban: we can only observe that none does.

The array in Figure 3 suggests that all the calendars had a common origin and, further, that they have been altered rarely, grudgingly, and minimally, principally in intervals of either 1 day or 20 days. Perhaps there is some constraint on calendrical variation that we have not yet identified. At the very least, a deeply rooted historical conservatism has operated upon it.

The earliest evidence we have, probably dating to the seventh century B.C., indicates the presence among the Olmec of a fully developed calendar round that is obviously cognate with and almost certainly ancestral to all the other calendars considered here (see chapter 3). A number of speculations have been offered concerning its origins, all of them resting upon the numerological features of the system and pointing with greater or lesser statistical fit to possible correlates in physiology, climatology, and astronomy (see chapter 4). No historical validation of these proposals is possible: we simply do not have the data.

By contrast, a combination of archaeology and calendrics itself appears to document empirically the quasi-evolutionary development from a single Olmec original of the forty-eight calendars we can examine. The steps by which this differentiation took place are strongly suggested by the data at hand, and reconstructing them tells us something about both the calendric system and the matrix of cultural history in which it grew (chap. 3–4). The essential data for such a reconstruction are also summarized in Figure 3. The correlational dates from which they derive are the subject of the next chapter.

CHAPTER 2

CALENDRICAL ANNALS

This chapter is a survey of the evidence correlating the specific calendars of Anahuac with one another and with the European calendars. Arranged in chronological order from the earliest to the latest dates, it is composed of hieroglyphic and linguistic texts, translated and annotated so as to provide as much as possible of the raw material from which any reasonable chronology must be derived.

Each text or group of related texts is introduced by a calendrical heading, giving the focal date in the Julian (J) or Gregorian (G) calendar, in the Olmec-Mayan Long Count (e.g., 11.16.0.0.0), and in the Mayan Tikal (T) or Mayapan (M) calendar round. The European dates are normally cited as Julian before 1584 and Gregorian thereafter. The Mayan dates are normally given as Tikal before 1539 and Mayapan thereafter. The second and succeeding lines of the heading summarize the evidence provided by the given text: European date, Long Count, day count and year count, together with the identification of the native calendar or calendars involved (e.g., Tarascan, Yucuñudahui). The day count and year count are cited in the native language implied by the text or in the English translations of day names from Figure 4. The letter coding of month names follows Figure 3. Translations of the day and month names are to be found under the relevant language listings in Chapter 5.

A few citations are included in the date list that are reconstructive or interpretative rather than empirical. These are preceded by an asterisk (*). They are included because of their inherent interest or their utility in the interpretation of the remainder of the corpus.

Most of the native correlational dates are given only in the calendar round. Their placement in the Long Count or in European years is thus necessarily provisional. They are correct to within 52 years or some (often unknown) multiple thereof. Errors of placement in absolute time may be quite material to the interpretation of the events dated, but they are totally irrelevant to the correlational significance of the dates for the reconstruction of the calendar.

The placement of the Long Count itself in European time continues to be debated (see Long Count section in Chapter 5). The solution adopted here invokes the 584,283 correlation constant quite simply because it is the only one that fits the data that follow. Although I am prepared to concede that serious chronological problems remain in the archaeology and astronomy of Middle America, I am convinced that ethnohistorically the debate is over—or should be.

It is manifestly impossible and would be counterproductive to attempt to cite all of the correlational dates that have been advanced in the now voluminous literature touching upon the chronology of Middle America. Erroneous and aberrant correlations are useful to us only if they suggest systemic alternative construals of the calendars themselves. It is not always possible to reconstruct the reasons for a date that does not fit. I have tried systematically to be inclusive without being redundant and to include enough problematic dates to represent extensively the kinds of difficulties they present.

What follows, then, is a date list comprising the calendrical annals of Anahuac. Not surprisingly, it is not only parallel to, but even constitutes a kind of skeletal outline of, the general culture history of the region.

THE OLMEC CALENDAR

679 B.C.   2 IX J   6.3.10.9.0   2 Ahau 3 Ceh T
                                         2 Lord 19 F.   Olmec

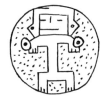

Al. 2 Lord (Olmec)

(Earspool, Museo de Cuicuilco)

If Caso and Bernal (1965:871) are right about the dating of Cuicuilco and if the inhabitants of that site used dot numerals instead of the digits of Oaxaca and Chiapas and if this figure is read as an Olmec year, then this is the earliest calendar-round date known. It may be the earliest day-count date in any case. The heading is to be read: reference date 679 B.C., 2 September Julian, 6.3.10.9.0 2 Ahua 3 Ceh Tikal, is documented by 2 Lord 19 F. (6th month or Ceh) Olmec. The origins of the more important calendars are indicated by superheadings, as in this case for the Olmec calendar.

667 B.C.   30 VIII J   6.4.2.12.0   1 Ahau 3 Ceh T
                                       1 Lord 19 F.   Olmec

A1. ?Water
A2. Jaw ?1
A3.
A4.
A5. ?New Fire
A6. year
A7. 1 Lord
A8. month G. (Olmec)
A9.
A10.

(Tapijulapa Ax, Simojovel;
Coe 1965a:747, Fig. 17;
Caso 1965b:932 Fig. 1)

This inscription documents the fact that the Olmec had Type V year bearers. The reading of A5 as New Fire makes 1 Lord the first year of the Olmec calendar round. The archaeological dating is vague and rests on the possibly Late Olmec style of the carving, but it is probably conservative to place it as earlier than Monte Alban I. The calendrical placement of the date rests on the correlational date at 563 B.C.

THE ZAPOTEC CALENDAR

594 B.C.   28 I J   6.7.16.2.9   8 Muluc 7 Uo T
                                    8 Niça 6 O.   Zapotec

A1. year
A2. Wind
A3. 4
A4.
A5.
A6.
A7. Water
A8. 8 (Zapotec)

(Stela 12, Monte Alban I;
Caso 1965b:933, Fig. 3)

The date is complete without a month designation, because 8 Water falls only once in a year 4 Wind. The archaeological dating is given by Caso as around 600 B.C. (see however chapter 4). The calendrical placement of the date depends on the correlational date at A.D. 1465 (below). The event commemorated is presumably civic or dynastic.

563 B.C.   4 VIII J   6.9.8.2.0   1 Ahau 3 Ceh T
                                            1 Làо 17 F.   Zapotec
                                            1 Lord 19 F.   Olmec

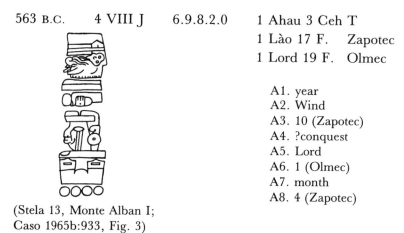

                                            A1. year
                                            A2. Wind
                                            A3. 10 (Zapotec)
                                            A4. ?conquest
                                            A5. Lord
                                            A6. 1 (Olmec)
                                            A7. month
                                            A8. 4 (Zapotec)

(Stela 13, Monte Alban I;
Caso 1965b:933, Fig. 3)

This date correlates the Olmec and Zapotec calendars. The glyphs at A2 and A5 are Olmec, and the date is the name day of the Olmec year and the beginning of the Olmec calendar round, but the date is given in the Zapotec calendar. The verb (A4) is read by Whittaker (1980:28, Fig. 6) as *conquest*. The identification of the Olmec calendar round is given in the date at 667 B.C. The archaeological and calendrical placement of the present date are the same as those discussed at 594 B.C.

528 B.C.   12 II J   6.11.3.2.11   2 Chuen 19 Zip T
                                            2 Pilloo 18 P.   Zapotec

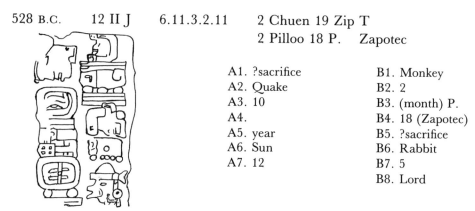

               A1. ?sacrifice   B1. Monkey
               A2. Quake   B2. 2
               A3. 10   B3. (month) P.
               A4.   B4. 18 (Zapotec)
               A5. year   B5. ?sacrifice
               A6. Sun   B6. Rabbit
               A7. 12   B7. 5
                                           B8. Lord

(Stela 17, Monte Alban I;
Whittaker 1980:50, Fig. 14)

A1. Wind	B1. Monkey
A2. 1 Jaguar	B2. 2
A3. ?Foot	B3. month
A4.	B4. 14 (Zapotec)
A5.	B5.
A6. ?capture	B6.
	B7.

(Stela 15, Monte Alban I;
Whittaker 1980:50, Fig. 14)

The two inscriptions refer to the same date in two different ways and to an event apparently involving the capture and sacrifice of 10 Quake, 5 Rabbit, 1 Jaguar, and perhaps (?) Wind and 5 Foot. Stela 17 gives the glyph name of the fourteenth Zapotec month (B3); Stela 15 gives only its number (B4). Whittaker (1980:28, Fig. 5) reads the former as *solar trecena*, but that requires relocating and renaming the days and invoking a different calendar. The month date has to be specified because the fourteenth month contains the same days as the first. The archaeological and calendrical placement of this date involves the same considerations as those discussed under 594 B.C.

THE LONG COUNT

*355 B.C.   13 VI J   6.19.19.0.0   1 Ahau 3 Ceh T

On this date, the ending of the Olmec calendar round and of the last tun in the sixth baktun coincide, making this a likely date for the inauguration of the Long Count. The co-occurrence of the beginning of the calendar round with the tun repeats each 936 years.

THE TIKAL CALENDAR

*236 B.C.   16 IX J   7.6.0.0.0   11 Ahau 8 Cumku T

This date was calculated by Teeple and published by Thompson (1932:370) as the probable date of the institution of the Tikal calendar, based on the Mayan mode of calculation of corrections to the solar year.

229 B.C.   3 VII J 7.6.6.16.3   11 Akbal 16 Kankin T
                                      11 Pée 10 I.   Zapotec

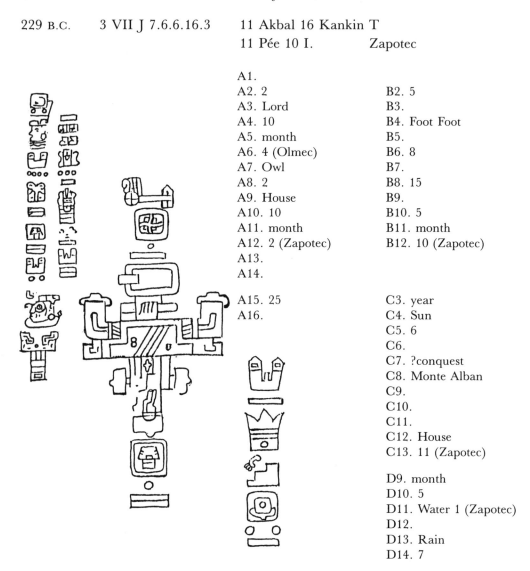

A1.
A2. 2   B2. 5
A3. Lord   B3.
A4. 10   B4. Foot Foot
A5. month   B5.
A6. 4 (Olmec)   B6. 8
A7. Owl   B7.
A8. 2   B8. 15
A9. House   B9.
A10. 10   B10. 5
A11. month   B11. month
A12. 2 (Zapotec)   B12. 10 (Zapotec)
A13.
A14.

A15. 25   C3. year
A16.   C4. Sun
                               C5. 6
                               C6.
                               C7. ?conquest
                               C8. Monte Alban
                               C9.
                               C10.
                               C11.
                               C12. House
                               C13. 11 (Zapotec)

                               D9. month
                               D10. 5
                               D11. Water 1 (Zapotec)
                               D12.
                               D13. Rain
                               D14. 7

(Tablet 14, Mound J,
Monte Alban II;
Caso 1965b:937, Fig. 12)

The central statement of the inscription (C3–C13) probably records the conquest of Monte Alban. The date is complete without the month designation, because 11 House occurs only once in a year 6 Sun. Whittaker's (1980:37–39) reading of the month sign as a trecena glyph requires changing the identification of the days and introducing new year bearers and a new calendar, none of which is necessary.

209 B.C.   18 X J   7.7.7.8.14   12 Ix 2 Uo   T
                                    12 Pêche 1 O. Zapotec

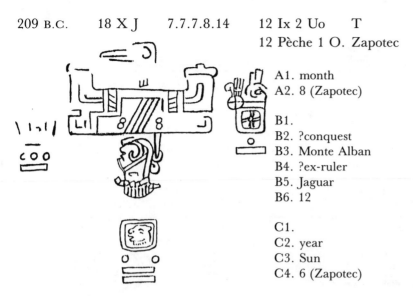

A1. month
A2. 8 (Zapotec)

B1.
B2. ?conquest
B3. Monte Alban
B4. ?ex-ruler
B5. Jaguar
B6. 12

C1.
C2. year
C3. Sun
C4. 6 (Zapotec)

(Tablet 10, Mound J,
Monte Alban II;
Caso 1965b:938, Fig. 13)

This date may record the conquest of Monte Alban and perhaps pictures the bearded conqueror. The date is complete without the month designation because 12 Jaguar occurs only once in a year 6 Sun.

THE KAMINALJUYU CALENDAR

147 B.C.   10 XI J   7.10.10.8.2   8 Ik 0 Zotz'   T
                                      8 Wind 0 Q. Teotihuacan
                                      8 Wind 16 P. Olmec
                                      8 Wind 15 P. Kaminaljuyu

                                      7 Anno? Wind Kaminaljuyu

                                      Anno? 1 Lord Olmec

                                      8 Anno? Wind Teotihuacan

(Stela 10, Kaminaljuyu; Ayala 1984:219)

Although there are a number of other day signs in this composition, I believe there are three dates. They are not accompanied by recognizable year signs, but I believe that years are intended. Two different writing systems and three different calendars are involved: Kaminaljuyu, Teotihuacan, and Olmec. The date at the upper left is probably 7 Wind Kaminaljuyu, and the glyph is probably the Teotihuacan glyph for that day (it is difficult to read). The central date is 1 Lord Olmec, written in Olmec style. If there were a year sign, it would be the headdress of the figure, which is unfortunately destroyed. The third date is 8 Wind Teotihuacan, written this time with the Olmec glyph for the day (cf. 563 B.C.). The result is a correlational date: 7 Wind Kaminaljuyu = 8 Wind Teotihuacan = 1 Lord Olmec. This confirms the placement of the Olmec year in 563 B.C. and of the Teotihuacan year in A.D. 861 (q.v.) and establishes the existence and placement of the Kaminaljuyu calendar (q.v., chapter 5) at II.B.0.

The text below begins (A1-B1) with the distance from the Olmec to the Teotihuacan New Year, 15 uinals. Though they are strikingly similar to Mayan glyphs, I believe these are the Olmec glyphs of this time. This was very close to the date of separation of the Olmec and Mayan calendars and perhaps of their writing systems as well. Although the suffix to the third date is commonly taken to be a day sign indicator, I suggest that its flamelike design may include the idea of the New Fire of the (Olmec) calendar round, and the double cartouches on the Wind glyphs may have been a year sign (see A.D. 47).

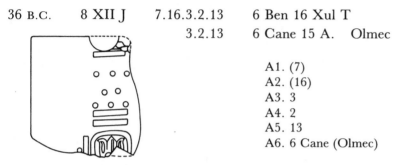

36 B.C.   8 XII J   7.16.3.2.13   6 Ben 16 Xul T

                     3.2.13   6 Cane 15 A.   Olmec

A1. (7)
A2. (16)
A3. 3
A4. 2
A5. 13
A6. 6 Cane (Olmec)

(Stela 2, Chiapa de Corzo;
Marcus 1976:51, Fig. 6)

Discovered by Lowe (1962:194) on the surface of Mound 5b at Chiapa de Corzo, Stela 2 contains what appears to be a fragmentary Long Count date, the earliest known. The glyph for the day Cane (A6) is neither Zapotec nor Mayan and may therefore be presumed to be Olmec.

32 B.C.   3 IX J   7.16.6.16.18   6 Etz'nab 1 Uo T
                          7.16.6.16.18   6 Flint 2 O.   Olmec

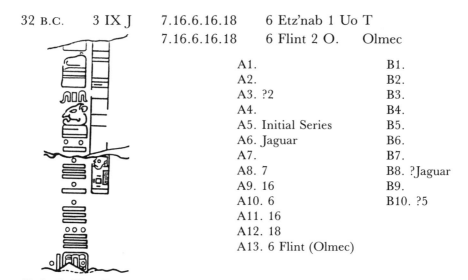

A1.   B1.
A2.   B2.
A3. ?2   B3.
A4.   B4.
A5. Initial Series   B5.
A6. Jaguar   B6.
A7.   B7.
A8. 7   B8. ?Jaguar
A9. 16   B9.
A10. 6   B10. ?5
A11. 16
A12. 18
A13. 6 Flint (Olmec)

(Stela C, Tres Zapotes;
Stirling 1943:14;
Marcus 1976:52, Fig. 7)

This is the earliest complete Long Count date. The glyph for the day Flint is neither Zapotec nor Mayan and hence is very likely Olmec.

(A.D. dates from here on.)

37   4 III J   7.19.15.7.12   12 Eb 0 Ceh T
                     7.19.15.7.12   12 Jaw   Olmec

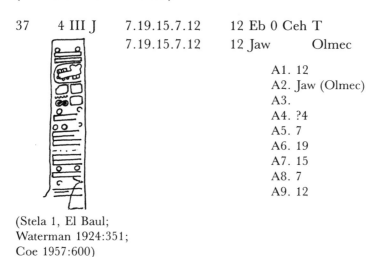

A1. 12
A2. Jaw (Olmec)
A3.
A4. ?4
A5. 7
A6. 19
A7. 15
A8. 7
A9. 12

(Stela 1, El Baul;
Waterman 1924:351;
Coe 1957:600)

The glyph for the day Jaw is Olmec (although it bears a close resemblance to the
corresponding Mayan glyph; see A.D. 320, A10). Another cycle 7 date on Stela 2,
Columba, reported by Lehmann (1926:176) and illustrated in Coe (1957:604, Fig.
6) supports the antiquity of this one, though only the baktun coefficient (7) can be
read. It is pictured here under its more recent designation as Abaj Takalik Stela 2.

A1. Initial Series
A2.
A3. 7

(Abaj Takalik, Stela 2;
Graham, Heizer and Shook
1978:103, Plate 2)

47   11 XII J   8.0.6.6.6   7 Cimi 19 Yaxkin T

7 Death   Izapa

A. Anno? 7 Death

(Izapa, Miscellaneous Monument
60; Norman 1976:282, Fig. 5.83)

If this date is read as a year name, it probably identifies Calendar B (Figure 3) as the calendar of Izapa. That, at least, is the only known calendar with Type I year bearers. It is not, however, accompanied by a known year sign.

126   4 VI J   8.4.5.17.11   7 Chuen 14 Kayab T
                      8.4.5.17.11   Olmec

A1.		B1.	
A2.		B2.	
A3.	8	B3.	8
A4.	4	B4.	3
A5.	5	B5.	2
A6.	17	B6.	10 (15?)
A7.	11	B7.	11
A8.		B8.	
A9.		B9.	
A10.		B10.	
A11.		B11.	

(Abaj Takalik, Stela 5;
Graham, Heizer and Shook
1978; 104, Plate 3)

This is not a correlational date. The lack of glyphs for baktun, katun, etc., suggests that it is Olmec.

162   13 III J   8.6.2.4.17   8 Caban 0 Kankin T
                       8.6.2.4.17   8 Quake (16 H.)   Olmec

A1. Initial Series
A2. month H.
A3.
A4.
A5. 8
A6. 6
A7. 2
A8. 4
A9. 17
A10. 8 Quake (Olmec)

(Tuxtla Statuette;
Marcus 1976:57, Fig. 9)

The glyph for the day Quake (A10.) is neither Zapotec nor Mayan and may thus be presumed to be Olmec.

*176   24 II J   8.6.16.7.14   9   Ix 7 Mac Tikal

Calculated from a ring number in *Dresden* (70), this may be the earliest date indicated in the codex that was contemporaneous when first recorded (Satterthwaite 1965:615). The month reached is not written.

199   7 X J   8.8.0.7.0   3 Ahau 13 Xul T
                                3 Ahau 12 Xul Campeche

A1. Initial Series
A2. 12 Xul
A3. G5 of the Lords of the Nights
A4. (?)
A5. 17 lunations had ended
A6. 3 Ahau
A7. he let
A8. his blood
A9. Bac T'ul
A10. title
A11. of or from
A12. Emblem Glyph

Hauberg Stela (Schele and Miller 1986:191)

The Long-Count position of this calendar-round date is fixed by its lunar notation and makes this the earliest known Mayan date. Such a placement requires reading A2 as 13 Xul Tikal, but, as written, the date is in the Campeche calendar. It is 466 earlier than the next Campeche date (see 665), and 369 years before I believe the Campeche calendar was invented (see Chapter 4). I believe the 12 is a scribal error for 13. The lack of provenience for the stela precludes a geographic assessment of the probability that it may be "Campeche." The atypical order of the dating elements associates it with Quirigua and Copan (Schele and Miller 1986:191). This inscription is the earliest representation of the *novena* (A3), the cycle of the Lords of the Night, which begins with G1 on every even tun.

292   6 VII J   8.12.14.8.15   13 Men 3 Zip T
                      8.12.14.8.15   Tikal

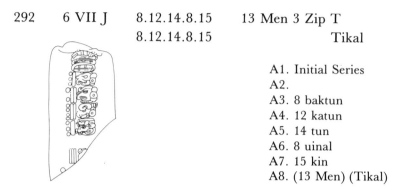

A1. Initial Series
A2.
A3. 8 baktun
A4. 12 katun
A5. 14 tun
A6. 8 uinal
A7. 15 kin
A8. (13 Men) (Tikal)

(Stela 29, Tikal;
Shook 1960;
Marcus 1976:58, Fig. 10)

This is the earliest provenienced and contemporaneous Mayan Long-Count date. The day and month are lacking, so it cannot be definitively assigned to the Tikal calendar by day glyphs, but the period glyphs are early Mayan.

320   14 IX J   8.14.3.1.12   1 Eb (G5) 0 Yaxkin T
                      8.14.3.1.12   1 Eb (G5) 0 Yaxkin Tikal

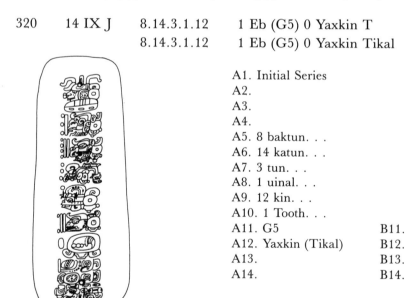

A1. Initial Series
A2.
A3.
A4.
A5. 8 baktun. . .
A6. 14 katun. . .
A7. 3 tun. . .
A8. 1 uinal. . .
A9. 12 kin. . .
A10. 1 Tooth. . .
A11. G5   B11.
A12. Yaxkin (Tikal)   B12.
A13.   B13.
A14.   B14.

(Leyden Plaque;
Marcus 1976:60, Fig. 11)

This is the earliest Long-Count date carrying an unequivocal designation of the Mayan month. The glyph for Tooth is notably similar to the corresponding Olmec glyph.

THE YUCUÑUDAHUI CALENDAR

426   29 I J   8.19.10.0.0   9 Ahau 3 Muan T
                                 9 Flower 7 J.   Yucuñudahui

	A1. year Cane 3	B1. year Cane 2
	A2. Flower 9	B2. ?Wind 7
		(Yucuñudahui)

(Stone 1, Tomb 1,
Yucuñudahui;
Moser 1977:145, Fig. 69)

The dates presumably refer to the accession (or birth) and death of the occupant of the tomb (A.D. 426–74). The year is marked with both the Ñuiñe and Teotihuacan year signs, and the numerals and day glyphs (A1, B1, A2) are also Ñuiñe. Paddock (1970:8) dates the tomb archaeologically to around A.D. 400. The calendrical position of the Yucuñudahui calendar is established by the correlational date at A.D. 768. It should be noted that 9 Flower occurs only once in an initially named Yucuñudahui year 3 Cane, so this is a complete calendar-round date, as is the second date, if it is correctly read.

468   8 IV J   9.1.12.14.10   1 Oc 3 Uayeb T
                 9.1.12.14.10   Olmec

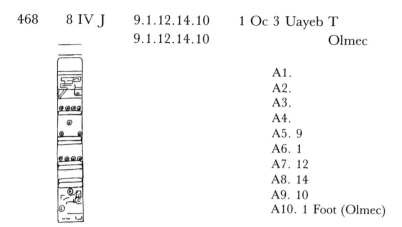

A1.
A2.
A3.
A4.
A5. 9
A6. 1
A7. 12
A8. 14
A9. 10
A10. 1 Foot (Olmec)

(Stela 6, Cerro de las
Mesas; Stirling 1943:
Fig. 11b; Coe 1965a:703,
Fig. 21)

The significance of the date is unknown, but whereas the day cited is an Olmec
year bearer, it is not the designation of a year.

533   4 VI J   9.4.18.16.8   9 Lamat 11 Zotz' T
                 9.4.18.16.8   9 Rabbit (12 Q.) Olmec

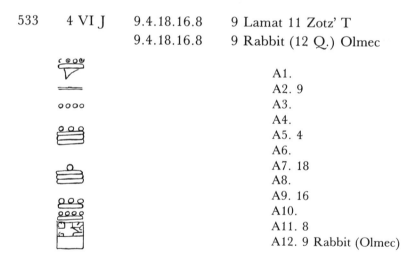

A1.
A2. 9
A3.
A4.
A5. 4
A6.
A7. 18
A8.
A9. 16
A10.
A11. 8
A12. 9 Rabbit (Olmec)

(Stela 8, Cerro de las
Mesas; Stirling 1965:
727, Fig. 15; Coe 1965a:
703, Fig. 21)

I believe the glyph for Rabbit to be an alternative form of that day sign in Olmec,
Maya, Zapotec, Teotihuacan, and Yucuñudahui.

565   26 X J   9.6.11.14.0   11 Ahau 3 Ceh T
                                11 Lord 19 F.   Olmec

A2.	
A3. House (Yucuñudahui)	
A4.	
B1. Eagle	
B2. Flint (Yucuñudahui)	
B3. Sun	
B4. 9 (Zapotec)	
B5. year	
B6. Lord	
B7. 11 (Olmec)	
C3. Wind	
C4. 6 (Zapotec)	

(Stone cover, outer face,
Tomb 104, Monte Alban IIIA–
IIIB; Whittaker 1980:35,
Fig. 8)

Three dates are specified and two are alluded to without coefficients. Three calendars and three writing systems are involved:

Date	Olmec	Zapotec	Yucuñudahui
515	(13 Foot)	9 Sun	(7 Rabbit)
525	(11 Serpent)	6 Wind	(4 Flint)
565	11 Lord	(7 Wind)	(5) Flint

Presumably the reference is to the birth, accession, and death of the occupant of the tomb. Because the order of reading is uncertain and these are floating calendar-round dates, it is impossible to specify which date refers to which event. The omission of coefficients falls short of giving us a direct correlation of the calendars involved.

665   9 III J   9.11.12.10.14   9 Ix 17 Pop T
                                      9 Ix 16 Pop Campeche

A1. 9? Ix
B1. 16 Pop
C1. T59.(?).60?
D1. 17 T528.116
E1. 12 (?)

(East ring, ballcourt, Uxmal;
Morley 1920:514, Fig. 75a)

A1. 9? Ix
B1. 17 Pop
C1. T59.(?).60?
D1. 12 (?)

(West ring, ballcourt, Uxmal;
Morley 1920:514, Fig. 75b)

The date on the west ring of the ballcourt is in the Tikal calendar, but that on the east ring is the earliest date in the Campeche calendar. We cannot rule out the possibility that this is in fact the earliest date in the Mayapan calendar, because (1) we do not know whether it was still using zero dating of the days of the month, and (2) we do not know whether it was a terminally or an initially named system. From the *Dresden Codex* 25–28b,c, we know that a Yucatecan calendar existed that used Type III name days, and it seems most likely that this was the Campeche calendar. If so, we may infer that it had abandoned zero dating (like the calendar of Palenque) and was identical to the Mayapan calendar except for its retention of terminal naming. This chain of inference is our only recourse, because the known dates in the Campeche calendar do not include an unambiguous one for the ending or beginning of a month. This would place the Campeche calendar as IV.N.20 and makes it a likely derivative of Palenque and ancestor to the calendar of Mayapan.

692   13 III J   9.13.0.0.0   8 Ahau 8 Uo T
                                 8 Akau   Tikal

uaxac ahau	[Katun] 8 Lord [Tikal]
uchci	It occurred.
chicanpahal chi ch'een ytza uchc	The revelation of Chichen Itza occurred.
u chicpahal tzucub te	It was the revealing of the Sacred Grove
sian can lae	That is Born of Heaven.
(*Tizimin* 21–26)	(Edmonson 1982:21–26)

This is the earliest date in the *First Chronicle* (Edmonson 1985), and it is the earliest possible historical date in any of the Colonial Yucatecan sources.

692   13 III J   9.13.0.0.0   8 Ahau 8 Uo T
                   0.0.1.9.2   13 Ik 20 Mol Palenque

C9. 13 Ik
D9. End Mol

(Lounsbury 1980:101, Fig. 1)

This mythological date from the Temple of the Cross Tablet C9–D9 marks the first appearance of the Palenque calendar. The tablet belongs to the katun cited, but the date corresponds either to 0.0.1.9.2 or 13.0.1.9.2, depending on how one wishes to refer to the beginning of the Mayan era. The expected or acceptable date in the Tikal calendar would be 13 Ik 0 Ch'en. This is the earliest documentable instance of the abandonment of zero counting of the days of the month anywhere in Middle America, though this is a feature of virtually all Postclassic calendars.

733   24 VI J   9.15.1.15.18   6 Etz'nab 1 Mol T
                                  6 Etz'nab 19 Yaxkin Campeche

A1. 6 Etz'nab
B1. completed
C1. u-hel-te (the changing)
D1. yax-kin(ne) (of Yaxkin)
E1. (?)
F1. (?)

Tonina Pediment, Fragment 35
(Becquelin and Baudez 1982:3:1275)

This is the only known Campeche date marking the end or beginning of a month. Victoria Bricker reads it as placing 6 Etz'nab on 0 Yaxkin; I believe it puts it on 19

Yaxkin, as in the heading. The calendrical difference is, of course, 19 days. Placement of the date in the Long Count is not certain: both the date given here and a date 52 years earlier would fit within the range of other dates from Tonina.

768   1 IV J   9.16.17.2.18   5 Etz'nab 6 Zotz' T
                                       5 Flint 19 Q. Yucuñudahui
                                       5 Flint 1 R. Zapotec

                        A1. Flint 5   B1. year Rabbit 4
                                                       (Yucuñudahui)
                        A2. Wind 7   B2. Sun 6 (Zapotec)

(Stone of the Four Glyphs,
Xochicalco; Davies 1977:253)

This date correlates the Zapotec and Yucuñudahui calendars. The inscription cites 2 years 12 years apart (A.D. 756–68), presumably relating to a reign at Xochicalco, and the glyphs are in the Xochicalco writing system. The inscription is written from right to left and places the Yucuñudahui calendar as IV.P.0.

768   14 VI J   9.16.17.6.12   1 Eb 0 Mol T
                     9.16.17.6.12   1 Eb 20 Yaxkin Palenque

                        A1. 1 Eb
                        A2.
                        A3. End Yaxkin
                        A4.

(Yaxchilan, Lintel 9,
Graham 1977:29)

This date demonstrates the use of the Palenque calendar at Yaxchilan.

790   22 IX J   9.17.19.17.7   11 Manik 5 Mac T

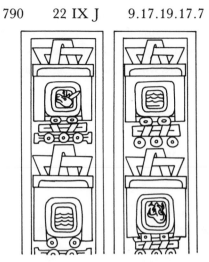

12 Deer 0 H.   Zapotec
11 Deer 19 H.   Olmec
11 Deer 1 H.   Tenango

A1. year	B1. year
A2. Water	B2. Deer
A3. 12 (Tenango)	B3. 12 (Zapotec)
A4. 3	B4. 8
A5. year	B5. year
A6. Serpent	B6. Water
A7. 12 (Olmec)	B7. 2 (Tenango)
A8. 3	B8. 7

(Tablets of Tenango;
Caso 1967:162, Fig. 19)

The two tablets are 16 years apart and presumably refer to the birth, accession, and death of a ruler of Tenango. The dates are recorded in three calendars in Xochicalco glyphs:

Date	Tenango	Olmec	Zapotec
774	12 Water	12 Serpent	(9 Deer)
790	2 Water	(2 Serpent)	12 Deer
838	(11 Water)	(11 Serpent)	(7 Deer)

The secondary numeral coefficients (A4, A8, B4, B8) are distance numbers in the trecena count, leading in each case to the following date. Thus, 12 Water + 3 = 2 (Water) and 12 Serpent + 3 = (2 Serpent). But 12 Deer + 8 = 7 (Deer), which is the latest date given in the inscription, and this provides the clue to the order of reading of the two tablets. The Olmec calendar is the only one known with a year Serpent (see 563 B.C.). From it, we can place the Tenango calendar as Type IV.L.0. It is thus identical to the Yucuñudahui calendar, except that it is initially named (100 days earlier). The remaining calendar uses Type II year bearers (Deer). It cannot be Tikal, which would be 1 Deer, nor Zapotec, which would be 11 Deer, so, by exclusion, it could be identified as the calendar of Teotihuacan, which is known from other evidence to have used Type II name days. No other Type II calendars are known for any time near this date or anyplace near Tenango. Ostensibly, this would put the Teotihuacan calendar in Type II.J.0, but we have other grounds for believing that it counted the day count from 2 to 14, omitting the coefficient 1. Thus its true New Year would appear to be two native months earlier: Type II.H.0. This, however, is the first name day of the Zapotec calendar, so what we actually have here is a Zapotec date given in the 2 to 14 day count of Teotihuacan. The

Teotihuacan New Year is fixed at II.Q.0 by the Cacaxtla correlation of 861. The archaeological dating of the Tablets of Tenango is uncertain. I have placed them in the eighth century because of their use of Xochicalco glyphs and the Teotihuacan numbers. Displacing them by some multiple of 52 years would not affect the calendrics of this correlation. The use of square dots in A7 and of flanged bars in B4 and B8 is of unknown significance.

848   5 IX J   10.0.18.13.15   2 Men 3 Ceh T
                                        2 Men 19 F. Olmec

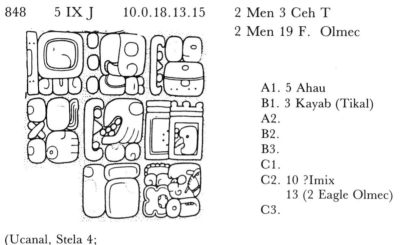

A1. 5 Ahau
B1. 3 Kayab (Tikal)
A2.
B2.
B3.
C1.
C2. 10 ?Imix
    13 (2 Eagle Olmec)
C3.

(Ucanal, Stela 4;
Graham 1967: Fig. 81)

The first date is a Tikal katun ending as follows: 10.1.0.0.0 5 Ahau 3 Kayab. Glyph C2, obviously in a foreign writing system, contains two signs with numeral prefixes. The second has an additional infixed coefficient, thus referring to the date of the Olmec year beginning 40 days before the Tikal date. Similar glyphs in square cartouches from Jimbal, Stelae 1 and 2, give dates of 12 Chicchan, 13 Cimi, 1 Manik, and Seibal Stela 13 has one of 7 Muluc. In none of these latter cases is there an explicit indication that a year is intended. The present inscription is the closest thing we have to a direct correlation of the Mayan calendar with any other native system. However, the Olmec year current at the Tikal date was 1 Ahau, and 2 Eagle was half a calendar round away: 25 years earlier or 27 years later. The significance of the inscription is thus uncertain.

861   8 III J   10.0.12.3.2   8 Ik 0 Zotz' T
                                    9 Wind 0 Q. Teotihuacan

A1. New Fire Anno Wind
A2. 9 (8) (Teotihuacan)
B1. Serpent
B2. 1
B3. Lord
B4. 2 (1) (Olmec)

(Cacaxtla, North Mural, Building A,
McVicker 1985:86, Fig. 3)

Although the Olmec date is not marked with a year sign, this date from the North Mural, Building A, at Cacaxtla is a direct correlation of the Olmec and Teotihuacan calendars. Glyph B3 is very likely Olmec, as is that of 1 Serpent (B1), which, if it is a year, was 13 years later. The glyph for Wind is Teotihuacan. The coefficient numeral dots are color coded with the extra unit required by the Teotihuacan calendar in white and marked by more than two axes, as can be seen in the illustration. The true coefficient in the general day count is given in blue dots (and in parentheses in the translation). This identifies the Teotihuacan New Year as II.Q.0.

889   28 IV J   10.3.0.0.0   1 Ahau 3 Yaxkin T
                                    1 Ahau   Tikal

hun ahau	[Katun] 1 Lord [Tikal].
uac kal haab,	Sixscore years
c u tepal ob ch ch'en itzaa	They ruled Chichen Itza.
ca paxi chi ch'en itza	They abandoned Chichen Itza.
ca bin ob cahtal chan putun	They went and settled Champoton
ti y anhi y otoch ob	Where there were the homes
ah itzaob	Of the Itzas,
kuyan uincob lae	The chosen people.
(*Mani* 70–78)	

Most of the inscribed dates from Chichen Itza fall just before this date (see Thompson, 1937:186).

928   30 IX J   10.5.0.0.0   10 Ahau 8 Muan T

This is the last Initial Series inscription on a stone monument; it is from San Lorenzo (Thompson 1965:343).

942   1 V J   10.5.13.14.1   5 Imix 19 Yaxkin T
                                        5 Imix 18 Yaxkin Campeche

A1. 5 Imix (T ?)
B1. 18 Yaxkin (T ?)
C1. T506.506.713a.(?).180
D1. T1.24.120
E1. T42.803

A2. 18 T528.116
B2. 13 T(?). 679c.(?)
C2. T506.506.713a.(?).180
D2. (?)
E2. T(?).146.501

(Capstone, East Wing, Monjas, Uxmal; Morley 1920:511, Fig. 74)

This is the last monumental use of the Campeche calendar, on the Capstone, East Wing, Monjas, Uxmal (Proskouriakoff and Thompson 1947). T numbers refer to Thompson (1962).

THE TILANTONGO CALENDAR

*973   1 V J   10.7.5.4.4   5 Kan 7 Mol T
                                        5 Kwu 1 C.   Tilantongo

This is the date of the inauguration of the Tilantongo calendar by King 5 Quevui "Tlachitonatiuh" (Caso 1965a:955).

1036   16 IX J   10.10.9.9.13   1 Ben 1 Pax T
                                        1 Acatl 20 Toxcatl   Tepexic

A1. 10 (Tilantongo)
B1. Anno Cane
C1. 1 (Tepexic)

(*Tlapitepec Map*, cited by Brotherston 1983:174)

This date correlates the Tilantongo and Tepexic calendars.

1124   7 II J   10.14.18.3.13   6 Ben 1 Xul T
                                    6 Acatl 20 Huei Pachtli Colhua

6 acatl.   6 Cane [Colhua].
Ypan yn xihuitl   In this year
mic yn colhuacan tlahtoani   Died the Colhuacan king
quahuitonal   Eagle Sun.
niman onmatlalli maçatzin   Thereupon Great Deer ruled
Y[n] colhuacan.   In Colhuacan.
(Lehmann 1938:118; see note to
   A.D. 1127)

1127   26 VI J   10.15.1.11.8   6 Lamat 1 Mac T
                                    6 Tochtli 20 Tlacaxipeualiztli Cuitlahuac

Este rey Huetzin murió en   This king Great Elder died in
el de 664, y asimismo en el   that of 664, and hence in that
que llaman chicuacen Tochtli.   they call 6 Rabbit [Cuitlahuac].
Sucedióle después Totepeuh,   There followed after him
que reinó otros tantos   Our King, who reigned as many
años y murió en el   more years and died in the
año llamado macuili Calli,   year called 5 House [Cuitlahuac],
que fué en el de 716 de la   which was in that of 716 of
Encarnación; y por su fin   the Incarnation; and at his
y muerte entró en la sucesión   end and death there entered
Nacazxoch, el cual   into the succession Flesh Flower,
reinó otros tanto 52 años   who reigned another 52
y acabó en el de 768, que   years more and ended in that
también se llamó macuili   of 768, which was also
Calli, a quien heredó el   called 5 House [Cuitlahuac],
imperio Tlacomihua.   from whom Owner of the Vat
                                            inherited the empire.

(Ixtlilxochitl 1952:2:29;1:35)

VI. tochtli xihuitl   Year 6 Rabbit [Cuitlahuac].
1186 años.   A.D. 1186.
nicanypanin motlahtocatlalli   This was when he ascended the throne
yn itoca Mallatzin.   Who was called Great Deer.
(Chimalpahin 1958:44)

Davies (1977:462) suggests that both references are to Mazatl, father of Quetzalcoatl, who was referred to as Our King (Totepeuh) and Great Deer (Mazatzin). The European dates given by Ixtlilxochitl and Chimalpahin are wrong because their calendrical assumptions were wrong (see *1629, *1632 below).

1127   5 VIII J   10.15.1.13.8   7 Lamat 1 Muan T
                                       7 Tochtli 20 Huei Tozoztli Colhuacan

7 tochtli =   7 Rabbit [Colhuacan]
ypan   Was when
in momicti yn vemac   There died Majesty
yn ocan   There
chapoltepec   In Grasshopper Mountain
çincalco =   At Plant House Cave.
ypan in 7 tochtli xihuitl   In the year 7 Rabbit
ypan tlamico yin inxiuh tolteca.   Was when the years of the Toltec ended.
(Lehmann 1938:109)

If the events described are indeed cognate, the dates given for 1124 and A.D. 1127
effectively correlate the Colhua, Cuitlahuac, and Colhuacan calendars, even despite
their disagreement about exactly when Mazatl became king.

1150   1 IV J   10.16.4.13.3   1 Akbal 1 Ch'en T
                                     1 Calli 20 Atemoztli Texcoco

1 Calli   1 House [Texcoco].
ypan in xihuitl   In this year
yn·momiquilli yn intlàtocauh   The king of the Toltec died.
yn quitzinti yn tlàtocayotl   He had begun the kingdom
yn itoca Mixcoamaçatzin.   And was called Celestial Great Deer.
(Lehmann 1938:69)

1151   1 IV J   10.16.5.13.8   2 Lamat 1 Ch'en T
                                     2 Tochtli 20 Atemoztli Texcoco
                                     2 Tochtli 20 Atemoztli Tepepulco

II. tochtli xihuitl   Year 2 Rabbit [Texcoco].
1026 años.   A.D. 1026
nican ipan in omomiquillico   This is when he died
yn Totepeuh.   Who was Our King.
(Chimalpahin 1958:9)

3 tochtli.   3 Rabbit [Tepepulco].
Ypã ynyn xihuitl   In this year
mic yn colhuacan tlàtohuani   Died the Colhuacan king
Maçatzin   Great Deer.
niman onmotlalli cuetzaltzin   Thereupon Great Iguana ruled
yn colhuacan.   In Colhuacan.
(Lehmann 1938:119)

1152	29 VII J	10.16.7.1.13	6 Ben 1 Muan T
			6 Acatl 20 Huei Tozoztli Colhuacan

6 acatl.   6 Cane [Colhuacan]
ypan   Was when
mic yn itatzin quetzalcoatl   Died the lord father of Quetzal Serpent
ytoca totepeuh.   Who was called Our King.
(Lehmann 1938:69)

Davies (1977:462) has suggested that the references under A.D. 1150, 1151 and 1152 are all to the death of Mazatl, father of Quetzalcoatl. Ixtlilxochitl gives the equivalent date for 1149 in the Cuitlahuac calendar (5 House) for the deaths of Our King (Totepeuh) and Flesh Flower (Nacaxoch) (see A.D. 1127). If these are indeed references to the accession and death of Mazatl, they have the effect of correlating the Cuitlahuac, Texcoco, Tepepulco, Colhuacan, and Colhua calendars. The disagreement of the sources on exactly which years were involved appear to be within acceptable limits for oral history transcribed several centuries later.

1171	27 III J	10.17.6.0.8	9 Lamat 1 Ch'en T
			9 Tochtli 20 Atemoztli Texcoco

yn ypan nin 9 tochtli   It was in 9 Rabbit [Texcoco].
ypan   That was when
momiquilli yn tollan tlatoani   There died the Tula king
tlilcoatzin   Great Blacksnake.
auh niman motlatocatlalli   Whereupon there ascended the throne
yn huemac   Majesty:
ytlatocatoca onmochiuh   His royal title
Atecpanecatl: Was Watercastle.
ca cenca ytolloca   Many, many tales of him
cecni amoxpan mocaquiz   Are to be heard from various books.
(Lehmann 1938:97)

Davies (1977:462) also cites a date of 8 Calli (1170) from the same source, but I am unable to locate it.

1176	25 III J	10.17.11.1.13	1 Ben 1 Ch'en T
			1 Acatl 20 Atemoztli Texcoco

1 acatl   1 Cane [Texcoco]
yn ipan   Was when
in xihuitl   It was the year
yn mic quetzalcoatl.   That Quetzal Serpent died.
(Lehmann 1938:93)

1178   20 X J   10.17.13.12.12   4 Eb 5 Pop Tikal
                          10.17.13.12.12

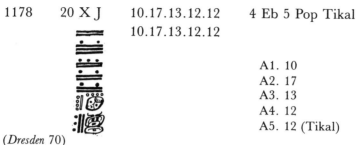

                                                A1. 10
                                                A2. 17
                                                A3. 13
                                                A4. 12
                                                A5. 12 (Tikal)

(*Dresden* 70)

This is the latest contemporaneous Long-Count date in the *Dresden Codex*, according to Satterthwaite (1965:615), who discounts the later 10.19.6.1.8 on *Dresden* 58.

1223   31 III J   10.19.18.14.5   13 Chicchan 18 Ch'en T
                          10.19.18.14.5   13 Serpent 15 O.   Olmec

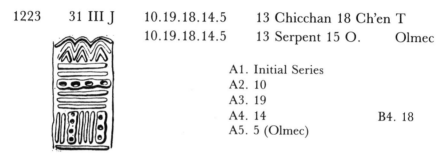

                                A1. Initial Series
                                A2. 10
                                A3. 19
                                A4. 14   B4. 18
                                A5. 5 (Olmec)

(San Andres Tuxtla, cylindrical seal;
García Payón 1949:403)

This date is from a cylindrical seal of red sand-tempered clay found in a house courtyard in San Andres Tuxtla prior to 1945. It is the latest known Olmec Long-Count date. A4 and B4 must be read from right to left because the uinal coefficient cannot exceed 17. This order of reading suggests that the inscription is Olmec rather than Maya.

1263   11 XI J   11.2.0.0.0   2 Ahau 8 Zip T
                                                2 Ahau   Tikal

cabil ahau   In [katun] 2 Ahau [Tikal]
u hetz'ci cab   The country was seated
ah cui tok   By Notch Flint,
tutul xiu   The Toltec Xiu
uxmal.   Of Uxmal.
(*Chumayel* 144–47)   (Edmonson 1986:144–47)

1296   24 II J   11.3.12.13.13   4 Ben 1 Ch'en T
                                          4 Acatl 20 Atemoztli Texcoco

4 acatl   4 Cane [Texcoco].
in yn xihuitl   In this year
mic yn colhuacan tlatohuani   There died the Colhuacan king
nonohualcatzin   Great Nonohualca.
niman onmotlalli achitometl   Whereupon Little Shanks ascended
yn colhuacan tlatocat.   To rule Colhuacan.
(Lehmann 1938:115–16)

1296   25 XII J   11.3.13.10.18   10 Etz'nab 1 Xul T
                                           10 Tecpatl 20 Tepeilhuitl Colhua

X. tecpatl xihuitl   Year 10 Flint [Colhua].
904 años.   A.D. 904
nican ypan inyn   This was when
onomiquillico   He died
yn itoca yohuallatonac   Who was called Night Steam,
ynic Ey tlahtoani culhuacan   Who was the third king of Colhuacan,
yn tlahtocal   Who reigned
epohual xihuitl   Threescore years.
auh ça niman ypan   And it was then
inyn omoteneuh xihuitl   In the year named
oncan hualmotlahtocatlalli   That there succeeded to the throne
yn itoca yn quetzallacxoyatzin   He who was called Great Quetzal Blight.
ye nahui tlahtohuani   He was the fourth king
mochiuh culhuacan.   To rule Colhuacan.
(Chimalpahin 1958:6)

1298   4 IV J   11.3.14.16.3   7 Akbal 1 Zac T
                                        7 Calli 20 Izcalli Tepepulco

VII. calli xihuitl   Year 7 Cane [Tepepulco].
953 años.   A.D. 953
nican ypan inyn   This was when
momiquillico yn quetzalacxoyatzin   Great Quetzal Blight died,
tlahtohuani culhuacã   The king of Colhuacan.
yc nahui tlahtocat   He was the fourth ruler
onpohual   For a score
on matlactli xihuitl   And ten years.
auh ca niman ipan   And it was then
inyn omoteneuh xihuitl   In the year named
oncan hualmotlahtocatlalli   He succeeded to the throne
yn itoca chalchiuhtlahtonac   Who was called Jade Hoard.
yc macuilli tlahtohuani   He was the fifth king
mochiuh yn culhuacan.   To rule Colhuacan.
(Chimalpahin 1958:6)

Davies (1977:456) suggests that Achitometl, Quetzalacxoyatzin, and Chalchiuh-
tlatonac were the same person, whose accession is chronicled in three successive
years in the Texcoco, Colhua, and Tepepulco calendars. This would reaffirm the
previously established correlation of these calendars (see A.D. 1152).

1302   2 II J   11.3.18.14.3   3 Akbal 1 Mol T
                                        3 Wa'i 20 B. Tilantongo

                                        3 House = 1300 años
(*Ñunaha* 1, cited by
Jansen 1981:1:413)

This date generally confirms the correlation of Tilantongo and Christian years. It is
the date of a journey by three chiefs of San Pedro Cántaros Ñunaha that can be
genealogically related to the life of 8 Deer "Tiger Claw," the conqueror–king of the
Mixteca, and would place the latter's death in 1167. However, the year 3 House
Tilantongo began in 1301, rather than 1300, as Jansen says.

1369   26 IV J   11.7.6.17.18   1 Etz'nab 1 Mac T
                                         1 Tecpatl 20 Tlacaxipeualiztli Cuitlahuac

contzintli	Then began
yn tlatocayotl acampichtli	The reign of Acampichtli.
ce tecpatl xihuitl	In the year 1 Flint [Cuitlahuac]
yc motlatocatlalli.	He ascended to the throne.
(*Aubin* 52)	

Después se sentó	Afterwards was seated
y fué soberano Acamápich	And was ruler, Acamapich.
En el año	In the year
1 Técpatl	1 Flint [Cuitlahuac]
se sentó	He was seated
como soberano.	As sovereign.
(Berlin 1948:15)	

Pasados casi veintiseis años	Nearly 26 years having passed
que ya era al año de ce Tepal,	So that it was already the year 1 Flint [Cuitlahuac],
y a la nuestra 1220 de la Encarnación . . .	And in our style 1220 of the Incarnation . . .
casi a los principios del año referido	Almost at the beginning of that year
los Mexicanos fueron a pedir	The Mexicans went to seek,
cada cabecera de por sí,	Each capital for itself,
señores	Lords
que los gobernasen	Who might govern them,
de lo cual Aculhua se holgó	With which Acolhua was pleased,
y les dió á sus dos hijos menores	And he gave them his two younger sons.

á los Tlatelulcos
les dió á su hijo el segundo,
llamado Mixcoatl,
y segun otros Coatecatl,
y los Tenuxcas
(ó Tenuchcas)
á su hijo el menor de los tres
llamado Acamapichtli,
que fueron
los primeros señores de México.

(Ixtlilxochitl 1958:1:118–19)

To the Tlatelolcans
He gave his second son,
Named Mixcoatl,
And according to others Coatecatl,
And to the Tenuxcas
(Or Tenochcas)
The youngest of his three sons,
Named Acamapichtli,
Who were
The first lords of Mexico.

In the year
1 Flint [Cuitlahuac]
Acampichtli
The Serpent Lord
Was seated
As Speaker.
(*Mendoza* 23)

Kirchhoff (1950:127) finds a date of 2 House in Mendoza, but I believe this is a misreading, perhaps of the preceding page. The limits of the reign of "Acampich" are clearly marked as 1 Tecpatl to 8 Tecpatl (1368–1375). His name glyph is clear and correctly annotated.

1369   5 VI J   11.7.7.1.18   2 Etz'nab 1 Muan T
                                           2 Tecpatl 20 Huei Tozoztli Colhuacan

Iniquac yn oacico yn chichimeca,
yn colhuacan chichimeca
yniquac yn otlachichiuhque
in tenochtli ytzintlan
ca çan oc tolxacaltzintli
yn inxacaltzin.
yn oncan mochantique,
Auh ca niman yc compehualtique,
yn tlalchuactli ypan
Auh yn quinyahuallotoque
yn tlalchuactli ypan
yn huallachia
ca tlapopotzticate.
Auh ca yuhquin tzoyac ehuatoc.
yn quihuallinecui.
ca miyec yc miquia.
yhuan popoçahuaya.
Auh miyecpa yn quimpehuanezquia,

ahuel mochivaya.

When the Chichimecs came to approach,
The Colhuacan Chichimecs,
When they had disposed them
At the base of the prickly pears,
Really they were just reed huts,
Their houses.
There they made their houses
And then they began.
They already fished with nets.
But those around them
On the mainland
Came to see
How they made smoke
And what a stench arose.
It began to reek.
Many died of this.
And many swelled up.
And many at times wanted to expel them
    for this,
But couldn't.

(2 Tecpatl)   2 Flint [Colhuacan].
(3 Calli)   3 House.
(*Aubin* 1893:49)   (León-Portilla, personal communication,
                                                  1984)

1369   25 VI J   11.7.7.2.18   9 Etz'nab 1 Pax T
                                        9 Tecpatl 20 Toxcatl Tepexic

El padre Torquemada pone su fundación   Father Torquemada puts its [Tenochtitlan's]
[de Tenochtitlán] en el año de 1341. Entre   founding in the year 1341. Among the
los indios D. Fernando de Alba la pone en   Indians, D. Fernando de Alva
una de sus relaciones en el año de 1140, en [Ixtlilxochitl] puts it in one of his accounts
otra el da 1142, y en otra el de 1220.   in the year 1140; in another he gives that
Muñoz Camargo en su historia de   of 1142, and in another that of 1220.
Tlaxcallan lo pone en el de 1131. Alvarado   Muñoz Camargo in his history of Tlaxcala
Tezozomoc da a entender que fué el año de   puts it in that of 1131. Alvarado
tres conejos, que puede referirse al de   Tezozomoc gives us to understand that it
1326. Chimalpain lo pone expressamente   was in the year 3 Rabbit [Tepepulco],
en el de 1325. D. Juan Ventura Zapata,   which may refer to that of 1326.
cacique de Tlaxcallan, la pone en el año de Chimalpahin puts it explicitly in that of
1321 que dice fué señalado con nueve   1325. D. Juan Ventura Zapata, the
pedernales, pero segun las tablas este año   Tlaxcala chief, puts it in the year of 1321,
no fué señalado sino con el signo de ocho   which he says was signaled by 9 Flint
cañas.   [Tepexic], but according to the tables this
                                             year was signaled only by the sign 8 Cane.

(Veytia 1836:2:137)

1370   27 II J   11.7.7.15.3   7 Akbal 1 Yax T
                                        7 Calli 20 Tititl Aztec

7 Calli xihuitl   Year 7 House [Aztec].
1369 años.   A.D. 1369
ipan in motlatocatlalli   In this year acceded to the throne
in Tlacatl Acampichtli   The Lord Acampichtli
in cenilhuitlapohualli   On the date
8 Ocelotl   8 Jaguar,
ic 20 de Febrero . . .   The 20th of February . . .
auh no ce matlaquilhuitl   Which was also the tenth day
on nahui mani   Plus four
huehue metztlapohualli   Of the old month
Itzcalli.   Izcalli.

(Glass 1975a, quoting León y Gama
1781–1784 unpublished, quoting
Chimalpalin's lost *Compendio de la Historia
Mexicana*)

The proposed European date is wrong because of Chimalpahin's calendrical assumptions (see A.D. 1629).

1370   17 III J   11.7.7.16.3   1 Akbal 1 Zac T
                                   1 Calli 20 Izcalli Tepepulco

Este Acolnauhtzin   This Acolnauhtzin
era soberano   Was king
cuando vinieron a llegar los mexica,   When the Mexica had just arrived,
ya tenochca.   Already Tenochca.
Y los mexica permanecieron juntos   And the Mexica stayed together
solamente 12 años en Tenochtitlan.   Only 12 years in Tenochtitlan.
Se separaron entonces en el 13º. año   They separated then in the 13th year
y se establecieron en el año 1 Calli   And settled in the year 1 House
                                           [Tepepulco]
en Tlatilolco   In Tlatelolco
Xaliyácac (nariz de arena).   Xaliyacac (sand spit).
(Berlin 1948:45)

Kirchhoff (1950:127) cites the same date from Chimalpahin. I do not find it.

1370   26 IV J   11.7.8.0.3   2 Akbal 1 Mac T
                                 2 Calli 20 Tlacaxipeualiztli Cuitlahuac

2 calli   2 House [Cuitlahuac].
ypan inyn xihuitl   In this year
quimiquanique in mexitin   The Mexitin were removed,
ynic oncan motlallico   So that they came to settle
Tlalcocomocco   Tlacocomocco [Tenochtitlan]
yn tencopa yn colhuaque.   On orders of the Colhuas.
(Lehmann 1938:138)

Después salieron   Afterwards they left
y se establecieron en Tenochtitlan,   And settled in Tenochtitlan
donde en medio de los tules   Where, in the middle of the reeds,
en medio de las cañas,   In the middle of the canes,
se levanta el nopal   The prickly pear stands.
Llegaron en el año 2 Calli   They arrived in the year 2 House
                                           (Cuitlahuac]
y luego levantaron un altar de césped.   And then they raised a sod altar.
(Berlin 1948:43)

La fundación de México   The founding of Mexico was
fué en el año 2 calli,   in the year 2 House [Cuitlahuac],
que corresponde al 1325 de   which corresponds to that of
la era vulgar, reinando el   1325 in the common era, in the
chichimeca Quinatzin, poco   reign of the Chichimec Quinatzin,
menos de dos siglos después   a little less than two
de la salida de los   centuries after the departure
nahuatlacas de Aztlán.   of the Nahua people from Aztlan.
(Clavijero 1964:72)

1371   5 II J   11.7.8.14.8   1 Lamat 1 Ch'en T
                                       1 Tochtli 20 Atemoztli Texcoco

1 tochtli	1 Rabbit [Texcoco].
yn ipan çe tochtli	It was in 1 Rabbit
ye ypã yn motlatocatique tenochca	That the Tenochca indeed took themselves a king
yniquac motlatocatlalli acamapichtli tenochtitlan.	When Acamapichtli ascended the throne of Tenochtitlan.
auh yuh motenehua	And it was so ordered,
mitoa ćan yehuatl ỹ quitlatocatlalli	Let it be told,
yn içihuauh	By her who was his wife,
catca ilancueytl	Ilancueitl.
oncan tzintic	There began
ỹ mexicatlatzcayotl =	The kingdom of the Mexica.
(Lehmann 1938:169)	

Los mexica permanecieron	The Mexica stayed for 42
42 años en Chapoltepec.	years in Chapultepec.
Fué en el 43 año cuando	It was in the 43rd year
fueron saqueados en el año	that they were sacked in the
1 Tochtli. En este fueron	year 1 Rabbit [Texcoco]. In
llevados a Colhuacan.	that [year] they were taken
(Berlin 1948:36)	to Colhuacan.

Kirchhoff (1950:127) cites the same date from Ixtlilxochitl and *Xolotl*.

1371   25 II J   11.7.8.15.8   8 Lamat 1 Yax T
                                        8 Tochtli 20 Tititl Aztec

8 tochtli	8 Rabbit [Aztec].
ypan xihuitl	In this year
ompeuh	Was the beginning
yn oncan	There
mexico	In Mexico
tenochtitlan.	Tenochtitlan.
(Lehmann 1938:152)	

1371   17 III J   11.7.8.16.8   2 Lamat 1 Zac T
                                         2 Tochtli 20 Izcalli Tepepulco

Kirchhoff (1950:127) cites Clavijero as giving a date of 2 Rabbit for the founding of Tenochtitlan. I am unable to locate the reference.

1371 26 IV J 11.7.9.0.8 3 Lamat 1 Mac T
3 Tochtli 20 Tlacaxipeualiztli Cuitlahuac

Veytia, in the passage cited previously, refers to Tezozomoc as giving a date of 3 Rabbit for the founding of Tenochtitlan. Neither Kirchhoff nor I can find it.

1371 5 VI J 11.7.9.2.8 4 Lamat 1 Muan T
4 Tochtli 20 Huei Tozoztli Colhuacan

3 House.
4 Rabbit [Colhuacan].
5 Cane.
Tenochtitlan.

(*Tira de Tepechpam*, Nóguez 1978:2:Apénd. 8.7, Lám. V)

Kirchhoff (1950:127) cites the same date from *Vaticanus A*.

1372 25 IV J 11.7.10.0.13 4 Ben 1 Mac T
4 Acatl 20 Tlacaxipeualiztli Cuitlahuac

Eligiéronlo [Acampich] por rey a tres de mayo de mil trescientos y sesenta y uno. (Sigüenza y Góngora 1960:294)

y así al tercer año de su fundación que corresponde al de 1330, señalado con el geroglífico de siete conejos, eligieron por capitan gobernador a un respetable anciano llamado Tenuhctzin, de cuyo nombre tomó despué su ciudad llamándola Mexico Tenochtitlan. (Veytia 1836:159)

They chose [Acampich] as king on the third of May of thirteen hundred and seventy one.

and thus in the third year of its founding, which corresponds to that of 1330, signed with the hieroglyphic of 7 Rabbit [Cuitlahuac], they chose as captain governor a respectable elder named Tenuchtzin, from whose name their city later took [its own] calling it Mexico Tenochtitlan.

This places the founding of Tenochtitlan in 4 Cane Cuitlahuac. Kirchhoff (1950:127) cites the same date from Boturini. It is startling that Sigüenza gives the correct Gregorian date.

1372   4 VI J   11.7.10.2.13   5 Ben 1 Muan T
                                   5 Acatl 20 Huei Tozoztli Colhuacan

Después se sentó Acamápich   Afterwards Acamapich was seated
(como soberano)   (As sovereign).
Fué en el año 5 Acatl   It was in the year 5 Cane [Colhuacan]
cuando se sentó.   When he was seated.
(Berlin 1948:51)

The series of dates for the founding of Tenochtitlan (1368–1372) intercorrelates the Cuitlahuac, Colhuacan, Tepexic, Aztec, Texcoco, and Tepepulco calendars.

1416   25 I J   11.9.14.7.13   7 Ben 1 Ch'en T
                                   7 Acatl 20 Atemoztli Texcoco

VII Acatl xihuitl.   Year 7 Cane [Texcoco].
ynic   On it
quimahmallque yn inteocalli   They made fire for their temple,
yn totomiuaque   The Totomihuacs
yn chiquiuhtepec   On Jade Mountain
yn chiautla.   Chiauhtla.
(*Historia Tolteca-Chichimeca*
1976:214, 349)

Jiménez (1961:144) sees this as cross-dating 1 Cane Aztec, 7 Cane Texcoco, and 13 Cane Tilantongo. It does not say so, but it does seem to imply that the Totomihuacs of southern Puebla used the Teotitlan calendar and began the calendar round on 2 Cane, whereas the *Historia Tolteca-Chichimeca* was using the Texcoco calendar.

1424   11 VI J   11.10.2.16.13   12 Ben 1 Pax T
                                     12 Acatl 20 Toxcatl Tepexic

                                     A1. 12 Cane [Tepexic]
                                     A2. 8 Rabbit [Aztec]

(*Rios 106*, cited by Brotherston
1983:174)

The assertion is that the year 12 Cane Tepexic was 9 Cane Aztec. Brotherston says that the alien name is "interjected," meaning that in a series of sequential years leading to 8 Rabbit, the expected 9 Cane is "ousted" in favor of 12 Cane.

1435   9 II J   11.10.13.13.8   7 Lamat 1 Yax T
                                          7 Tochtli 20 Tititl Aztec

1433 años,   A.D. 1433
6 An Ngū 7.   6 House [Tlaxcalan] 7 [Aztec]
Quequa ttzepātāthati acaanxïni   At that time Cutting Serpent [Itzcoatl]
coxoco nayābï nuquecquenqhuāy.   Conquered the principality of the Eagles
                                             [i.e., Cuauhtitlan].

                                         A.D. 1434
                                         7 Rabbit [Tlaxcalan] 8 [Aztec].

(Alvarado 1976:98–99)

It is curious that an Otomi document should use the Tlaxcalan and Aztec calendars, but not that of Otomi, for which the corresponding year would be 2 Rabbit.

1465   22 XII J   11.12.5.1.2   10 Ik 0 Mol T

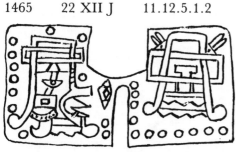

                                         10 Pée 19 B. Zapotec
                                         10 Chi 18 B. Tilantongo
                                         10 Chi 18 B. Tepexic

                                         A1. year 10 Wind [Zapotec]
                                         B1. 2 Flint [Tepexic]
                                         C1. year 11 Wind [Tilantongo]

(Jaguar Knight Pectoral,
Caso 1932:26; Jiménez
1940:74)

This date correlates the Zapotec and Tilantongo calendars. The specific day cited
in B1 could be that of the death of the Jaguar Knight. It falls on the sixteenth day
(Zapotec) or the fifteenth day (Tilantongo) of the tenth month in the year given. If
it is read as a year name (for which the inscription provides no direct warrant), it
implies a correlation with the calendar of Tepexic and places the death before the
month of L. (Pax).

1469   5 VIII J   11.12.8.13.4   6 Kan 2 Pop T
                                       6 Kan 9 Tun Mayapan

                                       A1.
                                       A2. 10?
                                       A3. 6 Kan
   A4. 9 Tun
                                       A5.
                                       A6.
                                       A7.

(Capstone, structure east
of Casa Principal, Chichen
Itza; Morley 1920:520)

As Morley points out, this calendar-round date cannot fall in the ninth tun in the
ninth or twelfth baktun. It can occur on the date cited or on 10.3.8.14.4. I believe
it shows that the Itzas had invented the Mayapan calendar this soon after the fall of
Mayapan, though it was not formally inaugurated until 1539. Like Morley, I do not
believe the earlier date makes historical sense, though it must be conceded that all
the other dates from Chichen Itza cluster just before 10.3 (see Thompson, 1937:186).

1480   12 IX J   11.13.0.0.0   6 Ahau 3 Zip T
                                      6 Ahau   Tikal

uac ahau   [Katun] 6 Lord [Tikal]
ca tac canlahun pix i.   And then was the fourteenth measure [of
                                          the baktun].
(Edmonson 1986:173–74)

The text provides no reference for what is being dated, but the apparently exceptional numeral classifier *pix* rather than *piz* or *ppiz* suggests something stacked, and
this is the fourteenth katun of this baktun.

1486   11 V J   11.13.5.13.7   6 Manik 5 Muan T
7 Deer 0 J. Tlapanec

Year 7 Deer [Tlapanec].
Picture: burning of the Temple of Tlachinolli
at Tlapa by the Mexicans.

(*Azoyú I* in Toscano 1943:129ff.)

En este mismo año	In this same year
(7 Tochtli)	(7 Rabbit [Aztec])
se destruyeron	Were destroyed
o acabaron	Or finished
los de Cozcacuauhtenanco	Those of Tenango
Tlapaneca.	Of the Tlapanec.

(*Cuauhtitlan*, quoted by Toscano 1943:129)

The equation of 7 Deer Tlapanec and 7 Rabbit Aztec places the Tlapanec calendar
as II.J.0, named initially and counting the day count from 2 to 14 instead of 1 to
13. The *Mendoza* (31) also records the first entry of the Mexicans into Tlapa in the
reign of Tizoc (1480–86), the last year of which was 7 Rabbit Aztec. The Tlapanec
captives were sacrificed at the dedication of the Great Temple in Tenochtitlan (Vega,
personal communication).

1486   12 VIII J   11.13.6.0.0   8 Ahau 13 Pop T
11.13.6.0.0   Tikal

u heklay	The account
t in tzi'btah	That I have written
utzcinnabal mul	Of the completion of the mounds
t u men heregesob	By the heretics.
ox kal katun utzcinnabci	Threescore katuns were completed;
ca tac holhun pis katun	Then it approached fifteen measured katuns
t u mentah ob	When they had done it,
nucuch uinicob (. . .)	The great people (. . .)
he x lik	Hence in fact
u mentic ob mull e	They built these mounds
ox lahun te katun	In thirteen [completed] katuns

ca tac uac p'el haab i	And just about six years [current].
lic	Thus
y utzcinnic ob e.	They finished the job.
(*Chumayel* 1367–1372, 1391–1393)	

This is one of the very few allusions to the Long Count in the *Books of Chilam Balam*. The assertion is that it took from 8.0.0.0.0 almost to 11.15.0.0.0 to build the pyramids and that they were actually finished in 11.13.6.0.0.

1493   20 V J   11.13.12.15.13   11 Ben 16 Muan T
                                         11 Ah 12 Ru Cab Tamuzuz Iximche

kicá ti pacataj	When dawn broke
ru xe caj	Over the land
chi julajuj Ah	On 11 Cane
ix boz	There began
pe	To come
Tukuchée	The Tukuches
chaká tinamit,	Across the town.
caní ixga ján ru zubak	At once were heard the flutes,
ru chabí tun	The war drums
ajauj	Of the lord
Caí Junajpú	2 Hunter,
güikital	Dressed up
chi tooj;	In state
chi tunatiuj gug,	In shining feathers,
chi tunatiuj cubul,	In shining garlands
chi calguach	With a crown
puak abaj.	Of silver.
(*Annals of the Cakchiquels* 248)	

This is the starting point of the Iximche calendar of the Cakchiquels: the Revolution of Iximche.

1500   30 V J   11.14.0.0.0   4 Ahau 8 Pax T
                                       4 Ahau   Tikal

can ahau uchci	[In katun] 4 Lord [Tikal] occurred
ma ya cimlal	Painless death,
oc nal	Entering the house,
kuchil ych paa.	Going into the Fort [Mayapan].
(*Tizimin* 202–4; *Chumayel* 112–14)	

Painless death is a euphemism for the Tikal calendar-round ceremonies of 1492. "Entering the House" is the name of any New Year ceremony, but perhaps particularly those that initiate important segments of the calendar round. This one was held at the entrance to ruined Mayapan on 11.13.12.4.17.

1507   13 XII J   11.14.7.11.13   1 Ben 1 Mol T
                                          1 Xo 20 B.   Chocho

año 1507 = 1. nca-ni-xo',   The year 1507 (was), first, the year 1 Cane
   acatl   (Acatl).
(*Quecholac*, quoted in Jiménez
1940:69)

This the first of a series of year-to-year correspondences in the *Quecholac*; hence the
word *first*. It equates the Chocho calendar with that of Tilantongo.

1513   23 III J   11.14.13.0.0   4 Ahau 3 Mac T
                                          4 Ahau   Tikal

cabil ahau   [Katun] 2 Lord [Tikal]
oxlahun tun   On the thirteenth tun
mani tz'ulob   The foreigners passed
u yax ilc ob   And first saw
u luumil yucatan   The lands of Yucatan,
tzucub te.   The sacred grove.
(*Mani* 206–13)

The thirteenth tun of katun 2 Lord landed on 4 Lord. The reference could be to the
sighting of the Yucatecan coast by Ponce de León in 1513. This is the earliest
contemporaneous correlational date relating to the European calendar.

*1517   Morley

t u y abil   in the year
1517 años,   A.D. 1517
lay y abil   That was the year
hauic cha katun lae;   They ended the taking of the katun;
lay hauic u uacuntabal   That stopped the erection
u tunil balcah   Of the world stones,
u oklal hun-hun-kal tun   Because each twenty tuns
u talel uatal   They had been setting up
u tunil balcah cuchi.   World stones.
(*Yaxkukul* 16)

Morley (1920:498) stresses this text as dating the end of katun 2 Ahau to support
the 12.9.0.0.0 (Spinden) correlation. The text says otherwise (i.e., the taking of
katun offices ended in 1517, not the katun) and is included here because of the
influence that Morley's argument continues to exercise in some quarters.

1519   9 XI J   11.14.19.13.2   8 Ik 10 Xul T
1519   8 XI J   8 Ehecatl 9 Quecholli Aztec

Auh in izqujhujtico	And these are all the days
in Mexico inic calaqujco in Españoles:	Since the Spaniards came to enter Mexico
ipã ce hecatl	On 1 Wind
in cemjlhujtlapoalli:	In the day count
auh in xiuhtonalli	And in the year count
ce acatl,	1 Cane [Aztec],
oc muztla	[When] yet next day
tlamatlactiz Quecholli:	Would be the tenth of Quecholli.
auh in cemjlhujtique	And when they had passed the day
ume calli:	2 House,
vel iquac	At that time indeed
in tlamatlacti quecholli.	It was the tenth of Quecholli.
(Sahagún 1975:12:80)	(Anderson and Dibble 1975:12:80)

y fue nra venturosa E	And it was our risky and
atrevida Entrada, En la gran	daring entry into the great city
çibdad de tenustitan	of Tenochtitlan Mexico on the
mex^co a ocho dias del	eighth day of the month of
mes de novienbre, año	November, the year of Our Savior
de nro saluador Jesuxpo,	Jesus Christ 1519, thanks be
de mill E quinientos, y	to Our Lord Jesus Christ for
diez, y nueve años	everything.
graçias a nro señor	
Jexuxpo por todo	
(Díaz 1904:1:272)	

Auh yn acico nican Mexico	When he arrived in Mexico
Tenuchtitlan	Tenochtitlan,
yn Capitan general	The Captain General
Hernando Cortés	Hernan Cortés
ynic connamicque tlacatl	Was received by the noble
Moteuhcçomatzin xocoyotl,	Montezuma the last,
ihuan Cacamatzin	By Cacamatzin,
tlahtohuani Tetzcuco,	Ruler of Tetzcuco,
yhuan Tetlepanquetzatzin,	And by Tetlepanquetzatzin
tlahtohuani Tlacopa,	King of Tlacopan,
ypan cemihuitlapoalli	On the day
chicuey ehcatl,	8 Wind,
auh yn ipan	And according to
yh inmetztapohual catca huehuetque	The count of months of the ancients
chiuhc nauiltia	It was the ninth day
quecholli.	Of Quecholli.
(Chimalpahin 1889:188–89)	

En las cuales dichas cosas,	In which things that have
y en otras no menos	been related,
útiles al servicio de vuestra	and in others no less
alteza, gasté de 8 de noviembre	useful to the service of your

de 1519, hasta entrante el
mes de mayo de este año
presente. . . .

(Cortés 1963:80)

highness, I spent from the 8th
of November of 1519 until the
beginning of the month of May of this
present year . . .

Sahagún's informants miscalculated by 1 month, citing 1 Ehecatl and 2 Calli instead of 8 Ehecatl and 9 Calli. Chimalpahin has the correct date. Bernal Díaz del Castillo was writing long after the event and probably converting back from Gregorian to Julian dating. In any case, he apparently entered Mexico the day before the Indians saw him do it. Cortés agrees with Díaz's date. Caso (1967:48–49) points out that Chimalpahin's native date is corroborated by Cristóbal de Castillo and the annals of Tula and Tlatelolco as well as the European date by Gómara, Cervantes de Salazar, Valadez, *Cozcatzin, Mazizcatzin, Actas de Cabildo 1528* and *Cédulas Reales.* Only two discrepant sources on the European date could be found, both placing it on 12 VIII (Torquemada and *Títulos de Santa Isabel Tola*). Caso quite rightly dismisses them. The dating discrepancy remains, but too much has been made of it. Like Caso, I cannot explain it, but the pattern of other correlations makes it clear that either the native or the Spanish sources were off by 1 day for reasons not yet explained.

1520   9 XII J   11.15.0.14.18   1 Etz'nab 1 Mol T
                                    1 Si 20 B.   Tilantongo

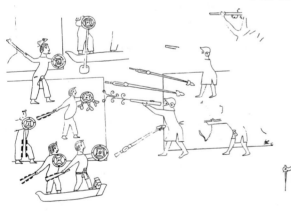

1 Flint [Tilantongo] was the year of the (pictured) battle [of the Noche Triste].

(*Yanhuitlan*, Pl. II, cited by
Jiménez 1940:70)

1521   2 V J   11.15.1.4.2   2 Ik 5 Muan T
                                      3 Wind 0 J. Tlapanec

                                          A1. 5 Grass
                                          B1. 4 Deer
                                          C1. 3 Wind (Tlapanec)
                                          D1. 2 Quake
                                          D2. 14 Grass
                                          D3. 13 Deer
                                          D4. 12 Wind

(*Azoyú I* in Toscano 1943:129ff.)

The text is written from bottom to top and from right to left. The day count is counted from 2 to 14. The picture represents the arrival of the Spaniards at Tenango (the toponym *Cozcacuautenango* is depicted) in 3 Wind Tlapanec. The relevant date is marked by a line from the year glyph into the picture. Toscano equates 3 Wind Tlapanec with 3 House Aztec (see A.D. 1486).

1520   14 VI J   11.15.0.6.0   5 Ahau 8 Kayab T
                                        5 Xochitl 2 Tecuilhuitontli Aztec

njmā qujvaltoqujla   Then followed after it
Tecujhujtontli,   Tecuilhuitontli.
ie uncan in qujzque in Españoles,   There was when the Spaniards went forth,
in moioalpoloque,   When they disappeared by night
amo tenemachpan,   Without one's knowledge;
amo iuh catca teiollo   It was not in one's heart
inic qujzque   That they would go forth
ioaltica.   By night.
(Sahagún 1975:12:80)

y fue nra Entrada En mexico,   And our entry into Mexico was
dia de señor san Juº   the day of San Juan in June
de Junio, de mill E   of 1520. Our departure in
quinientos y veynte años, fue flight was on the tenth of the
nra salida huyendo A diez   month of July of that year.
del mes de Jullio del dho año.
(Díaz 1904:1:437)

The day of San Juan is June 14 in the Julian calendar.

1521   13 VIII J   11.15.1.9.5   1 Chicchan 3 Uo T
1521   13 VIII J   1 Coatl 2 Tlaxochimaco Aztec

Auh in omomā chimalli;   And when the shields were laid down,
injc tixitinque,   When we fell,
in xiuhtonalli   It was in the year count

ei calli.
Auh in cemjlhujtlapoalli
ce coatl.
(Sahagún 1975:12:122)

3 House [Aztec].
And in the day count
It was 1 Serpent.

se prendio guatemuz y sus
capitanes En treze de agosto
A ora de bisperas En dia de
señor san ypolito año
de mill E quinientos y
veynte y un años graçias A
nro señor Jesuxpo y a nra
señora la virgen santa
maria su bendita madre Amen.
(Díaz 1904:2:129

Cuauhtemoc and his captains
were taken on the thirteenth
of August at the hour of
vespers on the day of the lord
San Hipolito in the year 1521,
thanks be to Our Lord Jesus
Christ and to Our Lady the
Virgin Santa Maria His
Blessed Mother, Amen.

Y plugo a Dios que un capitán de un
bergantín, que se dice Garci Holguín, llegó
en pos de una canoa en la cual le pareció
que iba gente de manera; y como llevaba
dos o tres ballesteros en la proa del
bergantín e iban encarando en los de la
canoa, hiciéronle señal que estaba allí el
señor, que no tirasen, y saltaron de presto,
y prendiéronle a él y a aquel Guatimucín,
y a aquel señor de Tacuba, y a otros
principales que con él estaban; y luego el
dicho captitán Garci Holguín me trajo allí
a la azotea donde estaba, que era junto al
lago, al señor de la ciudad y a los otros
principales presos; el cual, como le hice
sentar, no mostrándole riguridad ninguna,
llegóse a mí y díjome en su lengua que ya
él había hecho todo lo de su parte era
obligado para defenderse a sí y a los suyos
hasta venir en aquel estado, que ahora
hiciese de él lo que yo quisiese; y puso la
mano en un puñal que yo tenía,
diciéndome que le diese de puñaladas y le
matase. Y yo le animé y le dije que no
tuviese temor ninguno; y así, preso este
señor luego en ese punto cesó la guerra, a
la cual plugo a Dios Nuestro Señor dar
conclusión martes, día de San Hipólito,
que fueron 13 de agosto de 1521 años.

(Cortés 1963:189)

And it pleased God that a captain of a
barkentine who is called Garci Holguín
came up on a canoe in which it seemed to
him people of breeding were traveling; and
since he had two or three bowmen in the
bow of the barkentine and they were facing
those in the canoe, he signaled to them
that the lord was there and they were not
to shoot, and jumped quickly and seized
him and that Cuauhtemoctzin, and that
lord of Tacuba and other leaders who were
with him, and then this captain Garci
Holguin got me there to the roof where he
was which was next to the lake, to the lord
of the city and the other captured leaders;
and he, when I had him sit, showing him
no severity whatever, came up to me and
told me in his language that he had done
everything on his part that he was
obligated to do to defend himself and his
people until he had come to this state, that
now I should do with him as I wished; and
he put his hand on a dagger I had, telling
me that I should stab him and kill him. I
cheered him and told him to have no fear
at all; and thus, with this lord captured,
then at that point the war ended, which it
pleased Our Lord to put an end to
Tuesday, St. Hipólito's day, which was on
August 13 of the year 1521.

This date for the fall of Tenochtitlan is the decisive correlation of the Aztec and
Julian calendars.

| 1521 | 30 X J | 11.15.1.13.3 | 1 Akbal 1 Xul T |
| 1540 | 25 X J | | 1 Kuahta 20 Uapanscuaro Tarascan |

Y yera por la fiesta de	And it was around the fiesta
Uapansquaro a veynte e cinco	of Uapanscuaro on the 25th
de cotubre y salieron a	of October, and the women
cojer maçorcas de mayz las	went out to get ears of corn
mugeres para la fiesta.	for the feast.
(De la Coruña 1541:18)	

The Julian date is that of the *Relación*, so the date referred to is the preceding 25 X (1540), but the event was in 1521, 5 leap years earlier. This date correlates the Tarascan and Julian calendars on the assumption that Tarascan festivals fell on the last day of the native month. This and other Tarascan dates for month endings (*fiestas*) must be fitted into the year in some pattern of 20-day intervals.

| 1522 | 27 II J | 11.15.2.1.3 | 4 Akbal 1 Ceh T |
| 1541 | 23 II J | | 4 Kuahta 20 Purecoraqua Tarascan |

y bino nueva q abia benydo	And the news came that a
un español y havia llegado	Spaniard had come and that he had
a taximaroa en un caballo	arrived at Nuevo Hidalgo on a
blanco y hera la fiesta de	white horse and it was the
purecoraqua a veynte y tres	feast of Purecoraqua on the
de febrero y estubo dos dias	23rd of February, and he was
en taximaroa y tornose	in Nuevo Hidalgo for two days
a mexico.	and returned to Mexico.
(De la Coruña 1541:246)	

The Julian date refers to 1541 but the event was in 1522; the date is one day late in any case: the feast of Purecoraqua would have been on 22 II 1541 J.

1522 1 III J 11.15.2.1.5   6 Chicchan 3 Ceh T
6 Yo 17 F. Tilantongo
6 Serpent 19 F. Olmec

Anno Rabbit 3
Anno Serpent 6

Tlaxocopa

Lord 10 Flint and
Lady 7 Wind

Tlaxocopa

(*Dehesa*, Chavero 1892,
lam. 10)

The date records a year in the lives of 10 Flint and his wife 7 Wind at Tlaxocopa, but it records it in both the Tilantongo and Olmec calendars. The 2 day names are attached by separate lines to the year sign and equate 3 Rabbit Tilantongo with 6 Serpent Olmec, confirming the placement of the Olmec year in 563 B.C. (q.v.). The personages in question and the date referred to may be substantially earlier, but the *Dehesa* is post-Conquest.

1522 22 VII J 11.15.2.8.8   6 Lamat 1 Pop T
1541 17 VII J   6 Auani 20 Caheri Consquaro Tarascan

pues binyeron las nuebas al cazonci como los españoles abian llegado a taximaroa y cada dia le benyan mensajeros q benyan doscientos españoles y hera por la fiesta d cahyra cusquaro a diez y syete de jullio cuando lluebe mucho en esta tierra y benya por capitan un caballero llamado Xptobal de oli.	So the news came to the Cazonci how the Spaniards had arrived in Nuevo Hidalgo, and every day messengers came to him [saying] that two hundred Spaniards were coming, and it was around the feast of Caheri Consquaro on the seventeenth of July, when it rains a lot in this country, and there came as captain a knight called Cristobal de Olid.

(De la Coruña 1541:248)

The Julian date refers to 1541, but the event was in 1522.

| 1522 | 19 XI J | 11.15.2.14.8 | 9 Lamat 1 Yaxkin T |
| 1541 | 14 XI J | | 9 Auani 20 Caheri Uapansquaro Tarascan |

Y en otra fiesta llamada
caheri uapansquaro baylauā
con unas cañas de maíz
a las espaldas.
(De la Coruña 1541:10)

And in another feast called
Caheri Uapansquaro they danced
with cornstalks on their backs.

y ēbio el cazonci a don
pedro con aquellos ancoras
a zacatula que hera por la
fiesta a catorce de
nobyēbre del present año.
(De la Coruña 1541:262)

And the Cazonci sent don Pedro
with those anchors to
Zacatula, which was around the
feast of the fourteenth of
November of the present year.

The "present year" is 1541. The anchors were being sent to Cortes's fleet at Zacatula, which burned in 1523 (Prescott 1863:3:270), so the event must be placed in 1522.

| 1523 | 7 VI J | 11.15.3.6.8 | 1 Lamat 1 Kayab T |
| 1541 | 7 VI J | | 1 Auani 20 Mascuto Tarascan |

y [Cortés] mandolos traer para que se
esniasen la doctrina xpīana en san fran^{co} y
estubieron alla un año quince muchachos
que fueron por la fiesta de mazcuto a siete
de junjo y amonestoles el cazonci que
aprendieren que no estarian alla mas de un
año.

(De la Coruña 1541:264)

And [Cortes] ordered [the children of the
Cazonci and don Pedro] brought [to
Mexico] so they could be instructed [in]
Christian doctrine at San Francisco, and
fifteen boys were there for a year, who left
around the feast of Mazcuto on June 7,
and the Cazonci warned them that they
were to learn and would not be there more
than a year.

The feast of Mascuto should have been on 2 VI 1541 J, but the date is correct for 1523.

| 1524 | 14 IV J | 11.15.4.4.0 | 1 Ahau 8 Kankin T |
| 1524 | 14 IV J | | 1 Hunahpu 10 Nabe Tamuzuz Cakchiquel |

Jacá
chi jun Junajpú
tokí xul
Castilan güinak
pa tinamit
chi Izimchée
(*Annals of the Cakchiquels* 248)

On the day
1 Hunter
There arrived
The Castilians
At the city
Of Iximche.

Yo me parto para la ciudad
de Guatemala lunes onze de abril.

I'm leaving for the city of
Guatemala Monday the 11th of April.

Desta ciudad de Uclatan a	From this city of Utatlan on
onze de abril . . . Pedro de	the 11th of April . . . Pedro de
Alvarado.	Alvarado.
Que yo señor: parti d.	That I [my] lord, left the
la ciudad de Uclatan y vine	city of Utatlan and came in
en dos dias a esta ciudad	two days to this city of
de Guatemala donde fui	Guatemala where I was very well
muy bien recebidos de los	received by the lords of it,
señores della q. no	so that it could not have been
pudiera ser mas en casa de	more so in the house of our
ntros. padres;	fathers;
(Alvarado 1925:77–78)	

Alvarado apparently intended to leave Utatlan on April 11th, but stayed up late writing to Cortes and did not get off until the 12th, thus arriving on the 14th. The date cited is equivalent to 1.8.6.7 Iximche.

1530 22 XI J   11.15.10.16.13   9 Ben 6 Yaxkin T

                                  9 Mishi 0 B. Jicaque

Capítulo VI. Que prosigue lo de atrás	Chapter VI. Which continues what went
(1530). [Los naturales de Hibueras y	before (1530).[The natives of Hibueras and
Honduras] Contaban su año repartido en	Honduras] Counted their year divided into
diez y ocho meses, llamándole *Ioalar*, que	eighteen months, calling it *Ioalar*, which is
es cosa que va pasando; y ponían veinte	something that is passing; and they had
días al mes, aunque no contaban sino por	twenty days to the month, although they
noches; y así, ponían primero la noche que	only counted by nights; and thus they put
el día y contaban veinte noches o veinte	the night before the day and counted
alboradas. . . . En el primer día de su	twenty nights or dawns. . . . On the first
tercer mes cae nuestro Año Nuevo, 1.° de	day of their third month falls our New
enero.	Year, the 1st of January.
(Herrera 1952:9:121)	

Herrera's editor, Miguel Gómez del Campillo, footnotes the chapter heading to cite the *Relación de Honduras y costumbres de sus naturales* and *Descripción de Honduras e Higueras*, manuscripts in the Archivo de Indias, as Herrera's source for this information. It establishes the Jicaque New Year as III.B.0. Counting the days from noon to noon was also the Mixe custom, and this is 1 month after the Mixe New Year.

1531 2 XI J   11.15.11.15.18   3 Etz'nab 6 Xul T

                                  3 Tahp 0 Mih Kahpu'ut Mixe

Capítulo VII. De las costumbres	Chapter VII. On the customs
de las gentes de algunas	of the peoples of some
provincias de lo que hoy es	provinces of what today is the
distrito del Arzobispado de	district of the Archbishopric
México y algunos de sus	of Mexico and some of its
confines (1531) . . . En la provincia	borderlands (1531). . . . In the

de los Miges . . . entierran	province of the Mixes . . . they
a los que mueren, por la	bury those who die for the
mayor parte, en el campo, y	most part in the countryside,
cada año hacen aquella	and every year they make
memoria a los finados, lleván-	that a memorial for the
doles comida por ofrenda	deceased, taking them food as
sobre sus sepulturas, por el	an offering on their graves,
mes de noviembre, dos días	in the month of November, two
antes o después que nosotros	days before or after we
celebramos la memoria de los	celebrate the memory of the dead.
difuntos.	
(Herrera 1952:9:208)	

Herrera's editor, Gómez, attributes this material to a 1533 manuscript now in the Archivo de Indias entitled *Relación y descripción de la Nueva España con la pintura de la tierra*. If, as there is reason to believe, this aboriginal feast of the dead was actually the Mixe New Year, it would represent a frozen Julian date from 1532 on, placing the Mixe calendar as III.A.0. The 2-day leeway in the date probably relates to the distinctive Mixe way of coping with leap year in a frozen calendar (see chapter 5, Mixe).

1537   26 II J   11.15.17.5.1   9 Imix 4 Ceh T

                                        9 Imix   Mayapan

oxlahun ahau	[In katun] 13 Lord [Mayapan]
cimci ah pula	The Water Thrower died.
uac ppel hab u binel	Six years were to go
u xocel	To its counting.
haab	The year
ti lakin cuchie	Occurred in the east.
caanil kan	4 Iguana
cumlahci pop ti lakin	Was checked in Pop in the east.
he tun ten a	Now indeed this time
cici pahool	I make the best guess
katun	Of the katun
haab	Date:
hun hix	1 Jaguar
çip	In Zip.
ca tac ox ppel i	So then there were three [days].
bolon ymix	9 Alligator
hi u kinil	Was the day
lay cimci	That he died,
ah pula lae	This Water Thrower,
na pot xiu.	Na Pot Xiu.
(*Chumayel* 117–27; *Tizimin* 207–21;	
Mani 221–30)	

1 Ix can fall in Zip in an east year only in the Tikal calendar; 4 Kan can fall in Pop in an east year only in that of Mayapan. Having contemplated both calendars, the scribe gives the date only in the day count.

1538   1 III J   11.15.18.5.9   13 Muluc 7 Ceh T
        1 III J   1 Acatl 12 Huei Tozoztli Totonac

Testigo — Tlachinutl,	Witness — Tlachinutl, native
natural de una estancia del	of a farm of the aforesaid
dicho pueblo [Matlatlán],	town [Matlatlan], said . . .
dixo . . . que la Pascua de	that on Easter of last year . . .
flores del año pasado . . . que	that on that farm they
en la dicha estancia limpiaron	cleaned and swept the pyramid
é barrieron el cu del	of the devil so that they
demonio, que se ayuntaron á	got together to dance and get
bailar é hacer borrachera,	drunk, and they raised in
y levantaron en medio del pa-	the middle of the courtyard
tio un árbol, ó madero	a tree or large pole,
grande, sobre el cual en lo	on the top of which they put
alto pusieron ciertos insig-	certain symbols of the devil,
nias del demonio é el día	and the day they did
en que esto hicieron dicen	this they call in their
por su calendario *ce acatl*,	calendar 1 Cane, and
é que era fiesta de un	it was the feast of an
ídolo que ellos llaman	idol they call 8 Monkey.
Chicueyozumatli.	
(Olmos 1912:211)	

Easter Sunday fell on 17 III Julian 1538 in the year 7 Rabbit Aztec, and on 20 IV Julian 1539 in the year 8 Cane Aztec, or 1 Cane Texcoco and 7 Cane Tilantongo. The first name day of the Tilantongo calendar fell on 20 IV Julian in both years. The feast of 8 Monkey was a day-count festival, notable for the Flying Pole Dance, which is clearly alluded to. It would have fallen on 11 II and 20 XI in 1538 and on 7 VI in 1539. The first of these occurrences was 34 days before Easter but was only 18 days before the most likely date for the Totonac New Year (and hence only 12 days before its name day). It seems very likely that the festival was postponed to coincide with the year's end ceremonies and that the New Year fell on 1 III, as listed above. The plausibility of this construal rests on the fact that it would make the Totonac calendar congruent with the geographically adjacent Nahuatl calendar of Teotitlan. It would appear therefore that the reference to Easter is merely an approximation, implying "around Easter time." It may have been Olmos himself who calculated the year date as 1 Cane in the Texcocan calendar and placed it in the year of the deposition (1539) rather than in that of the event described (1538), which would have to be a year Rabbit. The coincidence of Easter with the Tilantongo (-Tlaxcalan) first name day of 1539 may then be dismissed as irrelevant. The place- ment of the Totonac year bearer is clearly fixed in the correlational date of 1539 (q.v.), but the identification of its initial month is constrained only by the present argument and must be considered tentative. The case rests on the likelihood that it was the New Year ceremonies that occasioned the sweeping of the pyramid some 3 weeks before Easter.

THE MAYAPAN CALENDAR

1539   14 VIII J   11.15.19.14.0   11 Ahau 7 Uo Mayapan
                                                  11 Ahau

buluc ahau	In [katun] 11 Lord [Mayapan]
hulciob	They arrived,
kul uinicob	The God people
ti lakin	From the east;
u y ah tal	The newcomers
ca hul ob	Arrived
u yax	On the new
chun	Base [year]
uay	Here
tac luumil	In the lands
c oon	Of us
maya uinic.	Maya people.
(*Tizimin* 223–30)	

This is the inauguration of the katun cycle (May) of the Mayapan calendar.

1539   16 XI J   11.16.0.0.14   1 Ix Yaxkin M
            16 XI J   1 Calcusot Totonac

En Matlallán, pueblo
encomendado por su Majestad al
Adelantado Montejo, en el
año de mill é quinientos
treinta é nueve, en el mes
de Noviembre, fueron tomados
los siguientes testigos sobre
la causa tocante á Don Juan,
cacique del dicho pueblo . . .
[quien] dixo: que es verdad
que el Domingo pasado que
era XVI de Noviembre del año
de mill é quinientos é
treinta é nueve, hicieron la
fiesta de Panquezaliztli todo
el pueblo como los otros
testigos han dicho.
(Olmos 1912:208–9, 214)

In Matlatlan, a town entrusted
by His Majesty to the
Adelantado Montejo, in the
year 1539,
in the month
of November, the following
depositions were taken on
the case relating to Don Juan,
chief of the said town . . .
[who] said: that it is true
that on last Sunday, which
was the 16th of November
of the year 1539,
they held the feast
of Panquetzaliztli, the
entire town, as the other
witnesses have said.

Luis, principal del dicho
pueblo, testigo . . . dixo . . .
que es verdad que el Domingo
próximo pasado, del
dicho mes é año,
hicieron una fiesta al demonio
que según su calen-
dario cayó en el dicho
Domingo, que se dice la
fiesta panquezaliztli, que
es una del las que tienen
como pascua, que en su
lengua Totonac se llama
calcusot, en la cual fiesta
toda la mayor parte del
pueblo celebraron.
(Olmos 1912:209)

Luis, an elder of the afore-
said town, a witness . . . said . . .
that it is true that on Sunday
last of the aforementioned
month and year, they held
a feast for the devil,
which according to their
calendar fell on the Sunday
in question, which is called
the feast of Panquetzaliztli,
which is one of those they
consider primary, which in
their Totonac language is
called Calcusot, in which
all the larger part of the
town celebrated.

The date given is the first day of Panquetzaliztli in Nahuatl or Calcusot in Totonac. It is noteworthy that the Totonac celebration was held on the first day rather than the last, implying the derivation of the Totonac calendar from an initial-dated system. This also establishes that the Totonac had Type IV New Years and makes it clear that the New Year fell later in the year than B. Yaxkin (Panquetzaliztli). The probable placement of the Totonac New Year is discussed in relation to the quasi-correlational date of 1538 (q.v.). *Calcusot* in Totonac means 'middle'; it appears indeed to be the tenth Totonac month (see Chapter 5, Totonac).

| 1541 | 10 IX J | 11.16.1.15.18 | 2 Etz'nab 15 Zip M |
| 1541 | 10 IX J | | 2 Tihax 17 Ru Kab Tokik Cakchiquel |

Chupam juná guaé,
ok ixbán niṁ ulalaj,
jaok xe cam Castilán güinak
Panchoy,
chi caí Tijax,
ix bokó tuj ul juyú
chi Junajpú
ya, ixpé chupam juyú
xeyaar Castilan güinak
chi camik
xeyaar,
ca rixjayil
Tunatiuj.

(*Annals of the Cakchiquels* 272)

During the year
There was a great landslide,
In which the Castilians died
In Ciudad Vieja.
On 2 Flint,
The volcano erupted
At Hunter [Volcán de Agua];
The water burst from inside the volcano;
The Castilians perished
And died.
They perished,
And then the wife
Of Tonatiuh [Doña Beatriz de la Cueva,
    Vda. de Alvarado].

Sábado a diez de
septiembre de mil y quinientos
y cuarenta y un años

Saturday on the tenth of
September of the year fifteen hundred
and forty one at two o'clock in

a dos horas de la noche,
habiendo l lovido jueves y
viernes no mucho ni mucha
agua el dicho sábado le
aseguro como dicho es; y
dos horas de la noche
hubo muy grande tormenta de
agua de lo alto del vol-
cán que está encima de
Guatemala y fué tan sú-
bita que no hubo lugar de
remediar las muertes y
daños que le recrecieron
fué tanta la tormenta de
la tierra que trajo por
delante del agua y piedras
y árboles que los que
lo vimos quedamos admirados
y entró por la casa
del Adelantado don Pedro
de Alvarado que haya
gloria.
. . . y todas las tomó
debajo donde dieron las ánimas
a su criador.
(Rodriguez 1948:94)

the morning, it having rained
on Thursday and Friday not
very much nor much water on
the Saturday in question,
I assure you as it is stated;
and at two o'clock in the
morning there was a great
tempest of water from the
heights of the volcano that
is over Guatemala and it was
so sudden that there was no
time to alleviate the deaths
and damages that grew upon it.
The flooding of the land was
so great that it brought
before the water also rocks and
trees so that those of us
who saw it were left astounded,
and it entered through the
house of the Adelantado don
Pedro de Alvarado, may he find
glory.
. . . and took them all down
where they gave their spirits
to their creator.

1541   9 XII J   11.16.2.2.8   5 Lamat 5 Mol Mayapan
1541   9 XII J

ti haab
del mil quinientos quarenta y uno.
181 tz'uul
at. 5:
dik: 9
2nhele.
(*Chumayel* 1361–66)

In the year
A.D. 1541
The 181st [day] of the foreigners
At Merida
[Was like] Dec 9th
Of the Four Changers.

Partially in rebus writing, the text asserts that the end of June in the Christian calendar is like the midpoint of the Mayan year at the winter solstice. The assertion is not intended to be numerological: 5 Mol is the 145th day of the Mayan year.

1544   2 VI J   11.16.4.11.14   10 Ix 1 Kayab M
                                      10 'Idzo 16 K. Tilantongo

A1. Jaguar	B1. year Flint
A2. 10	B2. 12 (Tilantongo)

(*Yanhuitlan*, pl. XX)

The date is presumably that of the codex. It is given by Jiménez (1940:73), who calculates that it falls on 16 K. Aztec or Mixtec. It is a complete calendar-round date, because 10 Jaguar occurs only once in a year 12 Flint.

1544	17 VII J	11.16.4.13.19	3 Cauac 1 Pop M	
			13 Cauac 1 Pop Mayapan	

. . . pax ci cah tu men ma ya
cim lal lae
hoyl kan ah cuch hab tu h
[un te pop]
[i]chil hab 1534 años . . . hu
he tun tu uaxac la hun yax
kine
1535 años uac muluc ah
[cuch hab tu] hun te pop
[uuc] ahau he tun tu uac te
yax kinne
he tun tu bu luc te ceh

años uuc hix ah cuch hab tu
hun te pop
ox ahau he [tun] tu uuc te
yax kinne
1537 años uaxacil ca uac
tu hun te pop cin ci o[b ah]
pul haob te otz male
hek laob lae ah tz'un tu
tul xiu ye tel ah ci ya na
puc chi ye na may che ye
na may tun ye ah men e uan
ha . . . ui ni cob te ma nie
ah pul ha ob tu chi chen y
tza cu chi he u putz' [a ho]
be na hau uech na pot co uoh
tu la hun hi çip
lah ca a hau hi he tun tu
ca te yax kin e
bay bin ka he bal
1538 años bo lo kan ah
cuch hau tu hun te pop
uchci chac y kal u . . .
hin tah ci mil lae
ua xac a hau he tun tu uac
la hun xule
1539 años la hun mu luc tu
hun te pop
can a hau he tun tu bu luc
te xule
1540 años buluc hix tu hun

. . . abandoned the town because
of painless death there
[In 1534] 5 Kan was the year
bearer on the f[irst of Pop].
In the year 1534 it was . . . [?tun 14];
that tun was on 18 Yaxkin
[completed].
In 1535, 6 Muluc was the year
[bearer] on the first of Pop.
[7] Lord that tun was on the
sixth of Yaxkin [current].
That tun was on the eleventh of Ceh
[current].
The year 7 Ix was the year bearer on the
first of Pop [current].
3 Lord [was the tun] on the
seventh of Yaxkin [current].
The year 1537, 8 Cauac on the
first of Pop [current], the
water throwers died there at Otzmal.
They account for the Head of
the Toltec Xiu and Ah Ziyah, Na
Puc Chi and Na May Che and
Na May Tun and the curer Euan
Ha . . . people there at Mani who
were rain pilgrims to Chichen
Itza. There escaped
Na Ahau Pech and Na Pot Couoh
exactly on 10 Zip.
12 Lord indeed, that was the tun on the
second of Yaxkin [current].
Thus it will be remembered.
The year 1538, 9 Kan was the
year bearer on the first of Pop [current].
There was a hurricane [causing]
death here.
8 Lord, that was the tun on
the sixteenth of Xul [completed].
The year 1539, 10 Muluc was
on the first of Pop [current].
4 Lord, that was the tun on
the eleventh of Xul [current].
The year 1540, 11 Ix was on

te pop
ox la hun a hau he tun tu
uuc
1541 años lah ca bil ca
uac tu hun te pop
bo lon a hau he tun tu ca
te xule
1542 años ox la hun kan tu
hun te pop u hetz' ci cah
españoreçob ti hoo
cah ci ob yax hopp ci pa ta
no be tu men ah ma ni ob y
et u pro bin ci a il
ho a hau tu uac lahun te
çeec
hun a hau hi tu bu luc te
çeec
ca hix u hun te pop 1544
años
la hun a hau tu ua te çeec
1545 años ox lahun ca uac
tu hun te pop
hoppci xp̄o ti anoil tu men
fray le çob uay ti cah lae
he u kaba u pa dre il lob lae
fra y luis ui lla pan do
fra y di e go de ue har
fray ju° de la puer ta
fray me chor de be na ben te
fra y ju° de he rre ra
fray an gel po cob tok
u he tz'a hob te ti cah ti
hoe
uac a hau he tun tu hun te
çeec
(*Oxkutzcab* 66)

the first of Pop [current].
13 Lord, that tun was on the
seventh [of Xul, current].
The year 1541, 12 Cauac was on
the first of Pop [current].
9 Lord, that tun was on the
second of Xul [current].
The year 1542, 13 Kan on
the first of Pop [current],
the Spaniards seated the town at Merida.
They settled it and first be-
gan their tribute from the
people of Mani and the provinces.
5 Lord was on the sixteenth
of Tzec [current].
1 Lord indeed was on the
eleventh of Tzec [current].
2 Ix was on the first of Pop
[current] in the year 1544.
10 Lord was on the sixth of Tzec [current].
The year 1545, 13 Cauac was
on the first of Pop [current].
Christianity was begun by the
friars here in the city.
Here are the names of the fatherships:
Fray Luis Villalpando,
Fray Diego de Béjar,
Fray Juan de la Puerta,
Fray Melchor de Benavente,
Fray Juan de Herrera,
Fray Angel scraped flint.
They seated it there in the
city of Merida.
On 6 Lord the tun was on the
first of Tzec [current].

This is the best native correlation of the tun and the haab with the Christian year. Its evidence may be summarized as follows:

Year	Haab	Tun
1532		11.15.13.0.0
		2 Ahau 3 Mol T
1533	11.15.13.11.4	11.15.14.0.0
	5 Kan 2 Pop T	11 Ahau 18 Yaxkin T
	5 Kan 1 (1)	18 Yaxkin (2)
1534	11.15.14.11.9	11.15.15.0.0
	6 Muluc 2 Pop T	7 Ahau 13 Yaxkin T
	6 Muluc 1 Pop (1)	Ahau 7 Yaxkin (1)

Year	Haab	Tun
1535	11.15.15.11.14	11.15.16.0.0
	7 Ix 2 Pop T	3 Ahau 8 Yaxkin T
	7 Ix 1 Pop (1)	3 Ahau 7 Yaxkin (1)
1536	11.15.16.11.19	11.15.17.0.0
	8 Cauac 2 Pop T	12 Ahau 3 Yaxkin T
	8 Cauac 1 Pop (1)	12 Ahau 2 Yaxkin (1)
1537	11.15.17.12.4	11.15.18.0.0
	9 Kan 2 Pop T	8 Ahau 18 Xul T
	9 Kan 1 Pop (1)	8 Ahau 17 Xul (2)
1538	11.15.18.12.9	11.15.19.0.0
	10 Muluc 2 Pop T	4 Ahau 11 Xul T
	10 Muluc 1 Pop (1)	4 Ahau 11 Xul (1)
1539	11.15.19.12.14	11.16.0.0.0
	11 Ix 1 Pop M	13 Ahau 7 Xul M
	11 Ix 1 Pop (1)	13 Ahau 7 (?)
1540	11.16.0.12.19	11.16.1.0.0
	12 Cauac 1 Pop M	9 Ahau 2 Xul M
	12 Cauac 1 Pop (1)	9 Ahau 2 Xul (1)
1541	11.16.1.13.4	11.16.2.0.0
	13 Kan 1 Pop M	5 Ahau 17 Tzec M
	13 Kan 1 Pop (1)	5 Ahau 16 Tzec (1)
1542	11.16.2.13.9	11.16.3.0.0
	1 Muluc 1 Pop M	1 Ahau 12 Tzec M
	(omitted)	1 Ahau 11 Tzec (1)
1543	11.16.3.13.14	11.16.4.0.0
	2 Ix 1 Pop M	10 Ahau 7 Tzec M
	2 Ix 1 Pop (1)	10 Ahau 6 Tzec (1)
1544	11.16.4.13.19	11.16.5.0.0
	3 Cauac 1 Pop M	6 Ahau 2 Tzec M
	13 Cauac 1 Pop (1)	6 Ahau 1 Tzec (1)

(1) = current; (2) = completed; T = Tikal; M = Mayapan.

The dates given for the beginning of the haab are all correct except for 1542 (omitted) and 1544 (miscopied?). The dates given for the endings of the tuns are correct in the Mayapan calendar for 1535, 1536, 1539 and 1540. The date given for 1537 is correct in the Tikal calendar. Because two of them are stated in completed time, it is of interest that reading all the tun ending dates that way would make the dates for 1541, 1542, 1543, and 1544 correct in the Mayapan count, which may have been the original author's intent. The remaining three dates (1533, 1534, and 1538) would still be wrong and appear to be transcription errors. The tun endings remain just that: there is no reference to the initial-dated katun 11 Ahau that begins the Colonial (Mayapan) katun count in 1539 (11.15.19.14.0).

It has been pointed out to me by Victoria Bricker that a simpler construal of the dates may be possible. Taken at face value (i.e., ignoring the nicety of completive counting and the inauguration of the Mayapan calendar), six of the tun ending dates use a constant "error." These six could be explained by invoking a hypothetical Oxkutzcab calendar of Type V.N.1. The suggestion is intriguing because it would also explain another aberrant date (see Mathews 1982). The fact that this pattern is confined to the tun ending and does not appear in the haab dates of the *Oxkutzcab* leaves me puzzled but unconvinced. The dates in question are the tun endings for 1537, 1538, 1541, 1542, 1543, and 1544. Alternative explanations have been offered above for all of these except the 1538 date. It may be noted that no other calendar exists with a Type V New Year.

1545   21 II J   11.16.5.6.18   1 Etz'nab 20 Zac M
                                          1 Flint 20 F. Teotitlan

çe tecpatl. . . . 1544. . . . ome acatl   (Year) 1 Flint (Teotitlan) = 1544 =
                                                                   2 Cane (Tepexic)
(*Book of Mimiahuapan*;
Brotherston 1983:174; fols. 1v, 1r, 17r);
Robertson and Robertson 1975:269)

The assertion is that the year 1 Flint Teotitlan was 2 Cane Tepexic. The notes must have been made between the Teotitlan New Year on 27 II 1544 and that of Tepexic on 12 VI 1544, when the Tepexic date would have become 3 Flint.

*1548   29 II J

After this date, several of the Indian calendars froze their correlations to the Julian calendar and stopped counting leap-year days. The dates are discussed in the summaries of the corresponding calendars. Their frozen New Year's dates are listed here:

17 I	1610	Aztec
15 V	1691	Chiapanec (Chiapa de Corzo)
4 VI	1691	Nimiyua Chiapanec (Suchiapa)
4 VI	1720	Pokom
14 VII	1931	Kekchi
16 VII	1553	Mayapan
27 XII	1936	Lachixola Chinantec
27 XII	1950	Lalana Chinantec
25 II	1553	Texcoco

For other frozen calendars see A.D. *1584 and *1617.

*1549   1 I J

exemplo ogaño de 1549	Example: the present year
tienen 4 calli xiuitl por año . . .	of 1549 they have Anno 4 House
el año de 1550	[Cuitlahuac] for the year . . . the
estarán en la 31 casa y	year 1550 they will be in the
será 6 tochtli xiuitl . . .	31st house and it will be Anno 6
este año presente es 6	Rabbit [Colhuacan] . . . this present
calli xiuitl, començó 1	year is Anno 6 House [Chalca]; it
de henero . . . es de notar que	began January 1st . . . it is to
el año de 1549 estava en	be noted that the year 1549 was
la 31 casa dela rueda mayor	in the 31st house of the larger
que es 5 calli xiuitl. . .	[52-year] cycle [Toltec] which is Anno 5
	House [Colhuacan].
exemplo: a 8º. de octubre	Example: on the 8th of October
deste año de 49 fue	of this year of [1548] it
el primer día de su quin-	was the [fifth] day of their
çeno mes que es 22 calli,	[fourteenth] month [Aztec]
oy a 20 de octubre, buscando	which is [13] House, searching
el caracol adelante dende 22	the wheel forward will land on
calli van a dar en 20 de octubre,	the 20th of October on [12]
en 22 quanhtli et sic de	Eagle, and so forth.
aliis.	
(Motolinía 1549 in Baudot 1983:	
429–30)	

The passage is badly garbled by an eighteenth-century copyist but is closely similar to the calendrical note for 1553 (q.v.) by Las Navas. This one manages to jumble five different calendars and yields no defensible correlational date in any of them.

1551   2 XII J   11.16.12.4.13   6 Ben 20 Yaxkin M
                                       6 Huiyo 20 "Flag" Tilantongo

1551 cuyya ñu-huiyo	1551 [began in] the year 6 Cane
	[Tilantongo]
(*Sierra*, quoted by Jiménez	
1940:70)	

This correlates the Tilantongo year with the Christian one.

1553   5 III J   11.16.13.9.12   10 Eb 14 Ceh M
                                       10 Malinalli 15 Tlacaxipeualiztli Texcoco

El año de mil y quinientos	The year 1552
y cinquenta y dos fue	was
su año de ocho tecpatl	their year 8 Flint Anno
xiuitl, fue bisiesto aquel	[Aztec]. There was leap year
año de los quince dias	that year on the fifteenth
del tercero mes sobre la	day of the third month [Texcoco]

figura diez malinalli, que	on the figure 10 Grass,
cayó a veynte y cuatro de	which fell on February 24th,
Hebrero y sobre esta figura	and on that figure they will
se haran dos dias, diziendo	make two days, saying
oy diez malinalli, mañana	today 10 Grass, tomorrow 10
diez malinalli y luego	Grass and then go on to the
proceder al día siguiente	next day 11 [Cane].
onze malinalli	
(Las Navas 1553:120)	

The date given for 1552 is 8 Flint Aztec (or 1 Flint Texcoco), the name day falling therefore on 1 Flint 20 Atemoztli Texcoco on 6 XII 1548 (frozen). Thus 10 Malinalli 15 Tlacaxipeualiztli would fall on 25 II J in 1553. This is one of many attempts to provide the native calendar with a leap year. There is no confirmation that it worked because this is the last date in the Texcoco calendar, but it would seem to be contradicted by the fact that the Julian date given is frozen.

1553   6 IV J   11.16.13.11.4   3 Kan 6 Kankin M

                                         3 In Xichari 1 In Thacani Matlatzinca

Caso (1946) reconstructs this date from a lost Boturini MS. published by León (1903), Ramírez (1600), and Chavero (1892) and commented upon by Veytia (1836). Eleven In Bani (= 11 Calli) was the name day of the year.

1553   15 VII J   11.16.13.16.4   12 Kan 1 Pop M
          16 VII J   12 Kan 1 Pop Mayapan

El primer día del año	The first day of this people's
desta gente era siempre a	year was always on the
16 días de nuestro mes	sixteenth day of our month
de julio, y el primero de sus	of July and the first of their
meses (era) de Popp.	months (was) Pop.
(Landa 1966:71)	(Cf. Tozzer 1941:150)

Landa's famous correlational statement was absolutely correct as far as the Maya were concerned: they had decided to stop counting leap years after 1548 and considered the Julian calendar to be frozen to their own from that date forward. Almost all Yucatecan correlational statements are premised on dating July 16th to 1 Pop (see Chapter 5, Julian).

1555   16 III J   11.16.15.10.13   10 Ben 5 Mac M

                                           10 Xo' 20 G. Tilantongo

porque de aquí comenzaban	Because from here they began
el año a los 24 días de	the year on the 24th day of
febrero en esta nación	February in this Mexican

mexicana; porque en la
Zapoteca y Mixteca comenzaban
a los 16 días de marzo,
y así variaban en estos
20 días, que es su mes,
aunque tenían el mismo
año de 365 días y 18
meses y las mismas 20 letras
o signos apropiados a ellos.
(*Vaticanus 3738*:42v.)

nation; because in the
Zapotec and Mixtec they began
on the 16th day of March,
and thus they differed by
these 20 days, which is their
month, although they had the
same year of 365 days and 18
months and the same 20 letters
or signs appropriate to them.

This note by Pedro de los Ríos (Jansen 1984) is generally correct but not quite accurately stated. The date given is the correct date for the Tilantongo name day of the year 1555: 10 Cane, but this is the first occurrence of that day in that year, rather than what was presumably its decisive occurrence 13 months later as the 360th day of the year. This was thus not the beginning of the Tilantongo year but the occurrence of its first name day. The date 20 days earlier would have occupied the same position in only one of the calendars of the "Mexican nation": that of Metztitlan. Neither date is directly relevant to Zapotec, though the first name day of the Zapotec calendar would be simply the day before Tilantongo's, hence 15 III.

1555   16 VI J   11.16.15.15.5   11 Chicchan 17 Kayab M
                                     11 Kòo 12 "Mace" Tilantongo

A1. 10	B1. 11	C1. 6
A2. year Cane	B2. Serpent	C2. Cane
	B3. month Mace	
D1. 10		F1. 11
D2. year Flint	E2. month Flag	F2. Death (Tilantongo)

(Cuilapa Cornerstone, quoted
by Jiménez 1940:69)

The assertion is that the construction of the convent at Cuilapa was begun between 11 Serpent 12 Mace and 6 Cane 20 M. Tilantongo, May 5 and June 5, 1555 Julian, and that it was completed on 11 Death 8 Flag Tilantongo, November 15, 1568 Julian. This identifies the glyphs for the Mixtec months L. (Mace) and B. (Flag). The latter is the equivalent of the Aztec Panquetzaliztli; the former (*pace* Jiménez) is Tecuilhuitontli.

1557   15 VII J   11.16.17.17.15   1 Men 12 Pop M
                                      1 Tz'ikin 16 Ru Kab Mam Cakchiquel

Oxlajú güinak
guaé juná

During the thirteenth month
Of this year,

tok xicó gij Sanctiago	The day of Santiago fell
Pangán,	In Antigua.
ja chi jun Tziquin	Right on 1 Bird
chupam ka gij	In our time.
aok ixban nimá quicotem	A great celebration was held
cumá Castilán güinak Pangań.	By the Castilians of Antigua
Jaok xoc cajaual chilá	Because our lord was proclaimed there
Castilán	In Castile.
jaok xoc ajauj don Feliphe	The lord Don Felipe was proclaimed
Emperador.	Emperor.
(*Annals of the Cakchiquels* 278)	

The date falls in the eighth month Cakchiquel (*Ru Kab Mam*) but in the thirteenth month Iximche (2.18.12.16 Iximche). The day of Santiago, however, is 2 days later.

1559   1 V J   11.16.19.14.0   9 Ahau 12 Muan M

                                        9 Ahau   Mayapan

bolon ahau	[In katun] 9 Lord [Mayapan]
hoppci	It began.
xp̄oil uchci	Christianity occurred,
ca put çihile	And rebirth.
lay tal	That was its arrival
ychil u katunil	In the cycle.
hulci obispo toral	There arrived Bishop Toral,
ua xan e hauci kuy tab e	But he did not end the sacrifices.
(*Chumayel* 141–46)	

Bishop Francisco de Toral arrived in Merida in 1562 (Barrera 1948:65), in the third year of katun 9 Ahau. The date cited is the beginning of that katun in the Mayapan count.

1562   29 XI J   11.17.3.7.8   4 Lamat 20 Yaxkin M

                                        4 Sayu 20 B. Tilantongo

Taquiyeni nicanañahandi	And then impatiently
Toho nisano	I called the elder,
Nani	An old man,
Franco Pérez,	Francisco Pérez,
Nisicatnuhundi dzina	And I asked him first
Nisiyndito:	Saying:
"Ñanasadzaha cuiya	"How many years is it
Nicacundo?"	Since you were born?"
Nidzaduidzoto,	He answered
Nicachito:	Saying:
"Cuiya qhu sayu	"In the year 4 Rabbit,
De 1562 años.	In 1562.
Huitna sa-uni dzico u hui cuiya	Now it is 62 years

Nicacundi."   Since I was born."
(*Probanza de Miltepec* 1622;
Jansen 1980:2)

Francisco Pérez seems to have a better memory for years than for elapsed time: he
must have been 60 in 1622. The year correlation is another independent confirmation of the Tilantongo year count.

*1567   2 II J   11.17.7.11.14   9 Ix 6 Zac M
                                          9 Ocelotl 1 Atlcaualo Tepepulco

El primero mes del año se   The first month of the year
llamaua entre los mexicanos   was called among the Mexicans
Atl caoalo: y en otras   Atl caoalo: and in other
partes quavitl eoa. Este   places Quauitl eoa. This
mes començaua en el segundo   month began on the second
día del mes de Hebrero   day of the month of February
quando nosotros celebramos,   when we celebrate the
la purificación de nuestra   Purification of Our
señora.   Lady.
(Sahagún 1975–81:2:1)   (Anderson and Dibble 1981:2:1)

D'Olwer and Cline (1973:192) place the writing of this passage in 1565–1568.
Along with Torquemada's date of 1611 (q.v.), this establishes the beginning of the
Tepepulco year.

tuvieron estos indios el   These Indians had the year
año dividido en diez y   divided into eighteen
ocho meses, cada mes de   months, each month of
veinte dias, que hacen nú-   twenty days, which make a
mero de trescientos sesenta   number of three hundred sixty
dias, y a los cinco que   days, and the five lacking
falta para el cumplimiento   for the completion of our
de nuestro año llaman en   year they call in the
la lengua otomí dupa   Otomi language *dupa*
(dias muertos), y a los   (dead days), and to the
mexicanos, dias aciagos,   Mexicans, fateful days,
siendo su primer día   their first day of the year
del año el 2 de febrero   being the 2nd of February,
por lo cual estos falsos   wherefore these false
sacerdotes cuidaban mucho   priests take great care of
de la fiesta de la Purifi-   the feast of the Purification,
cación, que ellos llaman   which they call the blessing
la bendición de las can-   of the candles.
delas.
(García 1918:301)

This datum appears to settle the equation of the Otomi and Tepepulco calendars, and placing García's otherwise undated correlation in *1567 fits all the evidence even though it is not independently attested (see Caso 1967).

*1573   31 I J   11.17.13.12.4   2 Kan 6 Zac M
                                          2 An Botaga 1 Am Buoentaxi Otomi

Reboltorio. Enero—Aquarios.   Report. January—Aquarius.
31. Ancãndehe.   31. An Candehe. An Buoendaxi.
anbuoendãxi. anttzâyo.   An Ttzayo.
(Santiago 1632:13)   (Alvarado 1976:61)

This is an unfrozen date placing the beginning of the Otomi year on 1 Am Buoendaxi and, by implication, dating the writing of this page of the *Huichapan Codex* to 1573, when that fell on 31 I Julian. A note on the same page equates the month with Xilomaniztli, and hence with Zac (see A.D. 1567).

1579   19 II J   11.17.19.15.14   2 Ix 6 Ceh M
          1 III G   2 Ocelotl 1 Cuahuitlehua Teotitlan

Primero día del mes de   On the first day of March this
marzo celebraban estas   [sic] nation formerly celebrated
naciones antiguamente el   its New Year,
año nuevo, como nosotros   just as today we
agora celebramos el primero   commemorate January 1.
de enero.   (Durán 1971:412)
(Durán 1967:1:239)

For reasons unexplained, this gives the Gregorian date for the Teotitlan New Year. It is dated to 1579 by Durán himself (1971:383). The year was 10 Cane Teotitlan or 9 Cane Aztec. Durán uses Teotitlan month names except for Ob. and Fa. (see Nahuatl).

THE GREGORIAN CALENDAR

1584	2 II G	11.18.4.15.13	7 Ben 20 Yax M
	2 II G		7 Ah 8 Izcal Cakchiquel

chu can gij	On the second day
ic ebrero	Of the month of February,
mi xicó ru gij	On which fell the day
Sancta Maria Purificación	Of Santa María de la Purificación,
x utzí ricax güi candelas	They blessed the candles
mix jel catij lajuj chi gij	And changed ten days
makuí xajilax	That should not be counted
can ca rulic chic rupí	Because the order came
xa kaní ma tatá	From our Great Father,
Sancto Padre de Roma	The Holy Father of Rome,
caí güi juná rodal	Of the year '82
juná rocal rubanic juyuj	On the fifth may of the Revolution
patinamit	In the city
chi Iximché.	Of Iximche.
(*Annals of the Cakchiquels* 296)	

This was the ninety-first year since the Revolution, 83 Iximche *juná* plus 15 days (4.3.0.15 Iximche). Recinos (1950:161) says eighty-sixth or eighty-first on different parts of the page, but the text specifies only that it was between 80 and 100. The reference to (15)82 is to the original promulgation of the Gregorian calendar (q.v., chapter 5). The news reached the Cakchiquel in 1584, and they dutifully adopted it at once (including the citation of this date), and used it consistently hereafter.

*1584   29 II G

After this date, the Tzotzil and Tzeltal froze their calendar correlations to the Gregorian year and stopped counting leap-year days. The dates are discussed in the summaries of the corresponding calendars (chapter 5). Their frozen New Year dates are listed here:

28 XII	1688	Tzotzil
17 I	1888	Tzeltal
3 III	1688	Guitiupa
23 III	1845	Istacostoc
12 IV	1845	Mitontic
18 V	1917	Cancuc

1589   2 XII G   11.18.10.14.3   5 Akbal 20 Yaxkin M
                                          5 Wa'i 20 Flag Tilantongo

                                          A1. Eagle
                                          A2. 5 House (Tilantongo)
                                          A3.
                                          B2. A.D. 1589 (Gregorian) it began;
                                          B3. A.D. 1590;
                                          B4. A.D. 1591 it has been finished.
                                          C2. 7 Cane (Tilantongo)

5 Wa'i
(?)
1589 años peuhqui
1590 años
1591 años (tl)amico
7 'uyo.

(Tecamachalco Convent Façade,
quoted by Jiménez 1940:69)

This correlates the Tilantongo and Christian years. The text is Nahuatl.

1593   15 VII G   11.18.14.8.4   13 Kan 1 Pop M
                                         13 Kan 1 Pop Mayapan

ox [la] hun kan   13 Iguana
t u hun te pop   On the first of Pop [current: Mayapan]
ch'ab u lac katun   Was taken the plate of the katun
ti ho ahau   Which was 5 Lord.
1593 cuchi   A.D. 1593 it occurred;
t u holhun seec   On 15 Tzec [completed]
y alkaba   It dawned.
heklai   The relation
u cuch   Of the events
lic u tal   As they occurred
u alic lae   May be said thus:
he uil   Indeed the moon
t u kinil   Was at the sun:
hi u ch'abal   That was the creation
katun lae   Of this katun.
Mayapan u u ich   Mayapan was the face
u kex katun.   Of the change of the katun.
(*Tizimin* 1549–62)

The ceremony of taking the plate of the katun took place 5 tuns before the beginning of katun 5 Ahau (1598), hence in the year 13 Kan (1593). The "dawn" (i.e.,
the day before) was 4 Cauac 16 Tzec Mayapan, but the text gives it as "15 Tzec
having passed." It goes on, however, to specify dating in the "changed" Mayapan
katun, with its moon at the sun, that is, its end at the beginning. The actual ceremony was presumably held on 5 Ahau 17 Tzec Mayapan (11.18.14.13.0), at the
end of a uinal of the tun having the right name, rather than waiting for the tun
ending 7 uinals later, which had the wrong one. Katun 5 Ahau was important
because it was the end of an even baktun (12.0.0.0.0 Tikal), but it is the beginning
of the Mayapan katun that is anticipated here.

| 1596 | 17 II G | 11.18.17.1.11 | 11 Chuen 18 Zac M |
| 1544 | 16 II J | | 11 Chuen 18 Zac Mayapan |

ti tz'oc in tz'aic	I have finished giving
uooh lae	These glyphs
t u uaxac lahun te sac	On the eighteenth of Zac [current: Mayapan]
ti buluc chuen	On 11 Monkey,
t u holhun pis kin	On the fifteenth measured day
febrero	Of February [completed: Julian],
1544	A.D.
hab.	1544.
(*Tizimin* 2975–80)	

There are three common ways of expressing dates in Yucatec (cf. 1593 text): (a) t u
holhun seec 'on the 15th of Tzec (completed)' = 16 Tzec; (b) t u uaxac lahun te sac
'on the 18th of Zac (current)' = 18 Zac; and (c) t u holhum pis kin febrero 'on the
15th measured day of February' = 16 February. The expression here is therefore
remarkably explicit and is correct as stated (i.e., for 1544), even though the context
places the event one calendar round later (and 13 leap year days earlier in the
Julian calendar). The earlier date is given to add authority to the prophecy it ter-
minates, the actual date of composition of which is probably 1618–23.

| 1596 | 3 VI G | 11.18.17.6.18 | 1 Etz'nab 5 Kayab M |
| | | | 1 Tecpatl   Teotitlan |

| Ze tecpatl, ome acatl . . . 1596 | 1 Flint (Teotitlan) = 2 Cane (Tepexic) = 1596 |

(*Tizayuca* fol. 17 r.;
Robertson and Robertson
1975:274)

This correlation in the Techialoyan *Codex Kaska* of Tizayuca repeats the identification of the Teotitlan calendar at Mimiahuapan in the previous calendar round (see 1544). The year corresponds to 13 Flint Aztec, but the note was made sometime between the Teotitlan New Year on 25 II and that of Tepexic on 24 VI, when the Tepexic date would have become 3 Flint.

1606	27 I G	11.19.7.3.3	3 Akbal 20 Yax M
	17 I J		3 Calli 15 Izcalli Aztec

Nican ompehua	Here begins
ontzinti	[and] commences
yn huehue Mexica Metztlapo-	the ancient Mexica month
hualli	count
yn iuh nican Metztlapohuaya	as in the past those counted the months
y ye huecauh	who were
huehuetque	the ancestors
catca Mexica tenuchca	of the Mexico Tenochca
yhuan tlatilulca	and Tlatelolco
yn onemico	when they lived
ipan oc ce cahuitl	in different times
Çan cecenpohualilhuitl	They arranged that they
quitlaliaya	would assign only 20 days
in cecen Metztli	to each month.
ynic caxtollomey metztli mochihuaya	so that there were 18 months
impan y contzonquiçaya ce	with which they completed
xihuitl	a year.
Auh in inyancuicxiuh	And the New Year
yn omoteneuhque huehuetque	Of the said ancestors
catca oncã conpehualtia	began on
yn ipan yc 18 mani	the eighteenth
metztli	of the month
Enero.	of January.
çan oqu iuhqui ayemo	But they did not yet then
niman yhuan quihuicalatiaya	make it go along with the count
in Tonalle cecemilhuitlapohualli	Of each of the day signs,
ca macuilpohualilhuitl ipan	for there were one hundred
macuilhuitl	and five days
yn amo tonalle cecemilhuitl	each of which had no day sign,
auh ca çan oc mixcahuava	and these months of the
yn inmetz huehuetque	ancestors were just still neglected
yn tlapohuaya	as they made their calculations.
auh ca quin ompa ipan conpehualtiaya	But later they began
yn incemilhuitonalpohualliz	their count of the day signs;
yn ipan yc cemilhuitl ompehua	it began on the first day
christiano Metzpohualli	of May in the Christian
Mayo.	month count.
(British and Foreign Bible	(Anderson 1985)
Society Manuscript 374,	
unpaginated)	

Calendrically, the passage is complex. It begins the Aztec year on January 18 and the day count on May 1. For both statements to be true, the first date must be read as a frozen Gregorian date of 1585 and the second as a current Julian date of 1605. A table later in the manuscript confirms this intention by listing a day count ending on January 15 with 13 Xuchitl (as it should if the day count began the preceding May 1). This leads to the reconstruction of the true unfrozen Aztec date listed in the heading, implying the beginning of the year 10 Rabbit on 2 Atl 1 Izcalli on 4 I J or 14 I G in 1606, and hence on 17 I J 1549 or 18 I G 1585.

The author has confused the matter in two respects. He cites the frozen Gregorian date for the New Year but applies it to the Julian calendar. The day count can begin on May 1 only once in 1,040 years (see Julian). It does so in 1605 in the Julian count. It did so in A.D. 1111 in the Gregorian count and will not do so again within the era. Second, he lists January 18 in an appended calendar as falling on the first of Tititl rather than Izcalli, implying that he is using the Texcoco rather than the Aztec calendar. (Many of the associated documents in the source manuscript are Texcocan.) However, the Texcoco New Year was 20 days earlier than Aztec. It should have fallen on January 18 Julian in 1633 or Gregorian in 1593, but neither date has any relation to the May 1 day count of 1605.

Although it documents the imminent collapse of the Aztec day count, the manuscript demonstrates that the Aztec calendar is among those that adopted frozen correlations, probably as of 1549, and certainly in 1585, and that it was still precariously alive in 1606.

1611	1 II G	11.19.12.4.14	1 Ix 6 Zac M
			1 Ocelotl 1 Atcahualo Tepepulco

El primer mes con que estos mexicanos comenzaban su año . . . llamaban atlacahualco o quahuitlehua, el cual corresponde al nuestro febrero y comenzaba en el primer día de él. (Torquemada 1976:3:364)	The first month with which these Mexicans began their year . . . they called Atlcahualco or Quahuitlehua, which corresponds to our February and began on the first day of it.

The citation is to Book 10 of Torquemada's *Monarquía Indiana*, probably written in 1611 (Alcina Franch 1973:260). It establishes the beginning of the Tepepulco year.

*1612	1 IV G	11.19.13.7.19	10 Cauac 6 Kankin M
			10 Muxi' 1 Ahit Huastec

This date, cited by Lehmann (1920:2:880) from marginal notes on the *Codex Mexicain 65–71* (fol. 94v.–102v.), establishes the beginning of the second month in the Huastec year if the notes were made between 1612 and 1615 in the Gregorian calendar. Any other assumption would make a definitive placement of the Huastec calendar highly problematic (see Caso, 1954, and chapter 5, Huastec).

```
*1613    11 I G      11.19.14.4.4    9 Kan 6 Yax M
         1 I J                       9 Iguana 1 Chạ Mê Mazatec
```

This frozen date was collected in 1936 by Weitlaner and Weitlaner (1946:195). If it was frozen in 1617 when it first landed on 1 I J, it makes the Mazatec calendar congruent with the Aztec.

```
1618     18 IX G     12.0.0.0.0      5 Ahau 12 Zotz' M
                                     5 Ahau        Tikal
```

Chapter 29 of the *Book of Chilam Balam of Chumayel* (Edmonson 1986) outlines the Baktun Ceremonial, a ritual program in twenty acts, dated to katun 3 Ahau Mayapan (11.19.19.14.0), implying that the ceremonial was held on the correct Tikal date.

```
*1629    28 I G                                  Chimalpahin
         18 I J
```

Studies by Glass (1975a) and Prem (1983) have discussed the ten correlational dates given by Chimalpahin in his *Compendio de la historia mexicana* (CH) and in the *Crónica mexicáyotl* (CM) attributed to him. The years are named and correlated in the Aztec calendar, but for correlating days, Chimalpahin assumed the year to begin 1 Tititl, as in the Texcoco calendar, and he assigned this to 18 I Julian, which was correct after 29 II 1628. In a second fiction, he began the year with 1 Cipactli for the purpose of calculating the month position of the days, thus placing them where they could not possibly land in any Middle American calendar except Olmec, and it is upon these fictions that he arrived at the European date. On the evidence, he continued to write and revise his calculations until 2 leap years had passed, thus introducing dates based on 17 I and 16 I New Years, as listed in column 2 in the next figure. Because we have no independent way of documenting when Chimalpahin wrote the passages from which these dates are taken, none of this constitutes a genuine calendar correlation, though if we did, it would be possible to reconstruct at least those dates for days occurring only once in their respective years.

```
1369 18 I Anno 7 Calli          8 Ocelotl 14 Izcalli = 20 II (CH)
1393 18 I Anno 5 Calli          5 Coatl 5 Tititl = 22 I (CH)
1415 17 I Anno 1 Acatl          3 Coatl = 21 VII (CM)
1427 17 I Anno 13 Acatl         13 Atl = 22 VII (CM)
1428 17 I Anno 1 Tecpatl        13 Atl = 22 VI (CH)
1440 18 I Anno 13 Tecpatl       3 Coatl = 22 V (CM)
```

1469 17 I Anno 3 Calli	11 Quiahuitl = 11 VIII (CM)
1481 16 I Anno 2 Calli	6 Cozcaquauhtli = 2 VI (CH)
1502 16 I Anno 10 Tochtli	9 Mazatl 7 Tozoztontli = 14 IV (CH)
1520 18 I Anno 2 Tecpatl	7 Cipactli = 16 IX (CH)

1631	4 VII G 12.0.12.17.12	10 Eb 4 Uayeb M
	4 VII G	10 Eb 0 Pop Chol

Todos estos del Manché hablan una misma lengua, que es la Chol, y tienen unos mismos ritos y ceremonias y se gobiernan por unos mismos meses, dividiendo el año en dieciocho de a veinte días cada uno, y todos los veinte días tienen su nombre como lo tienen los días de la semana. Llaman al mes uinal; los veinte dias de él dividen en cuatro divisiones, cada una de a cinco, y los cuatro primeros de estas cuatro divisiones se mudan cada año para iniciar los meses. Son, según ellos dicen, los que toman el camino y cargan el mes, andan en rueda. Constan estos dieciocho meses de trescientos sesenta días, al fin de los cuales dan cinco, que llaman de gran ayuno, días que no tienen nombre. Con estos cinco días se cumplen los 365. Y sólo un yerro les hallé en esta cuenta, que es por ignorar los bisiestos. Y no hay que admirar, pues tantos años lo erramos nosotros hasta que la Iglesia lo enmendó, añadiendo un día en cuatro años en el mes de febrero por las seis horas que cada año tiene más de los 365 días de la cuenta del sol.

All these of the Manche speak one and the same language, which is Chol, and they have the selfsame rites and ceremonies and are governed by the selfsame months, dividing the year into eighteen of them of twenty days each, and each of the twenty days has its own name just as the days of the week have. They call the month *uinal*; the twenty days of it they divide into four divisions of five each, and the four first ones in these four divisions change every year to begin the months. They are, as they say, the ones that take the road and carry the month; they rotate. These eighteen months consist of three hundred and sixty days, at the end of which they give five which they call of great fasting, days which do not have a name. With these five days they complete the 365. And I caught them in only one error in this count, which is from not knowing about leap years. And it's no wonder because we had it wrong for so many years until the Church corrected it by adding one day every four years in the month of February for the six hours that every year has over the 365 days of the count of the sun.

Cúmplense estos dieciocho	These eighteen months are completed
meses a 28 de junio, que el	on the 28th of June,
postrero día del mes, y	which [is] the last day of the
entonces entran los cinco	month, and then come the five
del gran ayuno. Dura hasta	of the great fast. It lasts
tres de julio, y esta vigi-	until the third of July, and
lia para ellos es de gran	this vigil is for them of
veneración, de forma que	great veneration, so that
a cuatro de julio entra el	on the fourth of July comes
primer día del año	the first day of the year
según su cuenta. Tienen	according to their count.
señalado lo que se ha de	They have it indicated what is
sembrar en cada mes, así	to be sown in each month, both
de semillas como de legum-	of seeds and vegetables,
bres, sin que discrepen un	without missing a day of it.
día en ello. Han tenido	These people of the Manche
continua guerra estos del	have had an ongoing war with
Manché con los de Ajiça	those of the Itza, but
mas siempre han salido des-	they have always come off
calabrados porque son pocos	damaged because they are few
y los de Ajiza muchos, y	and the Itza many, and so
así los más años	most years they come in
vienen en el yazquin, que es	Yaxkin, which is the summer,
el verano a llevar presa,	to take prisoners, as they did
como lo hicieron el pasado de	the past one of 1630
1630 que llevaron más de	when they took more than a
cien personas, y así volvían	hundred people, and so were
éste arregostados.	delighted to come back this one.

(Tovilla 1631:184)

This unfrozen Gregorian date places the Chols firmly on the Tikal calendar. The Guatemalan "summer" is the dry season from November to February.

*1632	30 III G	13 Ben 1 Ch'en M
	20 III J	13 Acatl 20 Tititl Ixtlilxochitl

These pseudocorrelational dates assembled by Prem (1983) from Iztlilxochitl's *Obras históricas* are based on Aztec year correlations. Column 2 in the following figure contains the reconstructed New Years' dates they imply. The month positions cannot be from authentic sources because they yield years beginning with their name days. It is curious nonetheless that Ixtlilxochitl's text omits redundant month dates, a mark of calendrical sophistication not consonant with the errors in the rest of his handling of the calendars. Dates 1 and 8 are particularly aberrant. Ixtlilxochitl started his calculations from an unfrozen but anachronistically Julian date for a year beginning on 1 Tlacaxipehualiztli (Teotitlan) on 20 III. This would have to be after 29 II 1632. He (or somebody else) apparently continued writing and revising through 5 leap years (i.e., until 1652), thus giving us a 5-day spread of New Years' dates back to 15 III (see column 2; Ixtlilxochitl, however, died in 1648). None of this gives us an acceptable correlational date, but it remains likely that the years and day-count days for the events chronicled were taken from authentic sources that probably used the Aztec and Teotitlan calendars.

1004	15 IV	Anno 1 Tecpatl	1 Ollin = 30 III
1359	19 III	Anno 1 Acatl	13 Tecpatl 6 Tozoztontli = 15 IV
1363	19 III	Anno 1 Acatl	13 Tecpatl 6 Atemoztli = 30 XII
1369	19 III	Anno 11 Calli	1 Mazatl = 30 X
1402	18 III	Anno 1 Tochtli	1 Mazatl 20 Tozoztontli = 28 IV
1418	15 III	Anno 4 Tochtli	5 Coatl = 10 VII
1418	19 III	Anno 4 Tochtli	5 Coatl = 24 VIII
1418	9 V	Anno 4 Tochtli	13 Cozcaquauhtli 18 Xilomaniztli = 22 IV
1418	20 III	Anno 4 Tochtli	10 Cozcaquauhtli = 16 VIII
1418	19 III	Anno 4 Tochtli	10 Cozcaquauhtli = 24 IX
1418	16 III	Anno 4 Tochtli	10 Cozcaquauhtli =21 IX
1427	16 III	Anno 13 Acatl	11 Cozcaquauhtli = 20 III
1427	20 III	Anno 13 Acatl	13 Acatl 1 Tlacaxipehualiztli = 20 III
1427	20 III	Anno 13 Acatl	4 Ollin 5 Tlacaxipehualiztli = 24 III
1427	17 III	Anno 13 Acatl	10 Xochitl = 23 VII
1427	10 III	Anno 13 Acatl	1 Cuetzpallin = 20 VII
1427	17 III	Anno 13 Acatl	1 Cuetzpallin = 27 VII
1427	19 III	Anno 13 Acatl	1 Ollin = 11 VIII
1465	19 III	Anno 11 Tecpatl	13 Coatl 8 Atemoztli = 1 I
1503	15 III	Anno 11 Acatl	1 Cipactli 9 Toxcatl = 24 V

1707	21 I G	12.4.9.11.6	5 Cimi 18 Zac M
1707	21 I G		20 Pa ri Che Cakchiquel

D'aprés l'auteur anonyme de la Cronica de la prov. de Goattemala, MS, l'année cak-chiquèle aurait commencé avec le 1er jour Tacaxepual, au 31 Janvier. Une note marginale, écrite d'une autre main, dit que le 1er du mois Pariché se trouva être en 1707, au 21 Janvier; ce que nous paraît plus en accord avec le reste, en mettant le premier jour de 1er Tumuzuz au 22 ou 23 Mars.

According to the anonymous author of the Chronicle of the Province of Guatemala, Manuscript, the Cakchiquel year would have begun with the first day of Tacaxepual, on the 31st of January. A marginal note, written in another hand, says that the 1st of the month Pariche was found to be on the 21st of January in 1707; which appears to us more in accord with the rest, in placing the first day of 1st Tamuzuz on the 22nd or 23rd of March.

(Brasseur 1858:3:466, Fn. 3)

The Cakchiquel year began on 1 Tacaxepual, which fell on 31 I J in 1645 and on 31 I G in 1685. In 1707, it fell on 27 I G. Thus the marginal note is actually dating the last day of Pa ri Che and the 360th day of the Cakchiquel year. If 21 I were the date of 1 Tacaxepual, 1 Nabe Tamuzuz would fall 20 days later, on 10 II, so Brasseur's final calculation remains mysterious. This is the last documentation of the Cakchiquel calendar.

1722	3 V G	12.5.5.2.7	9 Manik 4 Kayab M
1722	3 V G		9 Keh 1 Nabe Mam Quiche

9 Keh gut mixcha 20 qih   And 9 Deer took the 20 days
nabe mam d mayo 3   of First Elder, Thursday, May 3, 1722.

(*Chol*:1)

This date, the earliest in Quiche, establishes the beginning of the Quiche year.

1752	6 VI G	12.6.15.11.19	3 Cauac 1 Pop M
			3 Cauac 1 Pop Valladolid

cauac   [3] Cauac
u hun te pop   Was the first of Pop
                                                    [current: Valladolid],
y ahal cab   The dawn.
t u ca te u kinil   On the second day [current]
hab   Of the year
cutal can ahau.   Is the seating of [katun] 4
                                                    Lord.

(*Tizimin* 4819–26)

This is the inaugural date of the Valladolid calendar.

1770	13 III G	12.7.13.12.8	5 Lamat 5 Muan M
1770	13 III G		5 Q'anel 7 Kak Quiche

This date is from *Chol*:38.

*1778   18 III G   Clavijero

El primer día del segundo   On the first day of the second
mes que en el principio de   month which at the beginning
su siglo correspondía al   of their century corresponded
18 de marzo, hacían una   to the 18th of March, they
fiesta solemnísima al dios   held a most solemn festival
Xipe.   for the god Xipe.

(Clavijero 1964:182)

Clavijero elsewhere identifies this festival as falling on the first day of Tlacaxi-pehualiztli. The date he gives is for the first day of Tozoztontli, but the festival belongs to the last day of Tlacaxipehualiztli. He is at some pains to refer the date to the beginning of the calendar round, but it would not matter: it is a frozen Julian date of 1548. If this were the beginning of the second month, it correctly places the beginning of the Teotitlan year; Tlacaxipehualiztli is the second month of the Tepepulco calendar, but the date given is 20 days late for that. Clavijero's day count, names, and order of the Tepepulco months and (frozen) Julian dates are independently correct, but he has combined them in the wrong way, putting the date in question on an impossible 2 Cipactli. He begins the native "century" with 1 Tochtli. In the Tepepulco calendar, that would have been 1745. In the Teotitlan calendar, it was 1778, which is the probable date of reference of this passage.

| 1854 | 1 IV G | 12.11.18.17.7 | 11 Manik 4 Kayab M |
| 1854 | 1 V G | | 2 Noh 1 Nabe Mam Quiche |

This date is from Hernández Spina (1854; see Bunting 1932). It is 1 European month late.

| 1854 | 12 VII G | 12.11.19.4.9 | 9 Muluc 1 Zotz' M |
| 1854 | 12 VIII G | | 1 Hunahpu 4 U Kab Pach Quiche |

This date is from Hernández Spina (1854; see Bunting 1932). It is 1 European month late.

| 1927 | 16 III G | 12.15.12.17.13 | 7 Ben 5 Kayab M |
| 1927 | 16 III G | | 7 Ah 1 L.   Jacaltec |

This date is reported by LaFarge and Beyers (1931:174). It correlates the Jacaltec and Gregorian calendars.

| 1930 | 1 XI G | 12.15.16.11.19 | 7 Cauac 11 Yax M |
| 1930 | 1 XI G | | 7 Kavok 3 U Kab Zih Quiche |

This date is given by Goubaud (1937) as falling on a Saturday. It did.

| 1931 | 12 I G | 12.15.16.15.11 | 1 Chuen 3 Kankin M |
| 1931 | 12 I G | | 1 Baatz' 5 Tz'api Q'ih Quiche |

This date is given by Schultze-Jena (1933:32) as the last day of the Quiche year. For some reason, it is 60 days too early.

1932	14 III G	12.15.18.0.18	12 Etz'nab 5 Kayab M	
1932	14 III G		12 Chinax 1 Uex	Kanhobal

This date is from LaFarge (1947:168) It correlates the Kanhobal and Gregorian calendars.

1939	21 III G	12.16.5.3.1	1 Imix 3 Kayab M	
1939	21 III G		1 I'mux	Tzutuhil

This date is from Rosales (1939:763). It proves that the Tzutuhil were still keeping the day count accurately.

1940	11 III G	12.16.6.2.17	6 Caban 4 Kayab M
1940	11 III G		6 Noh 1 Mech Ki Ixil

This is from Lincoln (1942:118). It correlates the Ixil and Gregorian calendars and demonstrates the congruence of the Ixil and Quiche calendars.

1946	10 III G	12.16.12.4.7	12 Manik 4 Kayab M
1946	10 III G		12 Tc'e 1 L. Mam

This date is from Oakes (1951:100–108). It correlates the Mam and Gregorian calendars and demonstrates the congruence of the Mam and Quiche calendars.

1949	10 IX G	12.16.15.14.7	5 Manik 4 Mol M
1949	10 IX G	38 Ze Blagay	12 Sgab Gabil 12 Beydo San Agustin Loxicha

This date is from Carrasco (1951).

1949	12 IX G	12.16.15.14.9	7 Muluc 6 Mol M
1949	12 IX G	23 Ze Blagay	10 Sgab Lodios 10 Mdi Candelaria Loxicha

This date is from Carrasco (1951).

1956	7 III G	12.17.2.6.17	9 Caban 4 Kayab M
1956	7 III G		9 No'j 1 L. Aguacatec

This date is from McArthur (1965:38). It correlates the Aguacatec and Gregorian calendars and demonstrates the congruence of the Aguacatec year with that of the Quiche.

| 1960 | 9 IV G | 12.17.6.9.11 | 8 Chuen 18 Cumku M |
| 1960 | 9 IV G | | 8 Chuen 15 U Kab Mam Quiche |

The date of the 8 Monkey festival at Momostenango is reported in Edmonson (1961:1:200).

| 1967 | 5 III G | 12.17.13.9.12 | 7 Eb 4 Kayab M |
| 1967 | 5 III G | | 7 'Ee 1 Mech Ki Ixil |

This date is reported by Colby and Colby (1981:47). It shows the Ixil calendar to be congruent with that of the Quiche.

| 1977 | 2 III G | 12.18.3.12.2 | 4 Ik 4 Kayab M |
| 1977 | 2 III G | | 4 Iq' 1 Nabe Mam Quiche |

This date is reported by Tedlock (1982:103).

| 1978 | 15 X G | 12.18.5.5.14 | 11 Ix 6 Yax M |
| 1978 | 15 X G | | 11 Haiim̃ 0 Mïh Kahpu'ut Mazatlan |

Lipp (1982:187, 270) reports this date for the Mazatlan Mixe New Year.

| 1981 | 1 III G | 12.18.7.13.2 | 8 Ik 4 Kayab M |
| 1981 | 1 III G | | 8 Ik' 5 Hoyeb' Ku Chuh |

This date is the last day of the year in San Mateo Ixtatán (Maxwell 1981). It correlates the Chuh and Gregorian calendars.

Is it possible that additional correlational dates may yet be discovered? I believe the answer is almost certainly yes. The several avenues of investigation that have given us the dates we already have all remain open for further exploration, albeit perhaps with somewhat differential promise. These may be summarized as ethnographic, ethnohistorical, epigraphic, and astronomical.

The relatively rapid disintegration of the remaining native religions in most of Middle America makes ethnography the least promising of these possibilities. Even so, as the following chapter indicates, many of the more isolated ethnic and linguistic groups of the area are still very poorly documented calendrically, and it is possible that the calendar round still exists among some of them, shielded from casual inquiry by its traditionally esoteric mystique and an understandable mistrust of outsiders. Oaxaca and perhaps the Huasteca may be especially worthy of further attention in this connection.

Ethnohistorical research is a more likely source of new data. Not only are we still discovering lost or unexamined manuscript materials, but the accumulation of past scholarship is making them more amenable to interpretation. A firm understanding of the reasons for erroneous construals in the past is a necessary and important part of progress in this area (see, e.g., *1629, *1632). But particularly in Central Mexico, the unraveling of the voluminous traditional native annals will certainly sharpen and correct the complex calendrics of the region at the same time that it clarifies the events purportedly dated. Much has already been done, but the task is still an enormous challenge.

Epigraphic materials are increasingly accessible and interpretable. I would hope that the emphasis of the present work on the pluralism of the writing systems would help to make them more so. The demonstration that such materials frequently include intercalendar correlations should facilitate the recognition of many more of them than I have been able to find. One obvious target for such research is an inscription linking some Mayan calendar to a non-Mayan one. The lack of any such correlation is a startling feature of the corpus I have been able to assemble. Still buried in the bodegas or the soil of Mexico and Guatemala and in codices we are still learning how to read, there are, I am confident, many calendrical surprises. The definitive placement of the Toltec calendar may be among them.

Astronomy has been, on the whole, a somewhat disappointing ally in our research to date on calendrical correlation. Once the constraints of the ethnohistorical evidence are accepted, it can be far more useful. It should now be possible to discipline the rich array of astronomical interpretation of iconography, mythology, and epigraphy and move it from the realm of the brilliantly speculative into that of fact. This endeavor will almost certainly uncover additional correlational data as well, while greatly enriching our comprehension of the astonishing astronomical achievements of native America.

CHAPTER 3

HISTORY OF THE CALENDAR

The most likely pattern of development of the Middle American calendar is depicted in Figure 8, showing a somewhat episodic proliferation that nonetheless approximates a geometric increase in number of calendars over the long run. Thus we go from one calendar in the seventh century B.C. to six in the fourth century B.C., to thirty-eight at the time of the Conquest. The principal changes are concentrated in the Late Preclassic, the Early Classic, the Early Postclassic, and the Early Colonial periods.

The most common change is the displacement of the New Year to an earlier or later month. This seems to be equally likely to move it backward or forward, and the better documented instances suggest that the displacement was almost always in 1-month intervals. Where the interval is larger, it is likely that there were intervening calendars for which we lack documentation. In exceptional cases (Tzeltal, Tlapanec, Cakchiquel), a kind of syncretism may cause a more radical relocation of the New Year. Ignoring this possibility, eighteen forward shifts and eighteen backward shifts of 1 month each will account for the data observed.

Somewhat less frequent are changes in year bearers. These appear to be confined to a forward progression by 1-day intervals. Such a change may have been made independently as many as ten times to account for the calendars we have.

The calendar has been changed three times by abandoning zero dating. There are no instances of readopting it.

The shift from initial to terminal naming of the year may have been only once, whereas the reverse change occurs four times.

The adoption of the 2- to 14-day count seems to have occurred only once, and it was abandoned only once.

Many of these changes cancel each other out as possible calendar "corrections." Those that remain are far too infrequent to have served the functions of a leap year.

Other calendrical features are more specialized and do not affect the calendar round. They include the use of novena and trecena calendars and the invention and modification of the Long Count and will be considered later.

The transformations necessary to account for the developments displayed in Figure 8 are enumerated in the alphabetized list that accompanies Figure 3.

The total number (*n*) of feature changes listed in any one transformation yields a fair estimate of the maxiumum number of intervening calendars (*n*-1). Any change involving 1 day or 1 month (e.g., Aztec to Otomi) presumably involves no intervening calendar and very little time. The change from Olmec to Zapotec appears to have involved two intervening calendars and may have taken longer, thus pushing the probable date of the invention of the Olmec calendar back by a century or so at least.

The history of the calendar of Anahuac may be conveniently summarized in relation to the generally accepted periodization of Middle American archaeology: Preclassic, Classic, Postclassic, and Colonial. It will require continuous reference to Figure 3 and Figure 8.

PRECLASSIC

The Preclassic period saw the invention of the day count, calendar round, and Long Count, and the differentiation of the calendar into four traditions: Olmec, Teotihuacan, Zapotec, and Tikal.

In considering the invention of the day count, we are limited to two speculations and one fact, and they do not fit together. Specuation 1 is that the day count was an Old World invention. Speculation 2 is that it was a Mayan invention. The fact is that its earliest manifestation is Olmec.

The idea that the day count, or a least the order of the named days, was a diffusion from the Old World has been systematically explored by Kelley (1960) and Moran and Kelley (1969). The theory is difficult to falsify and impossible to prove. There is, however, no Old World analog to the permutative count of 260 days and that forces us to look closer to home for the origin of the day count as such.

The vigesimal base of the day count appears to be particularly rooted in the counting system of the Mayan languages, in which higher numbers are generally expressed in multiples of 20s. The Maya also have a tradition linking the day count to the duration of pregnancy, which might provide a rationale for sacralizing the number 13 because it takes 13 months of 20 days to reach 260. But there seems to be no way of proving that this is the way it happened because we have no record of Mayan calendrics earlier than the first Long Count dates.

Our one fact is that the earliest recognizable calendrical glyphs are Olmec, possibly those carved in stone at Chalcatzingo, perhaps as early as 1150–900 B.C. They are not accompanied by numeral coefficients and so do not directly attest to the presence of a day count.

The earliest indisputable evidence of an Olmec day count occurs as part of an already fully developed calendar round—the earliest known—and constitutes

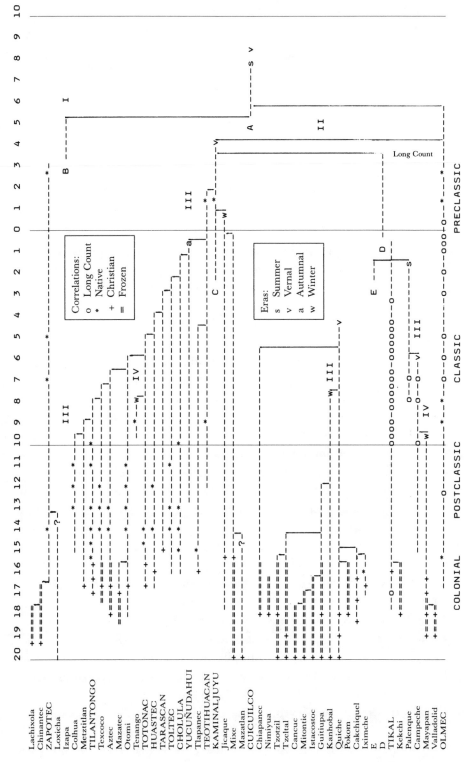

Figure 8. CALENDRICAL GENEALOGY (Year Bearers in Roman Numerals)

the first text in our date list in Chapter 2 (see 667 B.C.). It is almost certainly earlier than 600 B.C. and may be as much as 200 or 300 years earlier. It begins the year at the beginning: that is, on the universally acknowledged first day of the day count, and names it for the last day of the day count. From the glyphs used, it is likely that the Olmec name for the first day was Sun (compare Zapotec Day, Tequistlatec Light) rather than Alligator, and that the last day was Lord (compare Maya Lord, Mixe Eye) rather than Flower. The Olmec year bearer fell on V.G.0, and its calendar round began with 1 Lord.

The Olmec name day cited (see 563 B.C.) is the seventeenth day of the fourth Zapotec month (= 2 Ceh Mayapan). This would normally imply that its New Year fell on 8 Ceh Mayapan. However, the comparative calendrics of the earliest calendars strongly suggests that the date cited on Monte Alban Stela 13 is that of the first name day of the Olmec calendar, which would place its New Year 100 days earlier, on 3 Yaxkin Mayapan. Whether this is true or not, the existence of a calendar that was so placed would have to be postulated in order to explain the evolution of the other early calendars. It seem simplest to postulate that it was Olmec, thus interpreting the Olmec calendar as terminally named, with a New Year of Type I.B.19 falling on 3 Yaxkin Mayapan.

Sometime before the sixth century B.C., the calendar began to differentiate, giving rise to two intermediate calendars leading from Olmec to Zapotec: Calendar A (= Olmec + 1 month) and Calendar B (= Calendar A + 1 day). This sequence gave rise to the Zapotec calendar (= Calendar B + 1 day), documented by Monte Alban Stelae 12 and 13 (see 594, 563 B.C.). Like the Olmec calendar, that of the Zapotecs was terminally named and used zero counting. It was typologically III.C.19, and its New Year fell on 5 Mol Mayapan.

Sometime before the fourth century B.C., the existence of a third unknown calendar must be invoked. This calendar would have to be II.B.0 (= Olmec + 1d − t − 360*) in order to serve as the ancestor of the calendar of Teotihuacan. The correlational date of 147 B.C. identifies it as the calendar of Kaminaljuyu. The Teotihuacan calendar = Kaminaljuyu − 105d + 14*. The apparently simultaneous adoption of the 2 to 14 counting of the day count had the effect of shifting the resulting New Year forward by 2 months, so that in fact the Teotihuacan New Year is 65 days earlier than that of the Olmec. The elimination of the numeral 1 from the year count may also be responsible for the later preference of a number of Central Mexican calendars for beginning the calendar round on a year 2 rather than a year 1. The resulting Teotihuacan calendar was initially dated, using 0 to 19 counting of the days of the month and 2 to 14 counting of those of the day count, with a new year of type II.Q.0 on 19 Zip Mayapan (see A.D. 861).

In the fourth century B.C., the Olmec inaugurated the Long Count calendar. The date 6.19.19.0.0 1 Ahau 3 Ceh Tikal marks the end of the last tun of the sixth baktun. It also marks the end of the Olmec year and the inauguration of a

Figure 9. THE CALENDARS OF THE FIRST CENTURY A.D.

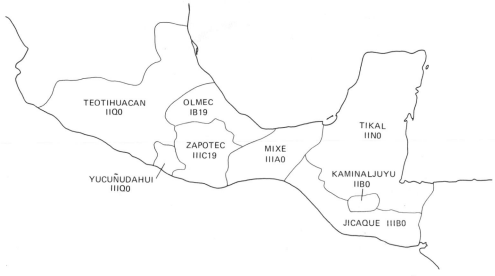

new calendar round. Such a calendrical coincidence can only occur every 936 years. It happened on June 13, 355 B.C. Julian. For the earliest Long Count date known, see 36 B.C., Chapter 2.

Sometime before the first century A.D. and possibly as early as the third century B.C. (see *236 B.C.), we may postulate a Calendar D (= Olmec – 1m), from which the same transformation that produced the calendar of Teotihuacan would have generated that of Tikal (= Calendar D – 105d – t). This would produce an initially named calendar with zero counting and the Long Count, which counted the days of the month from 0 to 19 and those of the day count from 1 to 13, with a New Year at II.N.0 on 4 Uayeb Mayapan (see A.D. 37, Chapter 2).

The similarity in the apparent derivation of the calendars of Tikal and Teotihuacan suggests a possible joint alternative derivation. From Calendar C, the transformation (– 105d – t) would produce a Calendar X at II.O.0, from which we could derive both Tikal (Calendar X – 1m) and Teotihuacan (Calendar X + 14* – 360*). On balance, it seems to me more likely that the initializing transformation (– 105d – t) was used twice independently than that Teotihuacan once had the Long Count (360*) and then forgot it.

CLASSIC

The seven known calendars of the Preclassic continued through the Classic period, and one of them, that of Teotihuacan, generated a whole family of additional calendars. Here again we seem to be missing some pieces of the puzzle. The Teotihuacan

Figure 10. THE YEAR 4 RABBIT (A.D. 870 TARASCAN) IN AMAPA
(Bell 1971:708, Fig. 10d.)

calendar produced one known direct descendant, Tlapanec (= Teotihuacan – 7m). The 7-month leap might be taken to imply as many as six intervening calendars, but it is more likely that it was motivated by the Olmec or Aztec calendar: it places the Tlapanec New Year 1 day before the Aztec New Year, resulting in an initial named calendar of Type II.G.0 that counted the day count from 2 to 14. Its New Year fell on 4 Muan Mayapan.

Teotihuacan also spawned the Yucuñudahui calendar (= Teotihuacan + 1d – 14*), which was the first in a series of six Classic calendars with Type III year bearers (see A.D. 426). The others were Cholula (= Yucuñudahui + 1m), Toltec (= Cholula + 1m), Tarascan (= Toltec + 1m), Huastec (= Tarascan + 1m), and Totonac (= Huastec + 1m). A separate development was that of Tenango (= Totonac + 1d).

All of these are initially named calendars using zero for month days and 1 to 13 for the day count. Their New Years' are indicated in Figure 3. Each has its own writing system, seemingly derived (but progressively divergent) from that of Teotihuacan. We have no correlational dates in any of them until the end of the Classic period, and their calendrical placement is partly conjectural because we lack even year dates in all of them except Huastec (Type III; see 1616) and Tenango (Type IV; see 798). To judge from the speed with which comparable strings of calendars were produced in later times, they could all have been generated in the Early Classic, and even the Totonac day glyphs, which are presumptively the latest, seem to belong solidly to the Middle if not Early Classic by archaeological context. (Archaeological provenience is given in the calendar index for specific glyph systems.)

The Yucuñudahui calendar itself can be dated to the early fifth century A.D. (see 426) and may not go back much earlier. But by the end of the Classic period, Type III year bearers are found all over Central and Northern Mexico, from Amapa in Nayarit (Figure 10) to Tajin in Veracruz and throughout Oaxaca. Differentiating one Type III calendar from another will take far more data than we now have, but we have at least a start on their glyphic systems, and it is likely that there were as many calendars as there were ways of writing them. The seemingly arbitrary calendrical placement of them is not purely speculative, as will become clearer in tracing their Postclassic histories.

The Zoquean languages of Oaxaca, Chiapas, Veracruz, Guatemala, Salvador, and Honduras are the most likely descendants of the language of the Olmec, but their calendars remain poorly documented. Only the Mixe calendar is adequately reported. Its history is clearly intertwined with that of the Jicaque, which was probably shared by the Lenca, and both calendars are almost certainly derived from Olmec via the intervening Kaminaljuyu calendar of the Late Preclassic, which was also the source of the calendar of Teotihuacan. Kaminaljuyu was II.B.0, probably initially dated and lacking the Long Count. It is possible that it was the calendar of the Xinca. Around the middle of the Classic period, it advanced 1 day to form the Jicaque calendar, which was probably shared by the Lenca and perhaps by the Xinca and Zoque as well. The Mixe calendar is 1 month earlier.

Toward the end of the Classic period, two new calendars put in a brief appearance in the western part of Mayan country: those of Palenque (see 692) and Campeche (see 665). The Palenque Calendar (= Tikal – 0 + t) abandoned the 0 to 19 counting of the month days in favor of 1 to 20. It was the earliest calendar to do so. It did not, however, change the year bearers, so the shift amounted to a return to terminal naming of the year. It was never more than a secondary calendar even in the cities where it is found, but it seems to have represented a tradition that survived, particularly by generating the Campeche calendar (= Palenque + 1d), which advanced the calendar by 1 day without making any other change. Historically, all of the Mayan peoples have clung stubbornly to initial dating of the year. It seems obvious that the introduction of terminal dating is an intrusive foreign idea, presumably attributable to Zapotec or Olmec influence. The occurrence of a few non-Mayan calendrical glyphs in the central Peten in the ninth century (see 848) seems to point to Olmec influence in that period and perhaps in the seventh century as well. There appear to be no Zapotec calendrical glyphs in Mayan sites at any date.

The Classic period thus expanded the number of calendars from seven to sixteen, and most of the expansion was traceable to Teotihuacan via the Mixtec calendar of Yucuñudahui. By the end of the Classic, only Olmec and Zapotec continued to use terminal naming of the year. The calendars of the end of the Classic period are mapped in Figure 11.

POSTCLASSIC

A great proliferation of calendars took place in the Early Postclassic period, traceable largely to Zapotec and Tikal. Most of the expansion can be attributed to the

Figure 11. THE CALENDARS OF THE TENTH CENTURY A.D.

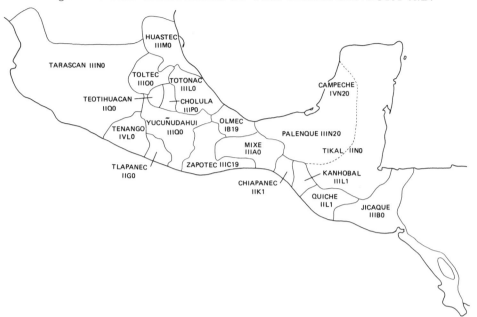

Tilantongo Mixtec calendar (= Zapotec + 1d − 0), which advanced 1 day and abandoned the use of zero without altering the terminal naming of the year. The result was a Type IV.C.20 New Year falling on 6 Mol Mayapan (see chapter 2, 1555).

The Tilantongo calendar promptly spawned a string of new calendars differing only in the month that begins the year. Thus Metztitlan = Tilantongo − 1m, and Colhua = Metztitlan − 1m. In the other direction, Texcoco = Tilantongo (Tlaxcalan) + 1m; Aztec = Texcoco + 1m; and Otomi (Tepepulco) = Aztec + 1m.

Then come the six calendars that were the initial dated calendars of the Classic period, now reinterpreted as terminally dated by the transformation (− 100d). Thus

Totonac	− 100d = Teotitlan	= Otomi	+ 1m
Huastec	− 100d = Cuitlahuac	= Teotitlan	+ 1m
Tarascan	− 100d = Tepanec	= Cuitlahuac	+ 1m
Toltec	− 100d = Colhuacan	= Tepanec	+ 1m
Cholula	− 100d = Tepexic	= Colhuacan	+ 1m
Yucuñudahui	− 100d = Chalca	= Tepexic	+ 1m

As the calendar names imply, these changes were accompanied by the Nahuatlization of the calendars as well as by the abandonment of zero and the adoption of the Tilantongo (−Aztec) writing system. Both Classic and Postclassic calendars are found at some sites, including Tula. Note that the final transformation on the preceding list makes the Yucuñudahui calendar congruent with that of Tenango, except that it now becomes terminally named: Tenango's *year bearers* were Type IV.

While the Tilantongo Mixtec were expanding in all directions (including the eventual occupation of Monte Alban), the Zapotec may have been doing the same, particularly to the north and east, giving rise to the Chinantec calendar (= Zapotec + 1m) and possibly motivating the shift of the western Mayan calendars from Type II to Type III year bearers (see later discussion). The calendars of the Zapotecan languages aside from Valley Zapotec itself remain largely undocumented, and perhaps they simply used the Zapotec calendar, although in one case, Loxicha (q.v.), they seem to have developed an alternative day count based on novenas and trecenas. This does not appear to be cognate with the distinctive trecena calendar of the Mixe.

It would appear to have been about the beginning of the Postclassic period (when zero counting was lost in so many calendars) that the Mayan calendar underwent fission. The Quiche calendar (= Tikal – 360* – 0 – 2m) replaced that of Tikal in the highlands of Guatemala and Chiapas. The Kekchi, the Cholans, and the Yucatecans remained on Tikal time. The genesis of the Chiapanec calendar (= Quiche – 1m) may have occurred at about the same period.

It was probably later in the Postclassic that the western Mayas, possibly under Zapotec influence, changed from Type II to Type III year bearers. The most likely sequence is

Kanhobal (Jacaltec)	= Quiche + 1d
Tzeltal	= Kanhobal – 7m
Tzotzil	= Tzeltal – 1m

The abrupt shift in the initial month of the Tzeltal calendar is probably explained by the invading Pochteca of the Late Postclassic: the Tzeltal adopted the initial month of the Aztec calendar as the beginning of their own year.

The Pipil invasion of Guatemala in the Late Postclassic had a demonstrable impact on the month names of both the Quiche and Cakchiquel, but in the latter case they actually changed the date of the New Year: Cakchiquel = Quiche – 5m. The Pipil brought with them the Teotitlan calendar of Puebla and Veracruz, and the Cakchiquel adopted its initial month (G. Ceh) as the beginning of their own year, retaining the other features of the Quiche calendar almost intact.

In 1493, the Cakchiquel also created the unique calendar of Iximche (q.v.); this is possibly an indication that a rudimentary knowledge of the Long Count persisted in the highlands until that date. The Iximche count was based on a 400-day "tun" that the Cakchiquel called a year (*juná*), counting such "years" in 20-"tun" units not unlike katuns, which the Cakchiquel called cycles (*may*).

In 1539, at the very close of the Postclassic, the Yucatecan Maya inaugurated the Mayapan calendar. It was derived from Tikal but probably not directly. It is more likely that it came from the Campeche calendar.

The Campeche calendar itself is problematic because we do not know for certain whether it was III.N.1 or IV.N.20. The most likely sequence is

Tikal	(II.N.0)	
Palenque	(III.N.20)	= Tikal + 1d + t – 0
Campeche	(IV.N.20)	= Palenque + 1d
Mayapan	(IV.N.1)	= Campeche + i

In this construal, the Campeche calendar kept the terminal dating (t) of Palenque, and the Mayapan calendar simply reverted to initial dating (i). The likelihood of this interpretation is enhanced by the fact that Mayapan switched to initial dating of the katun and the may at the same time. The alternative guess that the Campeche calendar was III.N.1 would involve more transformations to arrive at the known result.

The "most likely" construal of the Campeche calendar (i.e., as IV.N.20) would make it a seventh-century precursor to the whole family of terminally dated Type IV systems generated in the tenth century by the Tilantongo Mixtec. A Tabasco-Tlaxcala linkage of the sort suggested by ninth-century Cacaxtla might easily have provided the connection because the later Postclassic calendar of Tlaxcala was Tilantongo. The calendar situation at the time of the Conquest is illustrated in Figure 12.

By the Conquest, the sixteen known calendars of the end of the Classic had grown to thirty-eight. Perhaps only five of them (Kaminaljuyu, Teotihuacan,

Figure 12. THE CALENDARS OF THE SIXTEENTH CENTURY A.D.

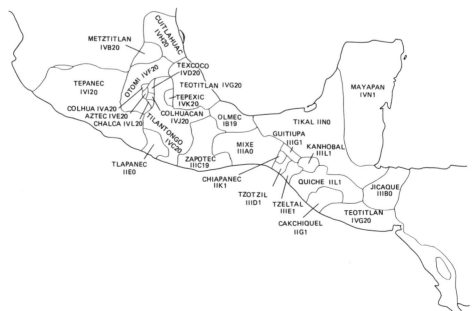

Tenango, Campeche, and Palenque) had become extinct. An additional six, although not technically extinct, had been replaced by their Postclassic equivalents (Totonac, Huastec, Tarascan, Toltec, Cholula, and Yucuñudahui). Many of the remaining twenty-seven were used by more than one ethnic or linguistic group, but they are not known to have been calendrically different on that account. My "best guess" about who used what calendar is registered in the calendar index (Chapter 5) for all the language groups that are likely to have had calendar-round dating.

COLONIAL

Although they are never to have been in use in the sixteenth century, seven calendars were never recorded with their correlational positions in the European calendar: Olmec, Colhua, Metztitlan, Toltec (Colhuacan), Cholula (Tepexic, Huexotzinco), Yucuñudahui (Chalca), and Zapotec. The remaining twenty calendars were thus correlated, many of them several times over and independently, and the seven that were not are firmly placed in relation to the rest by pre-Conquest correlations. It is absurd to pretend, as some scholars continue to do, that the "correlation question" hinges precariously on one mistake by Landa and another by Sahagún: what we have is an extremely tight web of mutually supportive correlations that prove beyond any reasonable doubt the relation between the European calendars and both the day count and the year count of virtually every known or suspected calendar in Middle American history.

Sixteen of the twenty correlated calendars are linked directly to acceptable contemporary dates in Julian or Gregorian time. The remaining four are more complicated: they involve "frozen" correlations.

The principal obstacle to satisfactory calendrical correlation has proved to be the European leap year. Numbers of Europeans have simply refused to believe that the native calendars did not contain a leap year, and a large number of published correlational dates are useless to us because they do not specify (or even allow us to reconstruct) the year to which they apply. The fact is, however, that no native calendar has ever had a leap year, so that there is a quadrennial loss of 1 day in the European date of any native New Year with the passing of time. (Thus the New Year date falls back a quarter of a year every 365 years, or nearly a month per century.)

The earliest frozen calendar was that of Mixe, which stopped counting leap years after February 29, 1532 Julian and retained a Julian New Year down to the present. Additional calendars were frozen in 1548, 1584, and 1616, and the Mixe calendar was refrozen as the Mazatlan calendar (q.v.) sometime between 1901 and 1923.

Leap year appears to have been an inconvenience, partly perhaps in relation to fixing the dates of Church festivals. The solution was to "freeze" the correlation to February 29, 1548 Julian and to ignore all subsequent leap years. The calendars known to have followed this procedure are Texcoco, Aztec, Chinantec, Tzeltal, Tzotzil, Mayapan, and Tikal (Kekchi but not Chol). The calendars known not to have done so are Tilantongo, Totonac, Quiche, Tarascan (Matlatzinca), Kanhobal, (Jacaltec, Chuh, Cakchiquel, Mazatec and (probably) Huastec. The scattered geographic pattern of calendar freezing suggests that it may have been a missionary invention, but it is one that has left no trace in the voluminous missionary records, and it seems to have become thoroughly rooted in the native calendrical traditions that adopted it. The frozen Julian date of New Year 1548 is still in occasional use among the Tzotzil despite their subsequent adoption of a frozen Gregorian date.

Four of the calendars using frozen correlations confirm thereby the direct and unfrozen correlations they had given us earlier (Tikal, Texcoco, Aztec, and Mayapan). In the other four cases, frozen correlations are all we have (Mixe, Chinantec, Tzeltal and Tzotzil). But a frozen correlation is just as valid as an unfrozen one, once it is recognized as such. Thus we may add these four calendars to our list of correlationally documented cases, bringing the total to twenty.

When the Gregorian calendar was promulgated in 1582 (1584 in New Spain), it was immediately implemented in almost all areas. It occasioned correlational confusion only among the Tzotzil and Tzeltal, who already had a frozen Julian correlation but who now decided to adopt a frozen Gregorian one. The calendar change from Julian to Gregorian meant adding 10 days to the European date, but the 9 leap years between 1548 and 1584 meant subtracting nine. So the Tzotzil have maintained to the present day a frozen Julian New Year (December 27) and a frozen Gregorian one (December 28). The Tzeltal maintain the same traditions for their new year, 20 days later.

One calendar may almost be said to have been created by calendar freezing. The Mazatec calendar was identical to the Aztec but seems not to have been frozen in 1548 when the Aztec calendar was. Instead, it appears to have waited until the precession of leap year brought its New Year date to rest on January 1 (1617) Julian and then froze it.

The Pokom calendar is a similar case. Structurally identical to Quiche, it froze its correlation in 1548, whereas Quiche did not.

The Tikal calendar survived in the Colonial period primarily among the Chol and Kekchi. The latter froze their correlation to 1548; the former did not. Both confirm the placement of the Tikal calendar with respect to other native calendars and to European dating.

Other late calendars were generated following the Conquest in Chiapas, Oaxaca, and Yucatan. Our documentation of the process may be incomplete, as there may be intermediate calendars we have not found.

17 c.	Nimiyua = Chiapanec	+ 1m
	Lalana = Chinantec	+ 2m
18 c.	Guitiupa = Tzotzil	+ 3m
	Istacostoc = Guitiupa	+ 1m
	Valladolid = Mayapan	+ 24-year katun
19 c.	Mitontic = Istacostoc	+ 1m
	Cancuc = Tzeltal	+ 5m
20 c.	Mazatlan = Mixe	+ 13* (f1901?)

Thus by the end of the nineteenth century, we can identify a total of fifty structurally distinct calendars known to have existed in Middle America (or sixty calendars that are distinct either calendrically or linguistically). By that date, however, the vast majority of them were extinct.

Only a handful of calendars are still extant today (1985). The Loxicha calendar preserves its novena (-trecena) day count, but that is only tangentially related to the calendar round. The Cancuc, Tzeltal, Mitontic, Istacostoc, and Tzotzil calendars still exist, but they are frozen *to* the European year rather than freezing the European year to their own leap-year-less system. The Tzotzil have altogether lost the day count and the month names, and though the Tzeltal retain theirs, they seem to have lost track of them. The Lachixola and Lalana Chinantec calendars and the Mazatec (-Aztec) calendar survive with frozen correlations. Independent unfrozen correlational dates may still be obtained in the Kanhobal, Quiche, and Mixe calendars but only the last two maintain a named month count and clear calendar-round dating, and the Mixe day count has almost vanished in the confusion produced by leap years and two European calendars. The Quiche Maya are the last refuge of a 3,000-year calendrical tradition.

CHAPTER 4

THE ROAD OF THE SUN

The structure of the year count of Anahuac taken as a whole shows clearly that it is based upon the observation of the sun. Certain of the details of this structure, in fact, may be read as markers of the historical evolution of astronomical knowledge among the various peoples who came to share the common solar calendar, preserving in their variation some of the logical steps by which their calendars were derived and some of the astronomical observations and calculations they incorporated.

What follows is an interpretation of the astronomical history of the calendar system. It is a complex history, and parts of it are inevitably inferential, but it parallels the general culture history of the region. It is consequently convenient to trace it in four principal phases that may reasonably be related to the Preclassic, Classic, Postclassic, and Colonial periods.

PRECLASSIC

The earliest model of the solar year in the astronomy of Anahuac was divided into quarters by observation of solstices and equinoxes. A first-order approximation to this calculation, confined to whole numbers, yields a calculational year of 364 days composed of 4 quarters of 91 days each or 13 zodiac signs of 28 days each. The calculation of 91 as 7×13 and of 364 as 28×13 may have been reason enough in itself to sacralize those numbers and to account for their subsequent prominence in the deified calendar.

A calculational year of 364 days is, of course, a very loose approximation. It could, however, be conveniently based on a simple attentive count of trecenas. It appears likely that such a count was invented and kept concurrently with an even looser count of yearlike cycles of 400 days, premised simply upon the arbitrary convention of vigesimal numeration. Even a brief period of observation would nonetheless demonstrate the superiority of a 365-day estimate of the year over both of these hypothetical cycles.

The Middle American mode of calculating the year is based upon the simultaneous use of both of these more primitive counts: the trecena and the veintena. This leads to the purely numerological construct of a 260-day cycle, the day count, and to the more astronomically satisfying 365-day cycle, the year count or vague year.

Close observation of the dates of recurrence of any one solstice or equinox would lead in a very short time to awareness of the shortcomings of the 365-day year because the tropical year is actually .2422 days longer. It would, however, require rather extensive records to establish the magnitude of this correction with any real precision. The ancient Middle Americans appear nonetheless to have reached the same accuracy as Western astronomers, not only independently but far earlier. In a sense, their discovery was as much numerological as it was astronomical.

The correct solution to the equation of the sun is that the tropical year exceeds the length of the vague year by 1 full day every 1,508 days. The occurrence of a particular solstice on a given date of the vague year will repeat after the passage of 1,508 vague years. The native (vague year) calendar will lose 1 day every 1,508 days with respect to the true (tropical) year, and it will take it 1,508 years to lose one full vague year (365 days): $365 \times 1,508 = 365.2422 \times 1,507$.

The Middle American astronomers arrived at their calculation of this fact through the trecena calendar. As they explored the numerology of the day count and the vague year, they discovered and mythologically underlined their properties:

$13 \times 4 = 52$	1 fifth day count
$13 \times 5 = 65$	1 quarter day count
$13 \times 7 = 91$	1 calculational year quarter
$13 \times 20 = 260$	1 day count
$13 \times 28 = 364$	1 calculational year
$13 \times 29 = 377$	1 leap year quarter
$13 \times 29 \times 4 = 1508$	1 leap year interval
$13 \times 365 \times 4 = 18,980$	1 calendar round
$13 \times 365 \times 29 = 137,605$	1 era quarter
$13 \times 365 \times 29 \times 4 = 550,420$	1 solar era

Having reached the crude approximation of a 91-day interval between solstices and equinoxes, the native astronomers might easily have extrapolated to an estimate of the era by observing how long it took for a given date in the vague year to regress 91 days. If, for example, the native New Year were set on a solstice, it would be expected to fall back to the next earlier equinox in one quarter of an era.

An initial crude estimate analogous to the Julian leap year could have supposed a regression of 1 day every 4 years (1,460 days). We have no evidence of such a speculation, which would presumably have been phrased as a correction of 91 days in 28 trecenas of years. What we do have is evidence for a rather more accurate estimate of 29 trecenas for the quarter of the leap-year interval, and hence of 29 calendar rounds for the era.

This solution is, in fact, correct, but it is not observably correct in these terms. The real tropical era is not divided into exact quarters. The unevenness of the earth's orbit around the sun dictates a somewhat irregular interval between solstice and equinox. Furthermore, the observation of equinoxes is more accurate than that of solstices, so that the observation may readily be reached that the distance between spring and fall is always 186 days, whereas that from fall to spring is either 179 or 180. In the Middle Preclassic, the intervals were approximately as follows:

Winter to spring	90–92 days (mean 91.09)
Spring to summer	94 days
Summer to fall	92 days
Fall to winter	88–89 days (mean 88.24)

The relative stability of the summer semester appears to have been an important aspect of the earliest calendar.

The structure and position of Calendar A, the earliest calendar for which there is any empirical warrant, is dictated by the data we have for its apparent calendrical derivatives and by the logic of scientific parsimony. It must have been a terminally named calendar using zero numeration for the uinal, counting the day count from 1 to 13, with a New Year falling on 4 Mol Tikal, and it began its calendar round on 1 Lord. I believe this structure to indicate a remarkably accurate estimate of the length of the tropical year, which was later improved upon. The following comparisons may be instructive:

Julian year	365.2500	Era 532,900 days
Calendar A year	365.2462	Era 550,426 days
Kaminaljuyu year	365.2422	Era 550,420 days
Yucuñudahui year	365.2422	Era 550,421 days
Gregorian year	365.2422	Era 550,421 days

Perhaps the most peculiar feature of Calendar A is that it named the year for the day name of its 360th day rather that its 1st day. The folk rationalization of this fact stresses the taboo character of the last 5 days of the vague year, variously described as extra, nameless, useless, or empty. That would, of course, rule out the 365th day as a day name, but it does not rule out the first. I believe that the real reason is astronomical, and relates to a calendar "correction" of – 6 days in comparison with initial naming. The need for such a correction is not apparent from Calendar A in isolation, however. It can be explained by the hypothesis that the calendar was designed so as to predict correctly the end of the solar era, 1,508 years later. The suggestion that this is so is contained in the structure, genealogy, and chronological placement of all of the calendars with respect to each other and to solar time. This chapter traces these features in detail.

It is not clear how the Olmec reached the erroneous conclusion that the era was 6 days longer than it was. It may be relevant to observe that the solar year implied (365.2462 days) would correspond to a precession rate of 93 days every 377 years, or 186 days every half era (754 years). Such a figure could have been reached by observation of the 186 days from the spring to the fall equinox and the arbitrary division of the 1,508-year era into equal halves. It would seem unlikely that a calendar would have been based on simply ignoring the shorter winter semester (of 179 or 180 days). This is, nonetheless, the nature of the error.

Some light is shed on the problem by calculation of the leap-year intervals generated by this estimate of the era. From the starting date of Calendar A, and counting in intervals of 116 (or 4 × 29) trecenas (1,508 days), the day-count dates reached would be 1 Ahau, 1 Lamat, 1 Cib, 1 Kan, 1 Eb. After 7,540 days, this sequence repeats: 29 day counts = 5 × 1,508 days. Squeezing 6 extra days into the era, however, implies a leap-year interval of 1,508.0189 days. The preceding calculation would still work for the first 52 iterations, but the .0189 fraction would then require the insertion of an additional day. The leap-year calculation would advance by 1.

Six such advances would occur within the era. The sequence is as follows:

1 Ahau	1 Lamat	1 Cib	1 Kan	1 Eb
2 Imix	2 Muluc	2 Caban	2 Chicchan	2 Ben
3 Ik	3 Oc	3 Etz'nab	3 Cimi	3 Ix
4 Akbal	4 Chuen	4 Cauac	4 Manik	4 Men
5 Kan	5 Eb	5 Ahau	5 Lamat	5 Cib
6 Chicchan	6 Ben	6 Imix	6 Muluc	6 Caban
7 Cimi	7 Ix	7 Ik	7 Oc	7 Etz'nab

The interval of 52 leap years between advances in the day-name coefficient is 79,925, or 10 days less than 219 years. Although this may seem complex, it would certainly have seemed quite simple to the diviner priests who were using the day count constantly for many other purposes, and for whom the number names of the veintena were as familiar numerologically as the names of the secular numbers.

Even though the solution is in error, it clearly embodies an approximation to the correct solution that would have had immediate appeal: 29 × 52 gives both the leap-year interval (in days) and the length of the era (in years). The reason for departing from this simple and elegant solution must have been compelling. Although I cannot reconstruct how it was reached, I believe it corresponded to the empirical discovery of the inequality of the quarters of the era. If, as seems to me possible, the original intuitive grasp of the era was a numerological rather than an astronomical discovery, the derivation of equal 377-year quarters of a 1,508-year era is both logical and mystically attractive (29 trecenas to a quarter and 29 calendar rounds to an era).

The insight that led to the empirical testing this model was probably the decision to use the year itself to measure the apparent movement of the sun. If, for example, the beginning of the year were placed on the autumnal equinox, New

Year's Day could be expected to fall back slowly toward the summer solstice until it fell exactly on the correct solsticial day after 377 years. The trouble is that it did not. In the event, it took 382 years. (The earliest date at which 4 Mol fell on the autumnal equinox before the eighth century B.C. was on 22 IX 1121 B.C. 5.1.2.2.11 1 Chuen 4 Mol.) The Olmec must have kept records of the recession long enough to generate the erroneous 6-day overcorrection.

It was apparently from this circumstance that the designers of Calendar A derived the necessity of lengthening the era by 6 days, apparently perceiving the necessary correction as equal to 3 days per half era. The mechanism used was terminal naming of the year, which added 6 days to the era at its beginning in the expectation that the New Year would then fall correctly on the summer solstice 29 calendar rounds later.

My reason for supposing that the New Year of Calendar A was fixed by the *summer* solstice depends importantly on the logic of the derivation of other calendars from it and upon archaeological constraints upon the latter. This will perhaps become clearer later. The alternatives would be to argue for placing this first calendar at the time of the congruence of its New Year with the spring equinox of 353 B.C. or the fall equinox of 1121 B.C. The later date would be far too late to permit the derivation of the other documented and dated Preclassic calendars. The earlier one is unnecessarily early and would leave a gap of a quarter era between the hypothetical dates of origin of the other calendars and their first documented manifestations.

The date selected for the genesis of Calendar A appears to be confirmed empirically by its name. It is not only the first year in which 4 Mol falls on an astronomically correct summer solstice: it is also the year 1 Lord. Because Calendar A used the same year bearers as the directly cognate Olmec calendar and because the latter began its calendar round on 1 Lord, it is reasonable to suppose that Calendar A did too. The date is 22 VI 739 B.C. Gregorian 6.0.9.8.1 6 Imix 4 Mol Tikal (Julian day number 1451684). It is noteworthy that the astronomical correctness of the name for this date is a tribute to the accuracy of observation of the summer solstice.

Calendar A was earlier derived (in Chapter 4) as a hypothetical calendar, the existence of which must be invoked to explain the later Olmec and Zapotec calendars. It is possible, however, that it is not completely hypothetical. Perhaps the date for 679 B.C. Olmec from Cuicuilco (in Chapter 3) is in fact a date of 691 B.C. in Calendar A. The most likely point of origin for Calendar A in any case would lie within the triangle of San Andres Tuxtla, Monte Alban, and Cuicuilco, and Cuicuilco is a reasonable provisional name for it.

That the establishment of Calendar A was based upon calculation of the solar era and not merely upon recognition of the summer solstice, is further indicated by its generation, less than a century after its creation, of the closely related Olmec Calendar. The two calendars are in fact identical except that the Olmec

New Year falls 20 days earlier, on 4 Yaxkin Tikal. Its founding is the earliest example of this transformation of – 20 days, which I believe to be astronomically
motivated. It is, in fact, a kind of anti-leap-year correction to the summer era established by the Cuicuilco calendar.

It will be remembered that the shortfall of .2422 days per vague year will
cause the summer solstice to occur later and later in the native calendar with the
passage of time. A leap-year correction analogous to that of the Julian calendar
could easily accommodate this by advancing the date of the New Year by 1 day
every 4 years or 20 days every 80 years. The Middle American calendars do almost
exactly the opposite. Recognizing that the 1,508 vague years of the solar era actually make up only 1,507 real (tropical) years of 365.2422 days, they set out to erase
the extra year by subtracting one 20–day month every 83 years.

The anti-leap-year "correction" was not in fact a correction at all. Any
given Middle American calendar was allowed to run its course without changing its
New Year. But when a new calendar was founded, it was timed to the ongoing
awareness of when the era began and when it was due to end. The Olmec calendar
was a summer-era calendar based on that of Cuicuilco, the second in a potential
series of eighteen such calendars to be expected at intervals of approximately 83
years. In fact, however, this summer calendar series ends with the Olmec. The
reason is that the invention of the Kaminaljuyu calendar rendered the series obso-
lete. This is discussed later.

The hypothetical date for the founding of the Olmec calendar would have
been 13 V 656 B.C. Gregorian 6.4.13.9.16 4 Cib 4 Yaxkin Tikal (Julian day number 1481959). It was probably preceded by observing the 5 days of Uayeb (1 month
earlier than they were observed in the Cuicuilco calendar). That is, the mechanism
of the change was to move Uayeb back 1 month. Because the year bearers were
unchanged, the effect was not a perfect adjustment of the new calendar to its astronomical position. Such an adjustment is due every 30,160 days, or 82 years plus
230 days. This one was presumably made after 82 years and 345 days. It is thus
115 days late. Such a discrepancy would have accumulated, so that, had the series
been continued, it would have amounted to more than a year after the third repe-
tition, and the fifth calendar would presumably have begun a year earlier.

The earliest archaeological evidence for the Olmec calendar is undoubtedly
the inscription listed in Chapter 2 at 667 B.C., giving an apparent calendar round
beginning on 1 Lord. The artifact bearing the inscription has no archaeological
context and may be dated only very approximately on grounds of artistic style to
the Late Olmec period. Given the expected date of 656 B.C. for the founding of the
Olmec calendar, we may perhaps place the date of the Tapijulapa ax one calendar
round later than 667 B.C. — at 615 B.C. It could, of course, have been later still (in
52-year intervals). But its inscription may be taken as documentation of the Olmec
calendar very probably within the first century of its existence. Olmec dates are

relatively well documented after about the third century B.C. (see Olmec, Chapter 5), but by that time the Olmec art style was ancient history.

It was the logic of the 6 extra leap year days of Calendar A that generated Calendar B in 520 B.C., 219 years after the inauguration of Calendar A (Cuicuilco), and the Zapotec calendar 219 years after that, in 305 B.C. It could have been expected to generate additional calendars at 219-year intervals for the remainder of the era. It did not do so, perhaps because it was surpassed by the more accurate solution of the Kaminaljuyu calendar.

Calendar B, with its New Year on 5 Mol Tikal, retains the terminal dating and all other structural features of Calendar A, merely advancing its New Year by 1 day. It remains almost entirely hypothetical. We do have one tantalizing possible use of Type I name days from a monument at Izapa (see Chapter 5), which therefore provides us with a highly provisional name for this calendar, but it is scarcely more than speculative. It was presumably inaugurated 1 V 520 B.C. Gregorian 6.11.11.8.17 5 Caban 5 Mol Tikal (Julian day number 1531620). The Izapa monument is half a millennium later.

In this reconstruction, the Zapotec calendar was installed on 10 III 305 B.C. Gregorian 7.2.9.8.13 13 Ben 6 Mol Tikal (Julian day number 1610096). This date requires substantial alteration of the dates of the early Zapotec stelae from Monte Alban I (594, 563 and 528 B.C. in the date list of Chapter 2). Placing them six calendar rounds later would relocate them at 282, 251, and 216 B.C., respectively. This construal would probably imply a corresponding reduction of some order in the presumed antiquity of some of the later Zapotec monuments as well (those listed at 229 and 209 B.C.), perhaps pushing forward the dating of the beginning of Monte Alban II. I cannot appraise the archaeological implications of these chronological shifts, but I know of no reason that the early stelae cannot belong to the end rather than to the beginning of Monte Alban I. It would be difficult to avoid the conclusion in any case that they document the Zapotec calendar immediately after its inauguration.

Calendar C was originally hypothesized as a necessary missing link between the Olmec calendar and those of Teotihuacan, Tikal, and Jicaque. In order to serve this purpose, it had to have a New Year located at 5 Yaxkin Tikal and use zero numeration of the uinal and 1 to 13 counting of the day count. At a later stage of my research, I found that this is precisely the placement of the calendar of Kaminaljuyu. It represents a 1-day advance from the Olmec calendar, with which it is otherwise structurally identical except in one feature: its names the year for its first day. This abandonment of terminal naming almost certainly reflects a recalculation of the length of the solar era, eliminating the extra 6 days previously thought to be necessary. It is at this point that the Middle American calendar attained the accuracy of modern astronomy in its estimate of the length of the tropical year: 365.2422 days. The date was 433 B.C.

The adoption of initial naming of the year at Kaminaljuyu implies awareness of a solar era of exactly 29 calendar rounds (550,420 days), which is astronomically correct. It has been pointed out that any 365-day calendar would have this potential, and initial naming does not in isolation imply awareness of the solar era. In the context of its derivation from a terminally named calendar, of the chronological pattern of the calendars derived from it, and of the numerological elegance and simplicity of its internal organization, however, the Kaminaljuyu calendar provides clear evidence that it was consciously designed to measure the solar era as well as the year. And it did so with stunning accuracy.

This conclusion is both explained and emphasized by the apparent date of its institution—on the spring equinox: 21 III 433 B.C. Gregorian 6.15.19.11.12 7 Eb 5 Yaxkin Tikal (Julian day number 1563355). The result was the inauguration of a new spring era and a new series of initially named Type II calendars, replacing the abortive summer era and terminally named series instituted with Calendar A. The new series remains incomplete but eventually included the calendars of Teotihuacan, Tikal, and the Tlapanec.

The Kaminaljuyu calendar itself remains poorly documented. It is found only on Stela 10 at Kaminaljuyu (see 147 B.C. in Chapter 2). Although this identifies Calendar C and confirms its existence, it is of no help in confirming its antiquity.

The same anti-leap-year mechanism that was responsible for generating the Olmec calendar from that of Cuicuilco generated Calendar D from that of Kaminaljuyu after the same 83-year interval. The transformation of – 20 days placed its New Year on 9 II 350 B.C. Gregorian 7.0.3.13.7 5 Manik 5 Xul Tikal (Julian day number 1593630). Calendar D is necessary to explaining the Tikal calendar as Calendar C (Kaminaljuyu) is to explaining that of Teotihuacan (*vide infra*).

The culmination of the development of the Preclassic calendar was the invention of the 13-baktun cycle—the Olmec (and Mayan) era, or Long Count. If it is true that the calculation of the calendar-round era of 1,508 years was initially derived from astronomical measurement of the 186-day summer semester (from spring to fall), it seems likely that the Long Count era of 8,000 tuns (7,890 years plus 300 days) was derived from the 180-day winter one (from fall to spring). It was, however, fundamentally a numerological rather than an astronomical invention, based upon the counting of 360-day tuns.

The tally of the tuns in the Long Count is structurally identical to the tally of the years in the Olmec calendar. Every fourth year of the latter is composed of uinals beginning with a day Imix and ending with a day Ahau. The uinals of the tun do the same. On this internal evidence alone, the Olmec invention of the Long Count seems self-evident. It seems reasonable to suppose, furthermore, that the Long Count was initiated at a time when the two cycles coincided. They did so dramatically on 8 VI 355 B.C. Gregorian 6.19.19.0.0 1 Ahau 3 Ceh Tikal (Julian day number 1591923).

The date is dramatic in that it is not only common to the year and tun cycles: it is the end of a calendar round and of the last katun of the sixth baktun. The co-ocurrence of an even calendar round and an even katun happens once in 936 years. It seems very likely that in this instance it was made the occasion for counting cycles of 13 and 20 katuns. The 13-katun cycle became the may and the 20-katun cycle became the baktun. Combining the two counts suggests a cycle of 13 baktuns (or 20 mays) and thus generates the Long Count. There was also a 20-baktun cycle (*Dresden* 61, 69), though it was not used in Long Count dates.

It is not self-evident that, having reached this point, the Olmec would have concluded that they were in the middle of the cycle thus generated. (The exact middle of the baktun cycle falls at the end of the thirteenth may, on the tenth katun of the seventh baktun: 7.9.0.0.0.) In any case, by naming the following katun the start of the seventh baktun, they generated the conventional and mythological starting point of 0.0.0.0.0 4 Ahau 8 Cumku Tikal 11 VIII 3113 B.C. Gregorian (Julian day number 584283).

There was, however, nothing arbitrary about the fixing of the end of the Long Count era. Victoria Bricker has pointed out to me that 13.0.0.0.0 4 Ahau 3 Kankin corresponds to an astronomically correct winter solstice: December 21, 2012 A.D. (Julian day number 2456283). Thus there appears to be a strong likelihood that the eral calendar, like the year calendar, was motivated by a long-range astronomical prediction, one that made a correct solsticial forecast 2,367 years into the future in 355 B.C.

It has been argued by Malström (1978), among others, that the beginning date of the Long Count may also be astronomically significant if it is read as a zenith passage date. Because the date of the zenith passage of the sun varies with latitude, the fact that 0.0.0.0.0 4 Ahau 8 Cumku fell on August 11 focuses attention on the vicinity of lat. 15° N. Because he was using a different correlation constant, Malström assumed that the date was August 13 and supposed that this date also marked the point at which the sun would have spent 260 days south of the zenith and would then pass to the north for 105 days, thus aligning the katun, tzol kin, and haab with a significant celestial event. Malström calculated that this made Izapa a likely locus for the invention of the Long Count.

The astronomy behind this brilliant speculation is not as solid as it sounds. The correct date of August 11 corresponds to a latitude of about 15° 39′ N — around Chama (but may vary as much as 10′ latitude from year to year), whereas the point at which the 105/260 division of the year occurs corresponds to the latitude of Izapa and Kaminaljuyu but a date of zenith passage on August 14. My following estimates are based on *The Astronomical Almanac for the Year 1986* (USGPO, Washington, Table C14, (cf. Aveni 1980, 1984), corrected for long. 90° W. It is doubtful that observations accurate to the day would have been feasible for the Preclassic Olmec, even sighting down a vertical well, because the variability amounts to about half the apparent width of the sun's face.

Zenith Passage	Latitude	Days N/S	Site
11 VIII	15° 39′ N	102/263	Chama (15° 38′)
12 VIII	15° 22′ N	103/262	Quirigua (15° 18′)
13 VIII	15° 04′ N	104/261	Utatlan (15° 04′)
14 VIII	14° 45′ N	105/260	Izapa (14° 42′)

Although the zenith-passage argument is intriguing, it does not seem to be a possible *determinant* of the establishment of the Long Count, even though that event is most likely to have occurred somewhere between the sites of the four earliest known Long Count dates—between lat. 18° 30′ N (Cerro de las Mesas and Tuxtla: zenith passage 31 VII) and lat. 14° 20′ N (Abaj Takalik and El Baul: zenith passage 15 VIII). But the starting point of the Long Count is determined by numerology once the ending point is fixed, and the correct prediction of the ending on a winter solstice would be a much stronger motive, particularly if it could be fitted to the numerological coincidence of the katun and haab cycles in 355 B.C. There is only one degree of freedom to fixing the course of the Long Count era. If it was done predictively, as I believe it was, its beginning is merely a numerological consequence of that fact and cannot therefore be indicative of anything else. And the predictive mechanism, of course, fits the pattern of the apparent development of the year calendar as well.

It is a matter of some theoretical interest that the Long Count is shared by the Olmec and Tikal but is absent from the Kaminaljuyu calendar that lies between them in the genetic model of derivation here presented. It is also absent at Teotihuacan, which derives from Kaminaljuyu. The count therefore diffused to the Maya directly from the Olmec quite separately from the genetic derivation of the Tikal calendar round from Kaminaljuyu and Calendar D. The Quiche calendar and other Mayan calendars derived from it also lack the Long Count, even though they derive from Tikal at a later date. Perhaps the undergraduate education in calendrics necessary for communication of the calendar round was inadequate for transmitting the more esoteric lore of the Long Count, which, in any case, remained confined to the Cholan and Yucatecan Maya and the Olmec. Whether the Kekchi, who used the Tikal calendar, also knew the Long Count has not been established.

All of the earliest archaeological evidence of the Long Count is Olmec (see 36, 32 B.C. and A.D. 37, 126, 162 in Chapter 2). It does not appear among the Maya until A.D. 292 (q.v.). The two cultures wrote the dates quite differently: Mayan dates used period glyphs for baktun, katun, tun, uinal, and kin; Olmec dates did not. They also used different glyphs for the 20 days of the day count—a less useful discrimination because these are not always legible in the early inscriptions.

It was after the genesis of the Long Count (355 B.C.) and Calendar D (350 B.C.) that the finishing touches were put on the anti-leap-year mechanism for calculating and predicting the solar era. They were incorporated into the calendar of Teotihuacan, which was instituted on 11 XI 165 B.C. Gregorian 7.9.12.3.12 3 Eb 0 Zotz' Tikal (Julian day number 1661475). Both the problem and the solution involved great ingenuity and subtlety.

The spring era of Kaminaljuyu presupposed that 1,508 vague years would exactly coincide with 1,507 tropical ones. To remain faithful to this calculation, any new calendar founded thereafter should have based its New Year on the same day of the month as the New Year of Kaminaljuyu (5 Yaxkin Tikal) but should have moved it back in 20-day steps at intervals of (approximately) 83 years. This was, as has been stated, the logic of Calendar D and of the derivation of the Olmec calendar from Calendar A (Cuicuilco). The system did not demand that a new calendar be founded—only that if one were, it would confine itself to foreordained placement within the era. If all the implied calendars had in fact been instituted, this would have obliterated 18 months of the full year by which the vague count exceeds the tropical one. But what about Uayeb? This was the problem addressed by the Teotihuacan calendar.

Why any particular people felt the need for instituting a calendar of its own is a separate question, presumably related to processes of religious, philosophical, and ideological change. On the record, in fact, two opportunities for doing so in the third and early second centuries B.C. were generally ignored. But the Teotihuacanos decided to do so and to solve the Uayeb problem at the same time.

Starting from the Kaminaljuyu New Year, they subtracted Uayeb, then, treating the date reached as a first name day, they counted down to 1 to reach the New Year of their new calendar (transformation − 100). By treating the day before the first of Uayeb as a first name day (100th day) rather than a name day (360th), they audaciously subtracted one full day count (260 days) from the calendar. The date reached (0 Uo Tikal) would have been correct calendrically, but it was premature astronomically. According to the eral calculation, it could not have been properly implemented until 1 B.C. Being unwilling to wait on the heavens for 164 years and having demonstrated a gift for calendrical legerdemain, they therefore took the still more extraordinary step of adding + 1 to the day count, henceforth counting the days from 2 to 14 rather than from 1 to 13. Advancing the day count coefficient by + 1 moves it forward 40 days (two uinals) in the year count (from 0 Uo to 0 Zotz' Tikal), thus making it possible to institute the new calendar on 3 Eb 0 Zotz' Tikal in 165 B.C. The Teotihuacanos called it 4 Eb 0 Zotz'.

The earliest archaeological appearance of the Teotihuacan calendar is on Kaminaljuyu Stela 10 in 147 B.C. (see Chapter 2). The Teotihuacan date on the

monument is given in the general day count in Olmec glyphs: the + 1 transformation of the day count is not directly attested, though the 0 Zotz' New Year can be
inferred from the associated Olmec date of 1 Lord. Both features are, however,
confirmed by the correlational inscription of A.D. 861 at Cacaxtla.

The anti-leap-year Uayeb correction that generated the Tikal calendar was
strictly parallel to the Teotihuacan mechanism, but the date of its institution made
alteration of the day-count coefficient unnecessary. It was achieved by a straightforward subtraction of 105 days from the New Year of Calendar D. It was inaugurated
on 13 VII 84 A.D. Gregorian 8.2.3.8.12 9 Eb 0 Pop Tikal (Julian day number
1751935).

Arithmetically, the subtraction of 105 days from the New Year's date seem
straightforward enough. Calendrically, it is much less so. In principle, the Tikal
calendar could have been derived directly from the calendar of Kaminaljuyu by
subtracting 125 days, which is also straightforward and which would eliminate the
necessity of supposing that there ever was a Calendar D. I believe, however, that
the transformation of – 105 days that created the calendars of Teotihuacan and
Tikal was perceived as a terminalization (– 1 – 5) and reinitialization (– 100 + 1)
of the existing (initial) calendar in both cases. It was a once-in-an-era procedure
that alone would justify the omission of one entire day count.

To be sure, such an omission is devoid of consequences because nobody
appears to have been counting day counts as such. Aside from the present
intercalendrical comparison, I do not see how such an omission could be detected
(*pace* Vollmaere 1984). The fact that the new calendars are tied back into the eral
calculation at the correct dates means that the days in question are not really lost:
they vanish into the abyss between two different calendrical perspectives of the
same reality like the interocular distance between two eyes focused on the same
object. The initializing transformation of – 100 days is separable from the anti-
leap-year transformation of – 20 days because of its different implication for the day
count: it is an initializing transformation.

The earliest archaeological documentation of the Tikal calendar is on Stela
29 at Tikal in A.D. 292 and may be considered to fall within the Early Classic
period. (The Hauberg Stela of A.D. 199 is aberrant and unprovenienced.) There
are, however, two other calendrical developments that appear to belong to the Late
Preclassic: the institution of the Jicaque and Yucuñudahui calendars. Both are actually earlier than the Tikal calendar, but they have been left for separate discussion because they were derived by a separate process: each involves the establishment of a new solar era. The mechanism in both cases was the advancement of the
calendar by 1 day.

The Jicaque New Year falls on 6 Yaxkin Tikal, 1 day later than that of
Kaminaljuyu, from which it is almost certainly derived. The occurrence of this
1-day correction (and a number of others like it) suggests that the Middle American

astronomers calculated correctly that the era was sometimes (albeit not always) 1 day more than 29 calendar rounds: 550,421 days. It is my belief that they accommodated this awareness as they did their other astronomical predictions — by precorrection. The only occasion for such a precorrection would be the founding of a new solar era, and it is reasonable to expect that any such event would be timed to a date on which their mnemonic marker, the day of New Year, coincided with a solstice or an equinox. It is my hypothesis that this is the sole motivation for changes in year bearers in the initial-dated calendars. Such a change was involved in the generation of the Kaminaljuyu calendar from the Olmec, establishing the spring era. I believe it occurred again in the derivation of the Jicaque calendar from that of Kaminaljuyu.

The Jicaque calendar was instituted on 21 XII 57 B.C. Gregorian 7.15.1.15.18 8 Etz'nab 6 Yaxkin Tikal (Julian day number 1700961). It thus established a new winter era, scheduled to end on the winter solstice in A.D. 1451. The case is peculiarly problematic. The earliest documentation of the Jicaque calendar is the identification of its New Year (as that of the natives of Las Hibueras) in 1530. Because of the total lack of archaeological documentation, we cannot altogether rule out the possibility that it was not a winter calendar at all. It could have been generated as an autumn calendar a quarter of an era later. It seems likely, however, that by that date (in the third century A.D.), the Kaminaljuyu calendar from which it derived may no longer have been in use. No later derivation can be seriously considered.

In the present reconstruction, I have systematically opted for the earliest opportunity in cases of this kind. Because they only become detectable when there is a change of year bearers, we have no way of determining at this point whether the native astronomers attempted to differentiate eras of 550,420 days from those of 550,421. Presumably an era instituted on the first premise would change nothing (see, however, the later discussion of the Palenque and Quiche calendars). The difference between the two calculations does not affect the estimate of the tropical year as 365.2422 days.

I have speculated that the Jicaque calendar may have been that of the Lenca and perhaps of the Xinca as well (see Chapter 3.) It is clear that it generated (83 years after it was founded) the calendar of the Mixe, presumably on 11 XI 26 A.D. Gregorian 7.19.4.17.8 5 Lamat 6 Xul Tikal (Julian day number 1730071). Certain of the intervening and related peoples, particularly the Zoque and Tapachultec, may well have employed either of these two calendars, because the two must have been in contact with each other someplace sometime in the late Preclassic. The mechanism of the formation of the Mixe calendar was the anti-leap-year transformation of − 20 days. The Mixe calendar is first documented in 1531.

A final calendrical development of the Preclassic was the genesis of the Mixtec calendar of Yucuñudahui from that of Teotihuacan. Like the Jicaque calendar, it involved a change of year bearers and the institution of a new era, but unlike the winter era of Jicaque, that of Yucuñudahui was autumnal. It was established on 23 IX 42 B.C. Gregorian 8.0.1.1.3 2 Akbal 1 Zotz' Tikal (Julian day number 1736666).

Although it derived from the Teotihuacan calendar, Yucuñudahui abandoned the 2 to 14 counting of the day count. This did not have any discernable influence on the counting of the year. It may be, however, that Yucuñudahui retained the Teotihuacan starting point for the calendar round with the coefficient 2 rather than 1. A number of later calendars (Aztec, Teotitlan, Tilantongo, Otomi, and perhaps Campeche) considered the calendar round to begin with 2 Cane. Yucuñudahui may have done so, too. Alternatively, the accident that it seems to have begun in a year 2 Akbal suggests that date as a possible starting point for its calendar round.

The first archaeological documentation of the Yucuñudahui calendar is dated to A.D. 426 (see Chapter 2). This is a calendar-round date, of course, rather than a Long Count date. The tomb from which it comes is guess dated to A.D. 400, so a date of a calendar round or so earlier is not impossible. Even so, it would appear likely that still earlier Yucuñudahui dates may yet be found. A correlational date of A.D. 768 (in Chapter 2) establishes the position of the Yucuñudahui calendar with respect to Zapotec.

Preclassic development of the calendar of Anahuac is marked by the invention of the calendar round of 52 years (739 B.C.), the solar era of 1,508 years (433 B.C.) and the Long Count of 7,891 years (355 B.C.) Because it lasted for half of a solar era, it generated four families of calendars that memorialize an equal number of calculations of the solar era, anchored to its quarters: the solstices and equinoxes.

Terminal Naming

Summer era	Type I New Year	Cuicuilco (739 B.C.)
		Olmec (656 B.C.)
	Type II New Year	Izapa (520 B.C.)
	Type III New Year	Zapotec (305 B.C.)

Initial Naming

Spring era	Type II New Year	Kaminaljuyu (433 B.C.)
		Calendar D (350 B.C.)
		Teotihuacan (165 B.C.)
		Tikal (A.D. 84)
Winter era	Type III New Year	Jicaque (57 B.C.)
		Mixe (A.D. 26)
Autumn era	Type III New Year	Yucuñudahui (A.D. 42)

From a technical point of view, the evolution proceeded from a solar era of 550,426 days, indicated by terminal naming, to one of 550,420 (or 550,421 days), marked by initial naming. The later calculation corresponds to that of a tropical year of 365.2422 days, anticipating by more than two millennia the comparable institution of the Gregorian calendar in Europe.

The mechanisms of calendrical change are somewhat different in the terminal and initial calendars. In the former case, the transformation of + 1 day is applied every 219 years in order to accommodate the extra 6 days of the era. In the latter case, the transformation of + 1 day is used only to generate a new era, and the transformation of – 20 days occurs as a kind of anti-leap-year correction every 83 years. A special transformation of – 105 days is found in the calendars of Teotihuacan and Tikal in order to accommodate Uayeb in the prediction of the end of the era. This last is not repeated at later dates, because it takes care of the problem for the whole era, at least for all calendars derived from these two. As we shall see, these comprise all of the calendars of the Classic period, and the end of the era is also the end of the Classic.

CLASSIC

Two of the series of calendars set up by the eral calculations of the Preclassic continued to be productive of new calendars during the Classic period: the spring era of Kaminaljuyu through the Teotihuacan and Tikal calendars, and the autumn era of Yucuñudahui. But the passage of time also provided the impetus for generating new eras as well.

Summer era	Type III New Year	Palenque (A.D. 177)
Spring era	Type II New Year	Quiche (A.D. 395)
	Type IV New Year	Campeche (A.D. 568)
Winter era	Type III New Year	Kanhobal (A.D. 768)
	Type IV New Year	Tenango (A.D. 772)
		Mayapan (A.D. 937)

The derivation of the Classic calendars involved no traceable changes of calendrical astronomy and almost no new calendrical mechanisms. It simply continued the tradition of the Preclassic — to its conclusion.

The only known calendars derived directly from that of Teotihuacan are those of Yucuñudahui, which abandoned the 2 to 14 day count, and Tlapanec, which retained it. The latter was inaugurated on 14 III 258 A.D. Gregorian 8.10.19.12.2 12 Ik 5 Muan Tikal (Julian day number 1815291). It is first documented in Postclassic codices (see 1487 in Chapter 2). It remains possible that some other calendars whose day counts are unknown or poorly documented may have a parallel origin. Chiapanec and Tequistlatec occur to me as possibilities; neither of them is archaeologically documented. The widespread distribution of the Teotihuacan calendar in early times makes it difficult to rule out any area of Middle America as the possible locus of an affiliated derivative.

Five of the six new eral calendars of the Classic period derived from the calendar of Tikal: Palenque, Quiche, Campeche, Kanhobal, and Mayapan. The transformation in each case was + 1 day, but the mechanisms were subject to some variation, including zero deletion, year-bearer advance, terminalization, and initialization. The Palenque, Campeche, and Mayapan calendars are lineal descendants of the Tikal calendar: they continued to begin the year with Pop. The Quiche and Kanhobal calendars begin it with Kayab.

The Palenque calendar appears to have been inaugurated on 22 VI 177 A.D. Gregorian 8.6.17.13.18 12 Etz'nab 1 Pop Tikal (Julian day number 1785881). The transformation was + 1 day, and the mechanism was terminalization. Dates in the Tikal calendar record the first day of the month with a glyph for "seating." The same Type II days are recorded in the Palenque calendar as the last day of the preceding month with a glyph for "ending." It is the seating glyph that has customarily been read as zero. The ending glyph must be read as 20. The result is therefore zero deletion, and it adds 1 day to the year count, advancing the New Years to Type III days.

The delicacy of the calendrical adjustment involved in the Palenque calendar suggests the possibility that it was intended to mark the inauguration of a new summer era *without* adding a day to the length of the era. In other words, it predicated an era of 550,420 days rather than of 550,421. This would have meant that the advance in year bearers was used to mark the fact of a new era, but the function of predicting its termination was moved to the name day and thus continued to be the same as in the Tikal calendar. This use of terminal naming, different from that of the early Preclassic calendars, may have been shared by the apparently derivative Campeche calendar. If this construal is correct, the inauguration of the Palenque calendar would have been 24 years earlier than A.D. 177, in 153, when the *name day* fell on the summer solstice. Given the minimalist character of the change and the fact that even Calendar A does not appear to have done this, I believe the A.D. 177 date is correct.

The earliest documentation of the Palenque calendar is dated to the katun ending in A.D. 692 (see Chapter 2), nearly half a millennium after its hypothesized origin. Its later appearance at Yaxchilan, Naranjo, and Piedras Negras suggests that it may have been a Cholan invention. It is only a secondary usage at these sites, but it may have been the primary Cholan calendar of the Early Classic.

The Campeche calendar was inaugurated on 20 III 568 A.D. Gregorian 9.6.14.3.14 1 Ix 2 Pop Tikal (Julian day number 1928597). It was derived from the

Palenque calendar by advancing the New Year by +1 day. It is thus the earliest Type IV calendar. It retained terminal naming, so the year bearers were Type III.

The Campeche calendar is documented earliest from Uxmal in A.D. 649, and later from Etzna, Jaina, Holactun, Yaxchilan, and Bonampak, a geographic distribution suggesting that it was a Cholan (perhaps Chontal) calendar. None of the documentation includes a clear date for the beginning or ending of a month, so we cannot be certain that it was terminally named. The existence of a calendar with Type III year bearers is, however, attested in the New Year pages of the *Dresden Codex* (25–28), whereas evidence of Type IV year bearers in Yucatan is found only in Colonial sources, where it relates to the Mayapan calendar. Because the inauguration of the latter is explicitly attributed in the *Tizimin* to a shift from terminal to initial dating, the most likely construal of the Campeche calendar is the one I have made—as a terminally named Type IV system.

The origin of the Mayapan calendar is harder to resolve. Astronomical considerations create two strong possibilities, whereas ethnohistorical documentation imposes a third. The first possibility is that it is a winter era calendar of A.D. 937. The second is that it is an autumn era calendar of A.D. 1300. The third is that it is a may-influenced calendar of A.D. 1539. The first two possibilities have the advantage of being astronomically motivated and the disadvantage of being totally undocumented. The last is clearly documented but is motivated by and timed to the end of the katun cycle of the Itza (13 Ahau) rather than to solar astronomy.

I believe the first solution to be correct: the Mayapan calendar was invented as of 21 XII 937 A.D. Gregorian 10.5.9.6.4. 7 Kan 2 Pop Tikal (Julian day number 2063647). It appears, however, that it was not successfully promulgated until A.D. 1539. At that time, it was made part of a calendrical reform that moved from terminal to initial dating of both the katun and the haab (Edmonson 1976). The 937 solstice has the advantage of being the earliest of the possibilities, and the only one that falls within the (terminal) Classic period. It is not clear to me that the sophistication of eral astronomy survived as late as 1300 (see the later discussion).

It is possible that the derivation of the Quiche calendar (New Year 5 Kayab Tikal) from that of Tikal (New Year 0 Pop) was mediated by an intermediate Calendar E (New Year 5 Cumku Tikal) in two anti-leap-year transformations of –20 days each. The hypothetical Calendar E would have been initiated on 4 VI 167 A.D. Gregorian 8.6.7.10.7 7 Manik 5 Cumku Tikal (Julian day number 1782210). We have no evidence that such a calendar ever existed, and I consider it unlikely.

The Quiche calendar should, by this logic, have been instituted on 25 IV 250 A.D. Gregorian 8.10.11.12.2 5 Ik 5 Kayab Tikal (Julian day number 1812485). At this point, it should also have retained the 0 to 19 numbering of the days of the month. The surviving Quiche calendar, however, numbers the days of the month from 1 to 20, indicating a transformation of +1 day and hence the establishment of a new era. The Quiche mechanism of this transformation is unique: it actually moved the month coefficients back 1, so that the first day of the month continued to land on the Tikal year bearers. This may imply that, like the authors of the Palenque calendar, the Quiche calculated the era at exactly 550,420 days, rather than 550,421.

The zero deletion in the Quiche calendar implies that it originated with a new spring era on 21 III 395 A.D. Gregorian 8.17.18.12.7 7 Manik 5 Kayab Tikal (Julian day number 1865410). Its inheritors are heirs to its zero deletion, although at the same time being impoverished by the failure of the Quiche to learn the low-land Long Count. The earliest documentation of the Quiche calendar is in 1722.

An additional anti-leap-year transformation of – 20 days generated the Chiapanec calendar from the earlier version of the Quiche calendar. Its founding date would have been 16 III 333 A.D. Gregorian 8.14.15.13.17 3 Caban 5 Pax Tikal (Julian day number 1842760). In this reconstruction, the Chiapanec system should have continued to use zero counting of the month. Unfortunately, it is not documented until 1691, and even then we have no hint of the numbering used. It is quite likely that in the 4th century A.D. the Quiche calendar was used throughout the Mayan highlands, which meant that it abutted directly on Chiapanec territory.

The later date for the origin of the Quiche calendar would place the Chiapanec calendar correspondingly later, and I believe this to be the correct derivation: 8 II 478 A.D. Gregorian 9.2.2.14.2 5 Ik 5 Pax Tikal (Julian day number 1895685). This could be proved if we find documentation of the Chiapanec numbering of the days of the month: I wager that they did not use zero.

The Maya of Chiapas and the Cuchumatanes seceded calendrically from Quiche with the establishment of the Kanhobal winter era on 21 XII 768 A.D. Gregorian 9.16.17.15.18 5 Etz'nab 6 Kayab Tikal (Julian day number 2001921). The transformation of + 1 day was achieved by a simple year-bearer advance. The Kanhobal calendar remained undocumented until 1932, even though it is shared by Jacaltec and Chuh.

In Central Mexico, the Classic period was marked by the generation of at least five, and perhaps as many as twelve new calendars derived sequentially from the Yucuñudahui calendar by the anti-leap-year correction of – 20 days. An additional calendar, that of Tenango, was generated in 772 as a new winter era by the transformation + 1 day. The first five are Cholula (A.D. 125), Toltec (208), Tarascan (291), Huastec (373), and Totonac (456). Documentation of the first two is sketchy; the others are reasonably solid. Only the Tenango calendar is directly attested by Classic archaeology.

The general contrast between the calendrics of the Classic and the Postclassic periods in Central Mexico is fairly clearly marked (wherever we have any evidence for calendrics at all) in the manner of writing both numerals and day signs. Postclassic dates are written in Tilantongo glyphs with dot numerals. Classic dates are written in a number of regional glyphic day-sign systems with bar and dot numerals. Each of the five Classic calendars just enumerated (omitting Tenango) gives evidence of having had its own Classic writing system, including at least one unique day sign and the use of bar numerals. In general it would seem reasonable to suppose that a different calendrical writing system implies a different calendar.

The Yucuñudahui calendar was widely known even beyond the borders of the Mixteca (426), including at Monte Alban (565), Xochicalco (762), and very likely Teotihuacan (Chapter 5). The presence of Type III year bearers at the latter site cannot be definitely ascribed to the Yucuñudahui calendar, because they could equally well be attributed to one or another of the five calendars under discussion — all of which are Type III. Unfortunately, they do not occur in correlational dates. All five of the calendars derived from Yucuñudahui's are plausibly geographically contiguous to the ancient Mixteca defined by this distribution.

The case for the Cholula calendar is the weakest. Puebla is the nearest region to the Mixteca, and it would therefore make sense for its calendar to be the first derivative from Yucuñudahui's. By the transformation of – 20 days, the Cholula calendar should have been established on 14 VIII 125 A.D. Gregorian 8.4.5.2.18 13 Etz'nab 1 Zip Tikal (Julian day number 1766941). We have, however, no independent warrant for this placement. Bar and dot notation occurs at Cholula: coefficients of 1 are probable. Year signs occur in various Classic period Puebla sites but apparently only in relation to Olmec, Teotihuacan, and Tenango dates.

The Toltec calendar, again on geographic grounds, is a good candidate for lying between Yucuñudahui-Cholula and the firmly placed Tarascan calendar. If so, it should have been instituted on 5 VII 208 A.D. Gregorian 8.8.9.4.13 11 Ben 1 Uo Tikal (Julian day number 1797216). Although we know that a Type III calendar was used at Tula with bar and dot notation and coefficients of 1, we have absolutely no evidence of the placement of its initial month, and the third century is over half a millennium before Tula was founded. It is perfectly possible that there was no 1 Uo calendar, and it is equally possible that the Toltec New Year was several months earlier and its invention several centuries later. If we were to place it on 6 Muan, for example, its date of origin would be displaced to the seventh century, and it would be the direct antecedent of the Aztec calendar, whereas putting it on 6 Mac would imply an origin in the eighth century and make it the precursor to the Tilantongo Mixtec calendar! A single Toltec correlational date would solve a host of problems, but no such date is known.

The Tarascan New Year is well established and its putative origin is therefore less hypothetical: 26 V 291 A.D. Gregorian 8.12.13.6.8 9 Lamat 1 Pop Tikal (Julian day number 1827491). The earliest Tarascan calendrical documentation is from 1521 and establishes the position of the Postclassic Tepanec rather than the Classic Tarascan calendar. This will be discussed later. Archaeological examples of Classic Tarascan dating are relatively abundant, mostly ascribed to the Middle to Late Classic and found as far afield as Nayarit. The Tarascan year sign has not been identified, but the disproportionate representation of the day Rabbit strongly supports a Type III calendar. Although we have Tarascan coefficients of 1, thus ruling out a Teotihuacan derivation, we have no coefficients higher than 4, so the Classic bar notation is not documented.

The Huastec calendar was derived from Tarascan by the anti-leap-year transformation of – 20 days and inaugurated on 16 IV 373 A.D. Gregorian 8.16.16.7.18 6 Etz'nab 6 Cumku Tikal (Julian day number 1857401). It is documented from the earliest times at Tajin together with bar and dot numeration, coefficients of 1, and year signs coupled with Type III days, but its initial month is not fixed until 1616 (q.v.), and by that time it is the Postclassic Cuitlahuac calendar rather than the Classic Huastec one that is in play (*vide infra*). The Classical month placement is inferential rather than directly attested.

The Totonac calendar was generated from Huastec by the same anti-leap-year transformation of – 20 days and was instituted on 6 III 456 A.D. Gregorian 9.1.0.9.13 4 Ben 6 Kayab Tikal (Julian day number 1887676). It is archaeologically documented at various sites of central Veracruz from the Middle and Late Classic periods. No year signs or coefficients higher than 4 occur, so bar and dot numeration is not attested, but the salience of the days Wind and Cane suggest the overlapping use of the Teotihuacan and Totonac calendars, and a date of 1 Flint presumably refers to the latter. The month placement of the Totonac calendar is confirmed by correlational dates of 1538 and 1539, though neither is entirely satisfactory. Its geographic proximity to Huastec and Teotitlan is a further confirmation because it derives from the former and is a precursor to the latter (*vide infra*).

The Totonac calendar would have to be postulated even if it were not documented at all in order to account for the calendar of Tenango. The latter was inaugurated as a winter era calendar on 21 XII 772 A.D. Gregorian 9.17.1.16.19 10 Cauac 7 Kayab Tikal (Julian day number 2003382). It is marked as an eral calendar by the transformation of + 1 day that derives it from Totonac. There are some striking similarities in both the names and glyphs of the days between Totonac and Tarascan, suggesting that they were at one time in direct contact, and enhancing the plausibility of deriving the Tenango calendar from the former. Tenango itself used the Xochicalco writing system. Its calendrical placement is assured by a correlational date of A.D. 790, and it is also documented by a date of anno 13 Iguana from Micaltepec, Puebla (Tschohl 1972:469, Fig. 2). This is almost certainly a Tenango date because Tenango had the only known Type IV year bearers outside of Yucatan.

The summer era of Cuicuilco (Calendar A) came to an end in A.D. 769. Its New Year fell on the summer solstice, exactly as had been predicted 1,507 tropical years (550,420 days) before. There was no celebration at Cuicuilco, which lay buried under a lava flow together with its calendar. The outcome of the astronomical calculations involved in all ten of the eral calendars is as follows:

Era	Begins	Ends	Duration
Cuicuilco (summer)	739 B.C.	A.D. 769	550,420 days
Kaminaljuyu (spring)	433 B.C.	A.D. 1075	550,420 days
Jicaque (winter)	57 B.C.	A.D. 1451	550,421 days
Yucuñudahui (autumn)	A.D. 42	A.D. 1549	550,420 days
Palenque (summer)	A.D. 177	A.D. 1684	550,421 days
Quiche (spring)	A.D. 395	A.D. 1902	550,420 days
Campeche (spring)	A.D. 568	A.D. 2075	550,420 days
Kanhobal (winter)	A.D. 768	A.D. 2275	550,421 days
Tenango (winter)	A.D. 772	A.D. 2279	550,421 days
Mayapan (winter)	A.D. 937	A.D. 2444	550,421 days

The contrary directions of the leap-year recession and the anti-leap-year transformation meant that there would have been time to generate two full sets of eral calendars, but no final autumn era has been identified.

In a calendrical sense, the Classic came to an end with the end of the first summer era of Cuicuilco. Somewhat paradoxically, only the Zapotec were in a position to celebrate it. All the other calendars then extant were keeping time in later cycles by A.D. 769. Monte Alban must have had a splendid celebration, but there was nothing else to be done: the calendar was correct as it stood and would remain so for the first 219 years of the new summer era. Then, at that point, the same calendrical logic that had generated Calendar B (Izapa) some 1,508 years before would impose the necessity of another year-bearer advance.

So it was that the Mixtec calendar of Tilantongo was inaugurated on 2 V 988 A.D. Gregorian 10.8.0.7.19 7 Cauac 7 Mol Tikal (Julian day number 2082042). The transformation was + 1 day, and the mechanism was zero deletion, thus creating a terminally named Type IV calendar with Type III name days, counting the days of the month from 1 to 20. The logic that created the Tilantongo calendar and the calendrical consequences it generated belong, however, to the second solar era and to the Postclassic period.

POSTCLASSIC

The genesis of the Tilantongo calendar in 988 proved to have epochal consequences throughout Central Mexico: it created a movement that resulted in a speedy conversion of the zero-based initial calendars of the Classic to terminal calendars counting from 1 to 20. All of the calendars that were ostensibly derived from that of Yucuñudahui were eventually affected by this development, which was, however, a syncretistic religious and political change rather than an astronomical one.

For the calendars affected, the change was a transformation of – 100 days in the date of the New Year. Thus the New Year's day of the Classic calendar became the first name day of the Postclassic one in each case. From the astronomical point of view, this would have meant no change at all, provided it was accompanied by recognition that the function of marking the era was retained by the first name day. We have no unequivocal proof of such awareness, but it is significant that a correlational date of 1555 identifies the first name day of the Tilantongo and Metztitlan calendars as the date of their New Years. The first name day at the end of Hueitozoztli was an important Aztec feast as well.

The earliest documentation of the twelve calendars affected is ethnohistorical rather that archaeological. All of the dates involved are calendar-round dates, and none is contemporary with the events dated, so that they can be placed in real time only through genealogical reconstruction of the dynastic histories to which they mainly refer. This has been most thoroughly accomplished for the Tilantongo calendar by Caso, whose comprehensive reconstruction of the dynastic genealogy is virtually complete and has withstood criticism with what may, for our purposes, be considered only minor revisions. Caso's date for the founding of the Tilantongo calendar is 973. Mine is 15 years later.

The "earliest dates" in the other calendars are partly accidents of my own sampling: I have chosen to focus on two historical points, the reign of Quetzalcoatl and the founding of Tenochtitlan. The first dates the Colhua, Colhuacan, Cuitlahuac, Tepepulco, and Texcoco calendars to the twelfth century, and the latter additionally dates the Aztec and Tepexic calendars to the fourteenth. The Teotitlan calendar is datable to the fifteenth century, and post-Conquest documentation places the Tepanec and Metztitlan calendars. The Chalca calendar is not adequately dated (see Chapter 5: Nahuatl).

The alignment of the Classic and Postclassic calendars and the earliest documentation of the latter is indicated next. They are listed in the order of their New Year's dates from earliest to latest, which is the reverse of the order of the dates of genesis of the corresponding Classic calendars.

Postclassic	Classic	Hypothetical	Documented
(?)	(6 Yax)	1036	(?)
Colhua	(6 Zac)	953	1124
Metztitlan	(6 Ceh)	870	1555
Tilantongo	(6 Mac)	788	973
Texcoco	(6 Kankin)	705	1150
Aztec	(6 Muan)	622	1370
Tepepulco	Otomi	539	1151
Teotitlan	Totonac	456	1416
Cuitlahuac	Huastec	373	1127

Tepanec	Tarascan	291	1521
Colhuacan	Toltec	208	1127
Tepexic	Cholula	125	1369
Chalca	Yucuñudahui	42	426

Although the dating of these calendars is not totally satisfactory, it is adequate to substantiate that the relationships among them are not explicable in terms of any systematic Postclassic use of a leap-year or anti-leap-year adjustment. They are simply ratifications of preexisting Classic calendars, regularized as zero-deleting terminal systems.

In addition to precise documentation of the Chalca calendar, we may also anticipate future documentation of the Classic predecessors of the Otomi, Aztec, Texcoco, Tilantongo, Metztitlan, and Colhua calendars and perhaps the eventual discovery of the postulated calendar with a 6 Yax New Year.

Outside the sphere of "Toltec" influence in Central Mexico, the Mayan calendars and six non-Mayan ones were unaffected by the Tilantongo calendrical reform, though the expansion of the Nahua guaranteed that virtually all of Middle America was eventually exposed to it. The six non-Mayan calendars are Olmec, Zapotec, Jicaque, Mixe, Chiapanec, and Tlapanec.

Having completed their summer era on the correct solstitial date in 769, Olmec and Zapotec slipped calmly into a second era without alteration. The same happened to the other calendars at various time: Tlapanec, as heir to the Kaminaljuyu spring era, in 1075; Jicaque and Mixe, when they ended their winter era in 1451. Chiapanec was party to the Quiche spring era, which ended in 1902.

In the Mayan area, the Postclassic brought different developments in the lowlands and the highlands. In Yucatan, the most important features are the apparent genesis of the autumnal era of Mayapan in 937 and the almost simultaneous (928) abandonment of the monumental use of the Long Count. The latter development cannot have any direct relationship to either the Cuicuilco–Tilantongo era or to solar astronomy, though it is perhaps the most important defining characteristic of the Postclassic in the Mayan lowlands. An Olmec Long Count date of 1223, and a Yucatecan baktun celebration of 1618 underline the point that knowledge of the Long Count persisted and that there was no formal or calendrical reason for its suspension.

In the highlands, the Kanhobal winter era generated the new calendars of Guitiupa (Tzeltalan), Tzeltal, and Tzotzil. A number of Mayan calendars came under substantial Nahuatl pressure, which may have been decisive in the placement of the Cakchiquel New Year and probably influenced other calendrical developments in less dramatic ways.

The Guitiupa calendar is known only from modern sources (Chapter 2:1845; Chapter 5:Tzotzil, date *e*). It is also the most plausible ancestor of the later Colonial

Istacostoc and Mitontic calendars (*vide infra*). The mechanism by which these latter calendars were generated (a leap-year advance rather than an anti-leap-year reces- sion) is one specific to the post-Conquest period. Such a mechanism could also be used to derive the Guitiupa calendar itself from Tzotzil. However, the 5-month difference between the Tzotzil and Mitontic New Years would require 5 x 83 or 415 years if it were astronomically motivated. This would imply pre-Conquest use of a post-Conquest calendrical calculation.

It is thus more reasonable to postulate that the Guitiupa, or northern, Tzotzil calendar was derived in pre-Conquest times from Kanhobal. The date would have been 4 VI 1182 A.D. Gregorian 10.17.17.6.8 7 Lamat 6 Ceh Tikal (Julian day number 2152931). The transformation used was the – 20-day anti-leap year.

The Tzeltal calendar was also derived from the Kanhobal winter era, pre- sumably via the calendar of Guitiupa, by the anti-leap-year transformation of – 20 days. It was instituted on 15 III 1348 A.D. Gregorian 11.6.5.9.18 3 Etz'nab 6 Yax Tikal (Julian day number 2213481). The first documentation of it is modern — in 1888 (see Chapter 5: Tzeltal) and is complicated by post-Conquest calendrical developments.

The Tzotzil calendar dates to 4 II 1431 A.D. Gregorian 11.10.9.11.13 1 Ben 6 Ch'en Tikal (Julian day number 2243756). It was generated by the – 20-day anti-leap-year transformation and is first documented in 1688 (see Chapter 5: Tzotzil). Its calendrical placement is somewhat obscured by calendar freezing, and by the Julian-Gregorian calendar reform (*vide infra*).

In common with a number of other systems, the Quiche calendar came under heavy Nahua influence in the Late Postclassic. The Pipil of Guatemala were apparently using the Teotitlan calendar, and some of their month names and calendri- cal mythology were adopted by both the Quiche and Cakchiquel. When the Cakchiquel broke away from the Quiche politically in the late fifteenth century, they seized the occasion to create their own independent calendar as well. They did so by adopting the Teotitlan New Year, but retained the year bearers, zero dele- tion, and other features of the Quiche calendar.

Because the establishment of the Cakchiquel calendar does not appear to have been astronomically motivated, the date is uncertain. It seems most likely that it coincided with the founding of Iximche, probably in 1480 (which was, perhaps not coincidentally, the beginning of the Quiche calendar round).

If we were to explain the genesis of the Cakchiquel calendar as an anti- leap-year displacement from the Quiche New Year, we would generate a seventh- century date for the event: 6 X 664 A.D. Gregorian 9.11.12.2.17 8 Caban 5 Ceh Tikal (Julian day number 1963860). This seems to me unlikely, as it is substan- tially earlier than the dates usually offered for the separation of the Quiche and Cakchiquel languages.

Cakchiquel separatism is emphasized by the genesis of the unique calendar of Iximche in 1493. Based upon a cycle of 400 days, this calendar bears no relation to solar astronomy, but the explicitly political rationale of its establishment (the revolt of the Tukuches) seems to corroborate a political rather than purely calendrical motivation for the founding of the general Cakchiquel calendar. The *Annals of the Cakchiquels* (Recinos 1950:101–2) dates the founding of Iximche to a day 2 Iguana (Qat) in some year before the death of the Quiche king Quicab (ca. 1490), at whose "advice" the Cakchiquels made the move. The beginning of the Quiche calendar round in 1480 might readily have crystallized good Quiche reasons for a major change in Quicab's unpopular pro-Cakchiquel policy. The Iximche calendar was begun exactly 13 years later.

If this surmise is correct, the Cakchiquel calendar was initiated on 22 III 1480 A.D. Gregorian 11.12.19.8.17 5 Caban 5 Ceh Tikal (Julian day number 2261700). Of course, they could have waited until 1488 and made it a new spring era calendar, but there is no year-bearer advance or other structural clue to indicate that they did.

Although the Spanish conquest was initiated in 1519 and Tenochtitlan fell in 1521, the most convenient calendrical watershed between the Postclassic and Colonial periods was the promulgation of the Mayapan calendar in 1539. As we have seen, there is some reason to suppose that this calendar had been devised for astronomical reasons as early as 937. But the primary calendar of the Chol, Kekchi, and Yucatecan Maya of the Postclassic remained that of Tikal. It was replaced in Yucatan by the Mayapan calendar on 12 XI 1539 Gregorian 11.16.0.0.0 13 Ahau 8 Xul Tikal (Julian day number 2283483), though it remained in use among the Chol and Kekchi.

The calendrical developments of the Postclassic period are substantially different from those of the Preclassic and Classic. No new eral calendars were originated, and only three new calendars were generated by the anti-leap-year process (Guitiupa, Tzeltal, and Tzotzil). A fourth new Mayan calendar (Cakchiquel) seems to have been founded for political rather than calendrical reasons, and something similar apparently underlay the sweeping conversion of all the Central Mexican calendars to the Tilantongo Mixtec model in the tenth and eleventh centuries and the Mayapan calendar reform on the eve of the Spanish Conquest of Yucatan. There is no evidence that the sophistication of eral astronomy of earlier times survived until this date. Certainly, it did not survive beyond it.

COLONIAL

The impact of the Spanish Conquest on the native calendars of Middle America was extensive, complex, and ultimately fatal. It was complicated by the existence of two different civil calendars, Julian and Gregorian, and an ecclesiastical Church calendar. The features of these that differ calendrically from the Middle American systems had the greatest impact: leap year, the 7-day week, the 12-month year, and movable feasts of the Church.

The European leap year was the most important of these influences and was responsible for putting an end to the generation of new native calendars through the anti-leap-year transformation, instituting at least six new calendars (Nimiyua, Chinantec, Lachixola, Cancuc, Mitontic, Istacostoc) through an opposite leap-year transformation and destroying at least one (Tzotzil) through disruption of its day count. Three other calendars (Mazatec, Kekchi, and Pokom) were generated through disagreements over how to handle the Spanish leap year. The only other new calendar after the Conquest (Valladolid) was only indirectly influenced by the leap-year problem.

The Julian calendar, with its regular leap year every 4 years, may have been only a minor inconvenience for sixteenth-century Indian converts. Nonetheless, it generated a prompt calendrical reaction: the practice of "freezing" the correlation. Ten native calendars simply stopped counting European leap years after that of 1548, thus effectively freezing the European date for their respective New Years' days to what it was in 1549–50 (see Chapter 1, Fig. 3; Chapter 2, 1548; Chapter 5, Leap Year). The ten calendars are Aztec, Chiapanec, Chinantec, Guitiupa, Kekchi, Mayapan, Pokom, Texcoco, Tzeltal, and Tzotzil. The Pokom calendar is actually identical with Quiche and differs from it structurally only in its adopting a frozen correlation date, which the Quiche did not do. Both the Chol and Kekchi were still using the Tikal calendar, but the Kekchi froze their correlation and the Chol did not. The Mixe calendar appears to have been frozen even earlier (see Chapter 2, 1531).

The Gregorian calendar was promulgated in Mexico in 1583 and reached Chiapas and Guatemala in 1584. The Aztec, Guitiupa, Tzeltal, and Tzotzil calendars reacted by refreezing their New Years' dates to the new calendar. Because 9 of the 10 days' difference between the Julian and Gregorian calendars had been canceled by the leap years between 1548 and 1584, the result was a 1-day advance in the frozen New Year date (see Chapter 2, 1606; Chapter 5, Tzeltal, Tzotzil). Modern Tzotzil and Tzeltal tradition maintains both the Julian and Gregorian correlations, 1 day apart. At least in Tzotzil, they have allowed the European date rather than the native one to govern, which is probably the reason for the disruption and loss of the Tzotzil day count.

Some calendars ignored the Gregorian reform. The Yucatecans continued to cite the frozen Julian date for the Mayapan New Year into the nineteenth century. The Mazatec remained on the unfrozen Aztec calendar even after the Aztec froze the correlation, remaining on Julian dating until 1617 when their New Year recessed to January 1 *Julian* and then froze their New Year to that date (see Chapter 2, *1617).

The anti-leap-year transformation of – 20 days vanished with the Spanish Conquest, to be replaced by the opposite leap-year transformation of + 20 days. It seems clear that this was a syncretism of native and Spanish practice (the direction of the correction was European, but the 20-day interval was native), and it is to be presumed that it was applied every 83 years. Six new calendars were generated in this fashion, all of them apparently based on calendars using the frozen Julian correlation of 1548. None of the Middle American New Years' dates was anywhere near a solstice or equinox in 1548 (see Chapter 1, Fig. 3), so the dating of these calendars does not appear to be astronomically motivated. Using 1548 as an origin point yields the following founding dates for these six calendars. Their earliest documented appearance is also listed.

	Hypothetical	Documented
Nimiyua	1631	1691
Chinantec	1631	1949
Istacostoc	1631	1845
Mitontic	1714	1845
Cancuc	1797	1917
Lachixola	1797	1949

The last three calendars are known primarily from frozen Gregorian correlations. If they also based their 83-year cycles on the promulgation of the Gregorian calendar, this would add 36 years to their hypothetical dates of origin.

The only other new calendar of the post-Conquest period is that of Valladolid, established in 1752. It was marked by changing the counting of the katun from a period of 20 tuns to one of 24 haabs (vague years), and of the may from a period of 13 katuns to a period of 24 of these new katuns. It made no change in the Mayapan year and is thus irrelevant to solar astronomy and to our present purpose.

The transformation + 20 days produced the Nimiyua calendar from that of Chiapanec on 25 V 1631 A.D. Gregorian 12.0.12.15.12 9 Eb 5 Kayab Tikal (Julian day number 2316915). This made the Nimiyua calendar congruent with the Quiche. An alternative construal would be that the Nimiyua calendar was simply the original Chiapanec calendar, borrowed directly from the Quiche in the early Classic and differing structurally only in its post-Conquest adoption of frozen Julian dating. Without further documentation, there is no way of choosing between these alternatives.

The same transformation of + 20 days produced the Chinantec calendar from the Zapotec on 17 XII 1631 A.D. Gregorian 12.0.13.7.18 7 Etz'nab 6 Ch'en Tikal (Julian day number 2317121). The implication would be that the Chinantec used the Zapotec calendar until the conquest, froze it to Julian dating (as the

Zapotec may not have), and then generated a new calendar by a 20-day advance in its New Year.

The Lachixola calendar represents a 2-month advance from the Lalana Chinantec calendar. It was presumably founded on 16 XII 1797 A.D. Gregorian 12.9.1.15.8 5 Lamat 6 Zac Tikal (Julian day number 2377751).

The Guitiupa calendar was responsible for generating the Istacostoc, Mitontic, and Cancuc calendars, by three successive transformation of + 20 days each. The Istacostoc calendar was established on 7 III 1631 A.D. Gregorian 12.0.12.11.13 8 Ben 6 Mac Tikal (Julian day number 2316836).

The Mitontic calendar was established on 7 III 1714 A.D. Gregorian 12.4.16.15.8 7 Lamat 6 Kankin Tikal (Julian day number 2347151).

The Cancuc Tzeltal calendar is known only from modern sources (1917; see Chapter 5, Tzeltal). It could be derived from the Tzeltal calendar by the post-Conquest mechanism of a leap-year advance (rather than the pre-Conquest anti-leap-year recession) but the 5-month difference between the two calendars implies 5 × 83 or 415 years for the derivation, if it were calendrically motivated. Rather than invoking the pre-Conquest use of a post-Conquest calendrical device, it seems less risky to derive a Tzeltal calendar from a Tzotzil one. This would generate the Cancuc calendar from Guitiupa on 6 III 1797 A.D. Gregorian 12.9.1.1.3 6 Akbal 6 Muan Tikal (Julian day number 2377466).

Aside from the leap-year problem, the native sages were particularly impressed by the Spanish week (*semana*) and month (*mes*), though these had no astronomical impact on their calendars. They recognized the days of the week as a new kind of year bearer, operating on a 28-year cycle (see Chapter 5, Church Calendar), but they did not alter their counting on this account. The European lore associating the week days with the heavenly bodies and the months with the zodiac was eagerly assimilated, along with the European explanation of eclipses. Although these teachings may have eroded allegiance to the native calendars, it did not change them.

The Church calendar was more influential but in a similar way. Saints' days quickly became as ubiquitous as the native day count—and rapidly replaced it in some of its functions, particularly in ritual programming and the naming of both persons and places. A certain amount of calendrical syncretism is found in these connections; actual calendrical change was not required. To some extent, the movable feasts of the ecclesiastical calendar presented real problems, as in the collision of Carnival with the Tzotzil year's end (see Church Calendar and Tzotzil in Chapter 5). Even here, however, the calendrical change induced was minor.

CHAPTER 5

CALENDRICAL INDEX

What we know about the native calendrical systems of Anahuac and about the European systems that have generally replaced them is summarized in this chapter in detail, system by system. Topics are listed alphabetically.

Most of the entries deal with individual native calendars. These are introduced by a coded heading that identifies the following:

1. Type of New Years' days (I, II, III, IV, or V, from Figure 4)
2. First month (A. through Q., from Figure 3)
3. Where the name day falls in the month and how it is numbered (0, 1, 19 or 20)
4. Beginning of the calendar round (CR, day-count date, year in the first half of the sixteenth century)
5. Historical span by century (c.) before Christ (B.C.) or after Christ (A.D.)
6. Derivation (=) from a preceding calendar (named), plus (+) or minus (–) a number of days (d) or native months (m), initial (i) or terminal naming (t), zero counting (0), or special features (*): novena (9), trecena (13), 2 to 14 counting of the day count (14), the 24-year katun (24), the tun and Long Count (360) or the 400-day cycle (400)

The information in the headings thus refers to the placement of the given calendar in Figures 3 and 8 of Chapters 1 and 3. In the ensuing description, an effort has been made to provide comparable information on each calendar if it is known, including:

1. The date of its New Year in the Mayapan calendar
2. How the day count was counted
3. How the days of the month were counted
4. When the calendar round began
5. The color associations of the directions
6. How the days were named and written
7. How the months were named and written
8. What calendrical units were recognized and their names
9. How the numerals were named (if different from general numbers) and written
10. The use of novena and trecena cycles

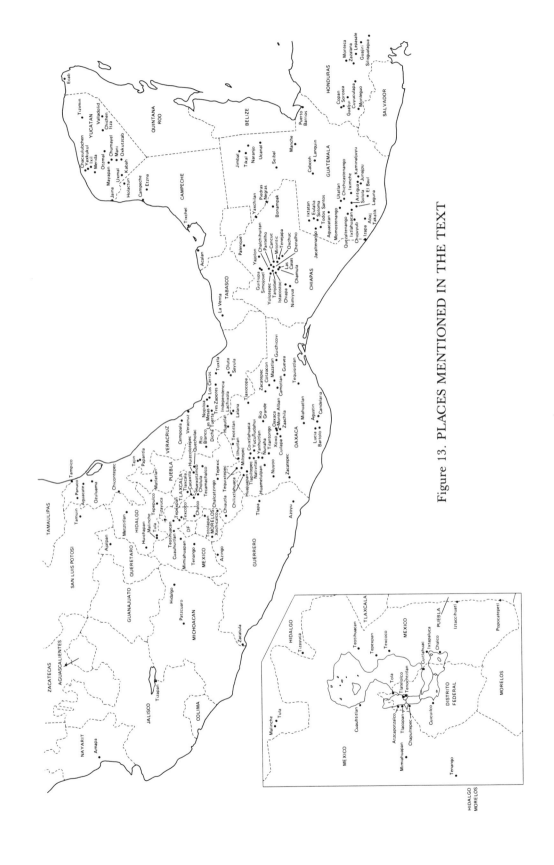

Figure 13. PLACES MENTIONED IN THE TEXT

11. How the correlational position of the calendar is known

12. When, where, and by whom the calendar was used

13. Dates relevant to the given calendar in the Calendrical Annals in Chapter 2

The index contains a listing for every language and every calendar in Anahuac. Often but not always, the language and the calendar bear the same name. Where this is not the case, they are cross-referenced. A few entries (Cacaopera, Iximche, Izapa, Loxicha, Mani) concern dubious or non-calendar-round calendars.

There are separate entries for the European calendars (Church, Gregorian, Julian) and for the comparative summary of calendrical features (calendar round, days, leap year, Long Count, months). The sections on the Long Count and correlation discuss the "correlation question." The entry on Days includes a discussion of writing systems.

The accompanying map (Figure 13) gives the location of places mentioned in the text — or at least of those I am able to locate.

AGUACATEC
(*II.L.1 CR 1 Noh 1532; XX c. A.D.= Quiche*)

The Aguacatec calendar begins its year on 4 Kayab Mayapan, naming the year for its first day, counting the days of the day count from 1 to 13, and presumably counting those of the month from 1 to 20 because in all other respects it appears to be congruent with the Quiche calendar. It presumably also followed the Quiche calendar round, though there is some indication that its senior year bearer may have been E'. Its directional colors are unknown, as are its month names. From McArthur (1965:34–37) the day names are:

a. —	f. Camey (death)	k. Batz (monkey)	p. AjMak (owl)
b. Ik (wind)	g. Chej (deer)	l. E' (tooth)	q. No'j (incense)
c. Ak'bal (night)	h. K'anil (ripe)	m. Aj (cane)	r. Chi'j
d. C'ach (net)	i. Choj (pay)	n. I'x (jaguar)	s. Cyok (storm)
e. Can (serpent)	j. Tx'i (dog)	o. Tz'ichin (quail)	t. AjPu' (hunter)

The units of the system were the day, month (*wink*), day count, year, and calendar round. The year bearers were called *alcal tetz munt, alcal tetz xe siwún* (rulers of the world, rulers of the bottom of the canyon).

McArthur (1965:38) gives the dates of three sequential new years as 9 III 1955 (8 E'), 7 III 1956 (9 No'j), and 6 III 1957 (10 Ik'), of which the middle one is correct (see 1956 A.D.).

AMUSGO
(=Tilantongo?)

From Mechling (1912) and Belmar (1901) a number of possible Amusgo day names
can be culled:

a. Cuchucuan, Ke   f. Tsoó (death)   k. —   p. Esten Staé
   Tsû (alligator)   (vulture)
b. Simanchudhe,   g. Sonducho,   l. Sondaa, Ndé   q. Sohot, Tuwá
   Ndié (wind)   Ke Tsoondé   (grass)   (earth)
                       (deer)
c. Huachioo, Waá   h. Tiusu (rabbit) m. Ndòhó (reed)   r. —
   (house)
d. —   i. Daatio,   n. Luichiayaa   s. Cobataa,
                       Da(teyó)   (jaguar)   Natuwa (rain)
                       (water)
e. Luechuchaa,   j. Luechee Ke   o. Ke Tchii   t. Ndàhá (flower)
   Ke Tsoondé   Tsué (dog)   (eagle)
   (serpent)

The units of the calendar included the sun (*yocunpat*) or day (*shué*), moon or
month (*tchii*), day count (?), year (*tchuu*), and calendar round (?), and the Amusgo
almost certainly used the Tilantongo calendar.

AZTEC
(IV.E.20 CR 2 Acatl 1507; XIII c. to XVII c. A.D.=Texcoco + 1m, f1548)

The calendar of the Aztecs of Tenochtitlan began the year on 6 Yax Mayapan. Its
glyphs, day names, month names, calendrical units, and modes of counting are
those of Nahuatl (q.v.). Its calendar round is fixed by Sahagún (1975-81:4:144). Its
directional colors are: east–yellow, north–red, west–white, south–blue/green.

In addition to the calendar round, the Aztecs also had a night cycle of 9
years and an air cycle of 13, both documented from the two Aztec pictorial manuscripts that are probably pre-Conquest: the *Codex Borbonicus* and the *Aubin Tonalamatl*. The night cycle is identified by the nine lords of the night, who are calendrically named for 9 sequential days of the day count, beginning with 1 Flint. Their
glyphic and Nahuatl names follow (from Caso 1967:22–23, Figs. 9a,b):

G1. Xiuhtecuhtli (lord   G4. Centeotl (corn god)   G7. Tlazolteotl (love
    of fire)   goddess)
G2. Itztli (obsidian)   G5. Mictlantecuhtli   G8. Tepeyolohtli
                               (lord of hell)   (mountain heart)
G3. Piltzintecuhtli   G6. Chalchiuhtlicue   G9. Tlaloc (rain god)
    (lord prince)   (jade skirt)

The calendrical placement of the Aztec night cycle is not clear.

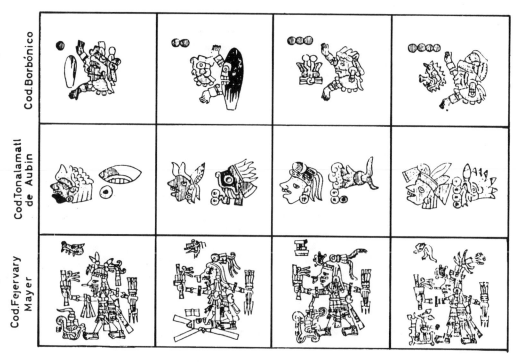

The air cycle is embodied in thirteen patron gods symbolized by as many flying creatures. The depiction of these and their identifications and god patrons are as follows (from Caso 1967:20, Fig. 8):

1. Blue hummingbird: Xiuhtecuhtli (lord of fire)
2. Green hummingbird: Tlaltecuhtli (earth lord)
3. Falcon: Chalchiuhtlicue (jade skirt)
4. Quail: Tonatiuh (sun god)
5. Eagle: Tlazolteotl (goddess of love)
6. Barn owl: Mictlantecuhtli (god of hell)
7. Butterfly: Centeotl (corn god)
8. Eagle: Tlaloc (rain god)

9. Turkey: Quetzalcoatl (feathered serpent)
10. Horned owl: Tezcatlipoca (smoking mirror)
11. Parakeet: Chalmecatecuhtli (lord of sacrifices)
12. Quetzal: Tlahuizcalpantecuhtli (god of dawn)
13. Parrot: Citlalinicue (goddess of heavens)

The Aztecs were certainly aware of the calendrical implications of these cycles for the counting of the year: $9 \times 40 + 5 = 13 \times 28 + 1 = 365$. The latter calculation is explicitly attested by Las Navas (1553) in Baudot (1983:429).

A seventeenth-century document (Anderson 1984) makes it clear that the Aztec calendar correlation was frozen to 1548, giving the date of January 17 (Julian) for the New Year. The subsequent history of the calendar is obscure, but its year bearers survive in twentieth-century Mazatec (q.v.) (see A.D. 1339, 1348, 1416, 1424, 1456, 1500, 1507, 1519, 1520, 1521, 1545, 1553, 1606).

BIXANAS
(=Zapotec?)

The Bixanas or Choapam Zapotec are calendrically unknown. Presumptively they used the Zapotec calendar.

CACAOPERA
(?)

From the vocabularies compiled by Lehmann (1920), the following are possible Cacaopera day names:

a. —	f. Kúla (dead)	k. —	p. Kúsma (buzzard)
b. Uin (wing)	g. Yán (deer)	l. Niní-ca (tooth)	q. —
c. Irrauta, U (night, house)	h. —	m. Naná (cane)	r. Súmu (flint)
d. Áluba (iguana)	i. Li (water)	n. Namá (jaguar)	s. Irra (rain)
e. Yárra (snake)	j. Álu (dog)	o. Guássirri (bird)	t. Báx-ka (flower)

The units of the calendar were the day or sun (*lan*), month or moon (*áicu*), and the year (?).

Cacaopera, a dialect of Matagalpa, is the only Chibchan language that may possibly have had calendar-round dating. The compilation of linguistic data in Lehmann (1920) does not encourage this speculation: it gives numerals from 1 to 5. I find no place names of the area that suggest calendrical derivation. The case for a Cacaopera calendar round looks extremely dubious.

CAKCHIQUEL
(II.G.1 CR 1 Iq 1541; XIV c. to XVII c. A.D.=Quiche – 5m)

The Cakchiquel year began on 4 Ceh Mayapan, counted the day count from 1 to 13 and the days of the month from 1 to 20, and named the year for its first day. Its calendar round began on 1 Iq (1541), and its direction colors are unknown. Its documented use is from the fourteenth to the seventeenth centuries, and it appears to have been known only to the Cakchiquel and their immediate neighbors. Its precolumbian writing system is unknown.

The *Annals of the Cakchiquels* makes use of the following day names in reference to events before 1521:

a. Imox (alligator) f. Camey (death) k. Batz (monkey) p. Ahmak (owl)
b. Iq (wind) g. Queh (deer) l. Ey (tooth) q. Noh (incense)
c. Akbal (night) h. Ganel m. Ah (cane) r. Tihax (flint)
d. Qat (net) i. Toh (rain) n. Ix, Balam (jaguar) s. Caok (storm)
e. Qan (snake) j. Tzíi (dog) o. Tziquín (bird) t. Hunahpu (one hunter)

The Quiche names are identical, and the Tzutuhil names are very nearly so. The *Annals* also cite a number of foreign day names in Pipil (e. *Cuat*, j. *Nacxit*, m. *Cat*, p. *Cakix*, t. *Xuch*), Mam (e. *Apak*, h. *Lamagi*, q. *Qikab*, t. *Ahauh*), and Pokomchi (g. *Quih*, l. *Iy*).

An anonymous Cakchiquel *Calendario de 1685* provides the Cakchiquel month names:

G. Tacaxepual (flaying) Q. Nabey Pach (1 moss)
H. Nabei Tamuzuz (1 termite) R. Ru Can Pach (2 moss)
 I. Ru Can Tamuzuz (2 termite) A. Tziquin Qih (bird time)
J. Cibixiq (smoke) B. Cakan (red cloud)
K. Uchum (reseed) C. Ybota (rolled mat)
L. Nabei Mam (1 elder) D. Katic (burning)
M. Ru Cab Mam (2 elder) E. Yzcal (rebirth)
N. Liquin Qa (soft earth) F. Pa Ri Che (in the trees)
O. Nabei Toqiq (1 damp)
P. Ru Cac Toqiq (2 damp) X. Tzapi Qih (extra days)

The names of the eleventh, twelfth, and seventeenth months (Q., R., and E.) are borrowed from Pipil. Those for G., L., M., N., Q., R., A., B., and X. are cognate with the Quiche names, though their calendrical correspondence is only partial. (L., M., N., A., and B. match.)

The units of the Cakchiquel calendar are the day (*q'ih*), moon or month (*iq'*), day count (*cholol q'ih*), year (*'a'*), old year of 400 days (*huna'*), cycle of 20 old years (*may*), and calendar round (also *huna'*). The use of "old years" and their cycle is unique to the calendar of Iximche (q.v.).

The *Calendario* already cited correlates the Cakchiquel and Gregorian calendars (see A.D. 1685). The year 1685 was 2 Noh Cakchiquel, but the *Calendario* presents instead what appears to be the "ideal year," beginning with 1 Iq, leading to the inference that 1 Iq was the beginning of the calendar round.

The Cakchiquel calendar has never been frozen. When they learned of the Gregorian count on February 2, 1584, the Cakchiquels immediately adopted it for European dates and kept to it thereafter (see A.D. 1584; see also A.D. 1493, 1524, 1541, 1557, 1584, 1585, 1685, 1707).

CALENDAR ROUND

Called *hunab* in Maya and *xiuhmolpilli* in Nahuatl, the calendar round is the cycle of 52 years named for the day counts of their initial (1st) or final (360th) days. Only 4 of the 20 days can begin (or end) the year, and they do so in a sequence of year bearers with coefficients that advance by 1 each year. Thus, in a Type I calendar (like that of the Olmec) the years would begin with 1a, 2f, 3k, 4p, 5a, and so on, to 13p, the 52nd year (see Figure 4). The calendar round normally begins with the senior year bearer of the set and the coefficient 1.

Invented by the Olmec before the seventh century B.C., calendar-round dating is a feature of all Middle American calendars, but it is possible to identify the beginning of the calendar round in only seventeen calendars, eight of them with initial, and nine with terminal year bearers. The beginnings (or endings) of the calendar round, of its 13-year quarters, and of the individual years were marked by the ceremony of new fire, euphemistically referred to in Yucatan as the "time stone year of painless death (*kin tun y aabil ma ya cimlal*)," though no stone monuments of the year are known.

Five of the terminally dated calendars begin the calendar round with the day 2 Cane, rather than the expected 1 Rabbit, a foible that would appear to have been introduced by the 2 to 14 day count of the calendar of Teotihuacan and that is known to occur in the Tilantongo, Aztec, Otomi, Teotitlan, and Campeche calendars. It also occurs in the initial dated Tlapanec calendar. The known starting points of the calendar round in the first half of the sixteenth century are given here in the Mayapan and Aztec calendar rounds with their year equivalents. Figure 14 gives the European year in which the native year of a given name *begins*.

Fig.14. CALENDAR ROUND BEGINS

Initial (Mayapan)		Christian	Terminal (Aztec)	
11 Cauac		1500		8 Tecpatl
12 Kan		1501		9 Calli
13 Muluc		1502		10 Tochtli
1 Ix	1 Tahp Mixe	1503		11 Acatl
2 Cauac		1504		12 Tecpatl
3 Kan		1505		13 Calli
4 Muluc	1 Che Ixil	1506		1 Tochtli
5 Ix	2 Grass Tlapanec**	1507	2 Acatl Aztec	2 Acatl
6 Cauac		1508		3 Tecpatl
7 Kan		1509		4 Calli
8 Muluc		1510		5 Tochtli
9 Ix		1511		6 Acatl
10 Cauac		1512		7 Tecpatl
11 Kan		1513		8 Calli
12 Muluc		1514		9 Tochtli
13 Ix		1515		10 Acatl
1 Cauac		1516	1 Lord Olmec	11 Tecpatl
2 Kan		1517		12 Calli
3 Muluc	1 K'anil Jacaltec	1518		13 Tochtli
4 Ix		1519	2 Acatl Teotitlan	1 Acatl
5 Cauac		1520		2 Tecpatl
6 Kan		1521		3 Calli
7 Muluc		1522		4 Tochtli
8 Ix		1523		5 Acatl
9 Cauac		1524		6 Tecpatl
10 Kan		1525		7 Calli
11 Muluc		1526		8 Tochtli
12 Ix		1527		9 Acatl
13 Cauac		1528		10 Tecpatl
1 Kan	1 Kan Mayapan	1529		11 Calli
2 Muluc		1530	1 Auani Tarascan	12 Tochtli
3 Ix		1531		13 Acatl
4 Cauac	1 Nooh Quiche	1532		1 Tecpatl
5 Kan		1533		2 Calli
6 Muluc		1534		3 Tochtli
7 Ix		1535	2 Ben Campeche	4 Acatl
8 Cauac		1536	1 Tecpatl Toltec	5 Tecpatl
9 Kan		1537		6 Calli
10 Muluc		1538		7 Tochtli
11 Ix		1539	2 An Xithi Otomi	8 Acatl
12 Cauac		1540		9 Tecpatl
13 Kan	1 Iq' Cakchiquel	1541		10 Calli
1 Muluc		1542		11 Tochtli
2 Ix		1543		12 Acatl
3 Cauac	1 Caban Tikal	1544		13 Tecpatl
4 Kan		1545		1 Calli
5 Muluc		1546	2 Huiyo Tilantongo*	2 Tochtli
6 Ix		1547		3 Acatl
7 Cauac		1548		4 Tecpatl
8 Kan		1549		5 Calli
9 Muluc		1550		6 Tochtli
10 Ix		1551		7 Acatl

* The Zapotec calendar round probably began the day before, on 1 Piy.

** The Tlapanec calendar round began the day before the Aztec first name day, on 2 Grass Tlapanec (=1 Malinalli Aztec).

CAMPECHE
(IV.N.20 CR 2 Ben 1535; VII c. to XV c. A.D.=Palenque + 1d)

The Campeche calendar, also called *Puuc-style dating*, was discovered first in the Campeche area and subsequently documented in the Usumacinta valley (Proskouriakoff and Thompson 1947). It uses Type III name days, and the *Dresden Codex* (25–28) suggests that its senior year bearer was 2 Ben. No unequivocal dates have been found giving the beginning or ending of a Campeche month, so its initial day is not established (see e.g., A.D. 733 in chapter 2); it either began with 0 and used initial dating or with 1 and used terminal dating. In the latter case, it would be identical to the Mayapan calendar except for terminal naming, and it is so coded in the heading. The geographic distribution of Campeche dates, especially if they are interpreted as having terminal year bearers, suggests an outside Zapotec or Mixtec influence on the Chol–Chontal area beginning in the seventh century. The terminally dated Palenque calendar (q.v.) may be a related phenomenon. Colonial Yucatecan sources reflect no awareness of either calendar, and both were secondary usages in sites that otherwise used Tikal dating. There are twelve relevant dates:

1. 9.11.12.10.14	9 Ix 16 Pop (9 III 665)	Uxmal, east wall, ball court
2. 9.12.0.0.0	10 Ahau 7 Yaxkin (26 VI 672)	Etzna, Stela 18
3. (*vide infra*)	8 Ahau Glyph G7	Etzna, Stela 19
4. 9.13.14.16.17	3 Caban 14 Muan (2 XII 706)	Jaina, ornamented shell
5. 9.15.14.16.17	6 Etz'nab 19 Yaxkin (24 VI 733)	Tonina, Fragment 35
6. 9.15.12.6.9	7 Muluc 1 (Kankin) (21 X 743)	Holactun (Xcalomkin)
7. 9.16.4.10.11	2 Chuen 3 Muan (9 XI 755)	Kabah, north doorjamb, Structure 1
8. 9.17.10.10.12	3 Eb 14 Mol (26 VI 781)	Yaxchilan, Stela 18
9. 9.18.13.17.14	1 Ix 1 Yax (17 VII 804)	Bonampak, lintel
10. 9.18.15.4.14	6 Ix 16 Kankin (20 X 805)	Yaxchilan, Stela 20
11. 9.19.0.0.0	9 Ahau 17 Mol (22 VI 810)	Etzna, Stela 9
12. 10.5.13.14.1	5 Imix 18 Yaxkin (1 V 942)	Uxmal, capstone, east wing, Monjas

Only dates 2 and 5 actually occur with Long Counts: the remainder are floating calendar-round dates, placed stylistically. Date 3 is read by Proskouriakoff and Thompson as 8 Ahau 7 Uo, but the 7 Uo is actually Glyph G7, and the expression is not a Campeche date. I reject the apparent Campeche date on the Hauberg Stela (see chap. 2, A.D. 199), considering it to be a simple error.

The units, day and month names, and writing system of the Campeche calendar are those of Tikal (see A.D. 199, 649, 733).

CANCUC
(III.J.1 CR?; XVI c. to XX c. A.D.=Tzeltal + 5m, f1584)

The modern Tzeltal calendar of Cancuc starts the year on 5 Muan Mayapan (see Tzeltal).

CAXONOS
(=Zapotec?)

The Caxonos or Ixtlan Zapotec are calendrically unknown. Presumptively they used the Zapotec calendar.

CHATINO
(=Zapotec?)

The units of the Chatino calendar are the day (*tsǫǫ*), month (*coo'*), and year (*yijǫ*), and presumably the day count and calendar round. From Pride and Pride (1970), the following day names are possible:

a. Cua'ñạ (alligator)
b. Cui'ị (wind)
c. Ngata (black)
d. Cuatsi' (iguana)
e. Cuañạ (serpent)
f. Nçujui (death)
g. Cuiñá (deer)
h.
i. Hitya (water)
j. Xne' (dog)
k. —
l. Cuicha (sun)
m. Lijya (cane)
n. Cuichi (jaguar)
o. Quiñi (eagle)
p. Cocụ (owl)
q. —
r. —
s. Tyoo (rain)
t. Quee (flower)

Nothing more is known of it.

CHIAPANEC
(II.K.1 CR?; XVII c. A.D.=Quiche – 1m, f1548)

The Chiapanec calendar begins its year on 4 Pax Mayapan, presumably counting the days of the day count from 1 to 13 and those of the month from 1 to 20. It presumably named the year for its first day (as does the Quiche calendar from which it apparently derives). Its calendar round and direction colors are unknown, as are its day names and pre-Conquest writing system.

From Lehmann's (1920) compilation of vocabularies, the possible day names of Chiapanec are:

a. Ico (alligator)   f. —   k. Ambi (monkey)   p. Ahau (crow)
b. Iho (wind)   g. —   l. Ihi (teeth)   q. —
c. Ango, Chi   h. Oko (rabbit)   m. Ama (corn)   r. —
   (house, night)
d. —   i. Imbo (water)   n. —   s. Amari (cloud)
e. Olo (snake)   j. Ombi, Laco   o. Aguagua   t. Eme (lord)
                             (dog, foot)   (eagle)

The units of the calendar were the day (*mindamo*), month or moon (*yoho*), day count (?), year (*butimi*), and calendar round (?).

Our only real knowledge of the Chiapanec calendar is limited to two lists of month names given by Albornoz (1691:51–52):

Name of the Months of the Chiapanec Year

1. Tumugûi (tamugûi), comienza á siembrar chile	15 de mayo	K.
2. Iatati (hatati), ya sale	4 de junio	L.
3. Ñumbi, siembra maguey	24 de junio	M.
4. Cutamé, muda el tiempo	14 de julio	N.
5. Iaumé (haumé), humedo	3 de agosto	O.
6.	23 de agosto	P.
7. Majua (mahua), helado	12 de setiembre	Q.
8.	2 de octubre	R.
9.	22 de octubre	A.
10. Mua, siembra camote	11 de noviembre	B.
11. Tupiu, sube la humedad	1 de deciembre	C.
12. Tuhu (tujiu), huba	21 de deciembre	D.
13. Muhu (mú-u), mosquitero	10 de enero	E.
14. Turi, maduro	30 de enero	F.
15. Manga, maduro	19 de febrero	G.
16. Puri, madurando el jocote	11 de marzo	H.
17. Cuturi, siembra jicalpestle	31 de marzo	I.
18. Cupané, madura el coyol	20 de abril	J.
Nbu, dias cinco intercalares para acabar el año.		X.

The Same, according to the Indians of Nimiyua or Tia Suchiapa.

1. Numana Yucu . . . . .principia á	4 de junio	L.
2. Numaha ñumbi, en que se siembra maguey	24 de junio	M.
3. Numaha muhu, mosquitero	14 de julio	N.
4. Numaha hatati, ya sale el viento	3 de agosto	O.
5. Numaha mundju, cuando se siembra chile	23 de agosto	P.
6. Numaha catani, fin del agua, principia el maís	12 de setiembre	Q.

7. Numaha manga, se cria el pescado	2 de octubre	R.
8. Numaha haomé, baja el rio y retorna pescado	22 de octubre	A.
9. Numaha mahua, principia el pico	11 de noviembre	B.
10. Numaha toho, ya no se siembra	1 de deciembre	C.
11. Numaha mua, siembra camote	21 de deciembre	D.
12. Numaha topia, sube la humedad	10 de enero	E.
13. Numaha tumuhu, ya no hay nada	30 de enero	F.
14. Numaha (?)	19 de febrero	G.
15. Numaha cupamę madura el coyol	11 de marzo	H.
16. Numaha puri, madura el jocote	31 de marzo	I.
17. Numaha puhuari (?)	20 de abril	J.
18. Numaha turi, maduridad	10 de mayo	K.
Numaha nbu, 5 dias complementares á.	30 de mayo	X.

This places the first Chiapanec calendar as stated before, whereas the Nimiyua calendar begins the year 20 days later. The months Puhuari, Yucu, and Mundju are missing from the first list, and Cuturi from the second, where it is presumably the fourteenth month. The dates given are frozen Julian dates after February 29, 1548 (see A.D. 1691).

CHICOMUCELTEC
(=Kanhobal?)

The units of the Chicomuceltec or Cotoque calendar were the sun or day (*k'ita*; cf. *k'ij* "fiesta"), month (*ich, itz*), day count (?), year (*equey, echél*), and calendar round (?). A number of possible day names are reported by Sapper (1897, 1912), Termer (1930), and Zimmermann (1960):

a. En (alligator)	f. Xemenejich (dead)	k. —	p. —
b. Ik (wind)	g. Vitím (deer)	l. Kaman (tooth)	q. Chauán (earth)
c. Akal (night)	h. Coi (rabbit)	m. Sin (cane)	r. Tuhil Ixlabon (flint)
d. Osou (iguana)	i. Yexjá (water)	n. Tsútúchó (jaguar)	s. Choki (storm)
e. Chan (serpent)	j. Sul (dog)	o. —	t. Vichil (flower)

Although its linguistic ties to Huastec are apparent, it seems geographically most likely that Chicomuceltec used the Kanhobal calendar.

CHINANTEC
(III.D.1 CR?; XX c. A.D.=Zapotec + 1m, f1548)

The Lalana Chinantec calendar begins the year on 5 Ch'en Mayapan, presumably counting the days of the day count from 1 to 13 and those of the months from 1 to 20, and presumably naming the year for its 360th day. Its calendar round and directional colors and its day count have not survived.

The names of the months have been published by I. Weitlaner (1936), Weitlaner and Weitlaner (1946), and Schulz (1955:234–38). They appear to differ only orthographically, and it is the latter list that is given here:

D. Tau Jaú (five, late)	I. Tou Jaú	N. Kûan (grow)	A. Moh (sow)
E. Ya (burning)	J. Huh (lie)	O. Lou (fall)	B. Mûî (fruit, grain)
F. Niu (depth)	K. Lhu (mosquito)	P. I (heavy)	C. Nyö (nausea, nine)
G. Luâ	L. Nô (stand, mountain)	Q. Jeh (hot)	
H. Jau (deep)	M. Lõ (mule)	R. Riu Kuîh (corn ear)	X. Ta Nyiu (seen?)

Each month name is preceded by Gi (Hị, Hî). Weitlaner (1936) says the name of month B. means *water*.

The units of the calendar are the day (*muiba*), moon or month (*zei*), day count (?), year (*gni*), and calendar round (?). A few of the possible daynames may be taken from Brinton (1893):

a. —	f. —	k. —	p. —
b. Tcì (wind)	g. Kwá-nùng (deer)	l. —	q. —
c. Nu (house)	h. Lù-kwù (rabbit)	m. —	r. —
d. —	i. —	n. —	s. —
e. Tcè (snake)	j. —	o. —	t. —

The date given by Schulz (1955) for the beginning of the Chinantec year is a frozen date of December 27, 1548. This places the Lalana Chinantec year as given before. The Lachixola Chinantec calendar is 2 native months (and 5 Uayeb days) later. Schulz (1955) gives the frozen date of 10 II (1549) for the Lachixola New Years of 1935 and 1936.

CHOCHO
(IV.C.20 CR?; XVI c. A.D.=Tilantongo)

There are sixteen known picture manuscripts from the Chocho-Popoluca area; none of them are pre-Conquest, and although six are said to be native in style and content (*Antonio de León*, *Coixtlahuaca I* and *II*, *Ilhuitlan*, *Nativitas*, and *Selden Roll*), none is primarily calendrical (Glass 1975b). Jiménez (1940) asserts that the Chocho calendar is congruent with the Tilantongo Mixtec calendar (q.v.) and gives its year bearers. Other possible day names are found in Starr (1900) and Mock (1977).

a. —

b. —

c. Chin, Nchà (house)

d. —

e. —

f. —

g. Xnià (deer)

h. Caha (rabbit)

i. Da (water)

j. Nià (dog)

k. —

l. Žoà (grass)

m. Xo (cane)

n. Ndúsù (jaguar)

o. Rxa (eagle)

p. Nde'šè (buzzard)

q. Che (earth)

r. The (flint)

s. Chiu (rain)

t. Sù (flower)

Nothing more is known of it.

CHOL
(II.N.0 CR?; I c. to XVII c. A.D.=Tikal)

The Chol used the Tikal calendar, beginning the year on 4 Uayeb Mayapan. The Chol day names can be partially reconstructed from their use as personal names in a number of unpublished sources, principally from Yajolón (Campbell 1979), Zacbalam (Feldman 1983), and Tamuctun (Ibid.). The most likely names are given below. Days a, c, p, r, s, and t are suggested by Campbell; e, g, and q are given by Feldman; b, d, f, h, i, j, k, l, m, n, and o are indicated by both. Campbell also supplies d (Nachan), g (Manich), and q (Tzanab), and Feldman gives r (Etz'nab).

a. Imux (alligator)

b. Ik (wind)

c. Votan (grain)

d. Canan (iguana)

e. Chacchan (serpent)

f. Tox (death)

g. Cuc (quetzal)

h. Lambat (rabbit)

i. Mulu' (rain)

j. Oc (foot)

k. Batz' (howler)

l. Eb (tooth)

m. Bin (cane)

n. Ix (jaguar)

o. Men (bird)

p. Chibin (spider)

q. Cabnal (quake)

r. Chaab (obsidian)

s. Chac (storm)

t. Ahau (lord)

The naming of different days for howler and spider monkeys is a noteworthy feature (k and p respectively). Most of the Nahuatl day names also occur in the Chol sources.

A good case can be made for the possibility that the Tikal month glyphs may match the Chol names better than the Yucatec ones. A comparison of the glyphs with the Chol-influenced Kekchi names (q.v.) suggests the following possibilities:

M. Pop (mat)	R. Cazeu	E. Yax Zih (blue flower)	J. Muan (owl)
N. Ik Kat (1-Kat)	A. Chichin (birds)	F. Zac Zih (white flower)	K. Canaazi (turtle)
O. Chac Kat (2-Kat)	B. Yaxkin (green time)	G. Chac Zih (red flower)	L. Ohl
P. Zotz' (bat)	C. Mol (gather)	H. Mac (cover)	
Q. Zec (skeleton)	D. Ik Zih (black flower)	I. Oneu	X. Mahi i Kaba (nameless)

Only the name of month B is independently attested as a Chol month name (see chapter 2, A.D. 1631). The displacement of Pop is forced by the preponderance of cognates among the other month names with those in the positions indicated in Yucatec. Either Pop was not the first Chol month, or an extra month has been inserted between it and Chichin. The Chol year cannot have begun a month earlier than the Yucatecan because of the dating of its New Year.

From their geographical distributions, it appears likely that the Late Classic Palenque and Campeche calendars (q.v.) were also Cholan, or even primarily Cholan.

The placement of the Chol calendar is given by Tovilla (1631). The units of the calendar are the day (*kin*), month (*uu*), day count (*tsic kin*), year (*hab*), and calendar round (?) (see A.D. 1631).

CHOLULA
(III.P.0 CR?; Classic; = Yucuñudahui – 1m)

The existence of a Cholula calendar is in some measure hypothetical. That there was a Classic period calendar at Cholula is beyond question. That it bore the same relationship to the Tepexic calendar that the Yucuñudahui calendar bore to that of Chalca is an inference based on geographic plausibility and the structure of adjacent calendars. Perhaps 7 of the 20 day glyphs can be documented in the Cholula-Tlaxcala area from the Classic period. The documentation is as follows:

a. Alligator?

b. Anno 9 Wind. Cacaxtla (McVicker 1985:86, Fig. 3).
7 Wind. Cholula? (Tschohl 1972:259, Pl. 10, Fig. 3).
7 Wind. Cacaxtla (McVicker 1985:87, Fig. 4).

c. House?

d. Anno 13 Iguana. Micaltepec (Tschohl 1972:469, Fig. 2).

e. 1 Serpent. Cacaxtla (McVicker 1985:86, Fig. 3).

f. Death?

g. 3 Deer. Cacaxtla (Ibid.:89, Fig. 6).

h. Rabbit?

i. Water?

j. Foot?

k. Monkey?

l. Sun. Cholula (Tschohl 1972:137, Pl. 7, no. 48; 161).

m. 3 Cane. Cacaxtla (McVicker 1985:88, Fig. 5).

n. Jaguar?

    o. Eagle?

    p. Owl?

    q. Quake?

    r. Flint?

    s. Rain?

    t. Anno 1? Lord. El Idolo (Tschohl 1972:269, Photo 1).

        2 Lord. Cacaxtla (McVicker 1985:86, Fig. 3).

        6 Lord. Huizcolotepec (Tschohl 1972:302). Postclassic?

The days marked with year signs are in other calendars: Olmec (t), Teotihuacan (b) and Tenango (d). The expected Type III calendar has not been found. I continue to believe that it will be.

CHONTAL

(=Tikal?)

The *Chronicle of Acalan-Tixchel* (Scholes and Roys 1968) suggests a number of possible Chontal day names; Feldman (1983) fills in others; and still others are derivable from Knowles (1984). Even so, the list is incomplete:

a. Imox (alligator)	f. Tox (death)	k. —	p. —
b. —	g. Chimay, Ceh (deer)	l. Eh (tooth)	q. Chaban (quake)
c. —	h. Lamat (rabbit)	m. Bin (cane)	r. —
d. Ache, Can (iguana)	i. Mulu (rain)	n. Hix, Balum (jaguar)	s. Chauac (storm)
e. Chan (snake)	j. Ok (foot)	o. —	t. Ahau (lord)

Most of the Nahuatl day names occur in the same sources and were apparently quite familiar to the Chontal. The units of the calendar were the day or sun (*kin*), month or moon (*uh*), day count, year, calendar round (?), and cycle (*may*).

CHUH

(III.L.1 CR?; XX c. A.D.=Kanhobal)

The Chuh calendar begins its year on 5 Kayab Mayapan. It counts the days of the day count from 1 to 13 and those of the month from 1 to 20 and names the year for its first day. The start of its calendar round and its directional colors are unknown, as are the units of the calendar.

The day names are reported by LaFarge and Beyers (1931:176), Termer (1930:385–86) and Maxwell (1981):

a. Himox (alligator)	f. Tox (death)	k. Ba'a'tz' (howler)	p. Chab'in (spider)
b. 'Ik' (wind)	g. Keh (deer)	l. 'Eyub' (tooth)	q. Kixkab' (quake)
c. Woton (grain)	h. Lañb'at (rabbit)	m. B'e'en (cane)	r. Chiñax (flint)
d. K'ana (iguana)	i. Mulu' (rain)	n. Hi'ix (jaguar)	s. Chawok (storm)
e. 'Ab'ak (soot)	j. 'Elab' (dog)	o. Tz'ikin (bird)	t. 'Ahaw (lord)

Termer also gives s. K'ak' (fire).

The Chuh month names are reported by Termer (1930:391) from Santa Eulalia:

K. Tap	P. Bac	C. Yaxul	H. Mak (cover)
L. Bex	Q. Tam	D. Savul	I. Oneu
M. Sacmay (white time)	R. Hua Tziquin (bird)	E. Xujim	J. Sívil
N. Nabich	A. Kanal (yellow)	F. —	
O. Mo	B. Yaxaquil (green)	G. Mol (gather)	X. Oyeb 'In

A correlational date for the last day of *H Oye' K'u* (=Oyeb 'In) establishes the beginning of the Chuh year (see A.D. 1981), and places it as congruent with that of Kanhobal (see A.D. 1981).

CHURCH CALENDAR

The ecclesiastical calendar of the Roman Catholic Church establishes both fixed and movable feasts. The movable feasts are determined by the date of Easter, which falls on the first Sunday after the Paschal full moon. This can be calculated by determining the date of the first Sunday of the year, given by the *dominical letter*, and by the *golden number*, which encodes the Paschal full moon (Ronan 1966).

The dominical letters indicate the date of the first Sunday of the year as January 1 (A), 2 (B), 3 (C), 4 (D), 5 (E), 6 (F), or 7 (G). Because of leap years, they follow a 28-year Julian cycle as follows:

1 FG	8 E	15 C	22 A
2 E	9 CD	16 B	23 G
3 D	10 B	17 AG	24 F
4 C	11 A	18 F	25 DE
5 AB	12 G	19 E	26 C
6 G	13 EF	20 D	27 B
7 F	14 D	21 CB	28 A

To find the dominical letter for any given year, add 9 to the year and divide by 28: the remainder will give the year of the 28-year "solar cycle" from 1 to 27; zero being the twenty-eighth year.

The golden numbers follow a 19-year cycle, giving the 19 possible (Gregorian) dates of the Paschal full moon as follows:

1	14 IV	7	8 IV	14	22 III
2	3 IV	8	28 III	15	10 IV
3	23 III	9	16 IV	16	30 III
4	11 IV	10	5 IV	17	17 IV
5	31 III	11	25 III	18	7 IV
6	18 IV	12	13 IV	19	27 III
		13	2 IV		

To find the golden number for any given year, add 1 to the year and divide by 19: the remainder will give the golden numbers of the lunar or "Metonic" cycle from 1 to 18, zero being the nineteenth year.

Of course the Julian table for the dominical letters must be converted to Gregorian for dates after 1582, and the Gregorian dates for the golden numbers must be converted to Julian for dates earlier than 1582. For example, 1688 had the dominical letter AG, indicating a leap year with Sunday falling on January 1 Julian, but that was January 11 Gregorian, so the Gregorian dominical letters would be EF, 1688 being a leap year in both calendars. The golden number is 17 (placing the Paschal Full Moon on Saturday, April 17 Gregorian or Friday, April 27 Julian. Easter would fall on the following Sunday, April 18 Gregorian or April 29 Julian.

The permutations of the solar and lunar cycles (the dominical letters and golden numbers) form a cycle of 532 years. It is a compromise cycle designed to keep Easter from falling on the date of the Jewish Passover or on the day of the Paschal full moon. The calculated date of the latter may or may not fall on a full moon because it is based on a fixed equinox of March 21.

The movable feasts are placed within the ecclesiastical year with reference to Easter but fitted between the fixed dates of Epiphany (January 6) and Christmas (December 25). This is achieved by varying the number of "Sundays after Epiphany" and the number of "Sundays after Pentecost." All other movable feasts bear an invariant relationship to Easter as follows:

Reyes Magos	Epiphany, January 6
Domingos después de Epifanía	Two to six Sundays
Septuagésima	Third Sunday before Lent
Sexagésima	Last Sunday before Lent
Quincuagésima	Last Sunday before Lent
Carnaval	Shrove Tuesday
Ceniza	Ash Wednesday: beginning of Lent
Cuaresma	Lent: the 40 days before Easter

Domingos de Cuaresma	Four Sundays
Domingo de Pasión	Second Sunday before Easter
Viernes de Dolores	Second Friday before Easter
Palmas y Ramos	Sunday before Easter
Jueves Santo	Thursday before Easter
Viernes Santo	Good Friday
Sábado de Gloria	Saturday before Easter
Pascua Florida	Easter Sunday
Domingos después de Pascua	Five Sundays
Ascensión	Sixth Thursday after Easter
Domingo de Ascensión	Sixth Sunday after Easter
Pentecostés	Seventh Sunday after Easter
Santísima Trinidad	Eighth Sunday after Easter
Corpus Christi	Thursday after Trinity
Domingos después de Pentecostés	Twenty-two to twenty-seven more Sundays
Adviento	Four Sundays before Christmas
Navidad	Christmas, December 25
Domingos después de Navidad	One or two Sundays

The 40 days of Lent do not include Sundays.

The fixed feasts of the Church are listed next in the Gregorian calendar. They are taken from two sources a century apart, the *Calendario Galván* of 1851 and the *Almanaque Guadalupana* of 1950, in order to illustrate something of the variation in the familiarity, accessibility, and calendrical placement of the saints' days.

Abraham 16 III
Abundio 11 VII
Acacio 8 V, 1851
Adalberto 23 IV, 1851
Adelaida 16 XII
Adrián 8 IX, 1851
Adulfo 27 IX
Agapito 20 IX
Agatón 7 XII, 1950
Agrícola 17 III
Agripina 23 VI
Agustín 28 VIII
Agustín C. 28 V, 1950
Alberto 7 IV
Alberto Carmelita 7 VIII, 1851
Albina 16 XII
Albino 1 III
Aldegunda 30 I
Alejandro 9 II, 1950
Alejandro 24 IV
Alejo 17 VII
Alfonso María de Ligorio 19 VIII, 1851
Alfonso Rodríguez 30 X, 1950
Amado 13 IX
Amador 30 IV, 1851
Amancio 8 IV
Amando 8 II

Ambrosio 7 XII
Amós 31 III, 1950
Ana 26 VII
Anacleto 13 VII
Anastasia 15 IV
Anastasio 2 V, 1851
Anastasio 22 I, 1851
Anastasio Papa 27 IV
Anatolio 3 VII
Andrés 30 XI
Andrés Avelino 10 XI
Andrés Corsino 4 II
Ángela 31 V, 1950
Ángeles 2 VIII, 1950
Ángeles Custodios 2 X
Aniano 11 XI, 1851
Aniceto 17 IV
Anselmo 21 IV
Antelmo 26 VI, 1851
Antemio 20 X, 1851
Antero 3 I
Antonino Arzobispo 10 V
Antonino 2 IX
Antonio Abad 17 I
Antonio de Padua 13 VI
Antonio del Águila 24 VII, 1851
Antonio María Zacaria 5 VII, 1950

Anunciación 25 III, 1950
Apolinar 23 VII, 1851
Apolinar Obispo 8 I, 1851
Apolonio 10 IV
Aquilino 4 I, 1950
Arcadio 12 I
Arelio 13 IX, 1851
Aristeo 3 IX, 1851
Arnulfo 15 VIII, 1851
Arnulfo 18 VII, 1851
Asteria 10 VIII, 1950
Asunción de Nuestra Señora 15 VIII
Atanasia 14 VIII
Atanasio 2 V, 1950
Atenedoro 18 X
Atenógenes 16 VII, 1851
Atilano 5 X, 1851
Aurea 24 VIII
Aureliano 16 VI
Aurelio 27 VII, 1851
Aurelio Obispo 12 XI, 1851
Ausencio 18 XII, 1950
Austacio 29 III, 1851
Austreberta 10 II, 1851
Balbina 31 III
Baldomero 27 II
Bárbara 4 XII

Bardomiano 25 IX, 1851
Bartolomé 24 VIII
Basilio 14 VI
Basilisa 15 IV
Basilisa 9 I, 1950
Beatriz 29 VII, 1950
Beda 27 V, 1950
Benigno 13 II
Bénito 21 III
Bénito 3 IV, 1950
Benjamin 31 III, 1851
Bernabé 11 VI, 1851
Bernardino de Sena 20 V
Bernardo 20 VIII, 1851
Bertoldo 29 III, 1950
Bibiana 2 XII, 1950
Blandina 2 VI, 1851
Blas 3 II
Blas 29 XI, 1950
Bonifacio 14 V
Bonifacio 5 VI
Braulio 26 III, 1851
Bricio 13 XI, 1851
Brígida 8 X, 1851
Bruno 6 X
Buenaventura 14 VII
Bulmaro 20 VII, 1851
Calixto 14 X
Camerino 21 VIII
Camilo de Lelis 15 VII
Cándido 2 II, 1851; 3 X 1950
Canuto 19 I
Caridad 1 VIII, 1950
Caritina 5 X, 1851
Carlos Borromeo 4 XI
Casiano 13 VIII
Casilda 9 IV, 1851
Casimiro 4 III
Casto 22 V, 1851
Cástulo 26 III, 1851
Catalina de Ricci 13 II
Catalina de Sena 30 IV, 1851
Catalina de Suecia 22 III
Catarina 25 XI
Cayetano 7 VIII
Cayo 22 IV, 1950
Cecilia 22 XI
Celedonio 3 III
Celerino 3 II
Celestino 29 V, 1851
Celestino 6 IV
Celso 6 IV
Celso Niño 28 VII
Cenobio 30 X, 1851
Cesáreo 27 VIII, 1851
Cesario Confesor 25 II
Cipriano 26 IX

Cipriano Doctor 16 IX
Cirenia 1 XI
Ciria 3 VIII
Ciriaco 18 VI, 1851; 8 VIII, 1950
Cirilo 9 VII
Cirilo 18 III, 1950
Cirilo 9 II, 1950
Cirino 10 V, 1851
Ciro 31 I
Clara 12 VIII
Claudio 30 X, 1851; 6 VI, 1950
Clemente 23 XI
Cleofas 25 IX, 1851
Cleto 26 IV
Clicerio 20 IX, 1851
Clotilde 3 VI, 1851
Cointa 8 II
Coleta 6 III
Columba 31 XII, 1851
Concepción 8 XII
Conrado 26 XI, 1851
Consorcia 22 VI, 1950
Constancia 18 II, 1851
Constancio 1 IX, 1851
Constantino 11 III
Cornelio 16 IX
Cosme 27 IX
Crescenciana 15 VI, 1851
Crescenciana Martir 5 V
Crescenciano 14 IX, 1851
Crescencio 19 IV
Crescencio 29 XII, 1851
Crisanto 25 X, 1851
Crisóforo 20 IV
Crisógono 24 XI, 1851
Crispín 25 X
Crispina 5 XII
Crispiniano 25 X, 1851
Cristeta 27 X
Cristiana 15 XII
Cristina 24 VII
Cristóbal 25 VII
Cutberto 20 III
Damaso 11 XII
Damián 27 IX
Daniel 21 VII, 1950
Daria 25 X
Darío 19 XII
Degollación de San Juan
   Bautista 29 VIII
Delfino 24 XII
Demetrio 22 XII
Desiderio 11 II, 1851
Desposorios de San José 26 XI, 1851
Diego de Alcalá 12 XI

Dimas 25 III
Dimna 15 V, 1851
Diodoro 3 V, 1851
Dionisia 6 XII, 1950
Dionisio 8 IV, 1851
Dionisio Areopagita 9 X
Divino Pastor 4 V, 1851; 9 VIII, 1950
Divino Redentor 20 VII, 1851
Divino Rostro 4 III, 1851
Domingo 20 XII, 1950
Domingo de Guzmán 4 VIII
Domitila 12 V
Donaciano 24 V
Donaciano Obispo 6 IX
Donato 22 X, 1851; 7 VIII, 1950
Dorotea 6 II
Doroteo 5 VI
Dulce Nombre de Jesús 19 I, 1851
Dulce Nombre de María 14 IX
Dulcino 24 XII, 1851
Dunstano 19 V, 1851
Edmundo 20 XI
Eduardo 13 X
Eduwigis 17 X
Efigenia 21 IX, 1950
Efrén 9 VII, 1851
Efrosina 1 I, 1851
Eladio 8 V, 1851
Elena 18 VIII
Eleucadio 14 II
Eleuterio 20 II
Eleuterio 26 V, 1950
Elfego 19 IV, 1851
Eligio 1 XII
Eliseo 14 VI, 1950
Elodia 23 X, 1851
Elpidio 10 XI
Elpidio Obispo 4 III
Emelia 30 V, 1950
Emerenciana 23 I, 1851
Emeterio 3 III
Emigdio 5 VIII, 1851
Emilia 30 V, 1950
Emiliano 8 VIII, 1851
Emilio 22 V, 1851
Emilo 8 VIII, 1950
Encarnación del Divino Verbo
   25 III, 1851
Enedina 14 V, 1851
Engracia 16 IV, 1851
Enrique 15 VII
Epifanía 6 I
Epifanio 7 IV
Epigmenio 24 III

Ireneo 3 VII
Isaac Monge 3 VI
Isabel 5 XI, 1851
Isabel Reina de Portugal 8 VII
Isabel Reina de Hungría 19 XI, 1851
Isauro 17 VI
Isidro Labrador 15 V, 1851
Ismael 17 VI, 1851
Isodoro 4 IV
Jacinto 11 IX
Jacinto 16 VIII
Jerónimo 20 VII, 1950
Jerónimo 30 IX, 1950
Joaquín 16 VIII, 1950
Joel 13 VII, 1950
Jonas 21 IX, 1950
Jorge 23 IV, 1851
Josafat 14 XI, 1950
José, Patrocinio de 11 V, 1851
José 19 III
José de Cupertino 18 IX, 1950
Jovita 15 II
Juan 13 V, 1950
Juan 6 V
Juan Apóstol 27 XII
Juan Bautista 24 VI
Juan Bautista Rossi 23 V, 1950
Juan Bautista de La Salle 15 V, 1950
Juan Cancio 20 X, 1950
Juan Capistrano 31 X, 1851; 28 III, 1950
Juan Climaco 30 III
Juan Crisóstomo 27 I
Juan Damaceno 23 V
Juan Damas 27 III, 1950
Juan Francisco Regis 16 VI, 1851
Juan Gualberto 12 VII
Juan Mártir 26 VI
Juan Monge 21 VII
Juan Nepomuceno 16 V
Juan Papa 27 V
Juan Sahagún 12 VI
Juan Silenciario 13 V
Juan de Dios 8 III
Juan de Mata 8 II
Juan de la Cruz 24 XI
Juana Francisca 21 VIII, 1950
Judas Tadeo 28 X
Julia 22 V, 1950
Julián 9 I
Julian Obispo 28 I
Juliana 16 II
Juliana de Falconeris 19 VI
Julio 20 XII

Julio Papa 12 IV
Julita 30 VII, 1851
Justa 19 VII
Justina 26 IX
Justino 13 IV
Justo 6 VIII
Juvencio 25 I, 1851
Ladislao 27 VI
Lamberto 15 IV, 1851
Lamberto Obispo 17 IX, 1851
Largo 8 VIII, 1950
Laureano 4 VII, 1851
Lauro 18 VIII
Lázaro 17 XII
Leandro 27 II
Ledegario 2 X, 1851
Leobardo 18 I, 1851
Leocadia 9 XII
Leo
León II 28 VI, 1950
Leonardo 6 XI
Leonicio 10 VII, 1851
Leonides 8 VIII, 1851
Leonila 17 I
Leovigildo 20 VIII, 1851
Liborio 23 VII, 1851
Librado 17 VIII
Lidia Tintorera 3 VIII
Lino 23 IX
Liova 28 IX
Llagas del Divino Redentor 7 III
Llagas de San Francisco 17 IX
Longinos 15 III
Lorenzo 10 VIII
Lorenzo Justiniano 5 IX
Lourdes 11 II, 1950
Lucano 30 X, 1851
Lucas 18 X
Lucía 25 VI, 1851
Lucía 13 XII
Luciano 7 I
Lucina 30 VI, 1851
Lucio 15 XII
Lugarda 16 VI
Luis Beltrán 9 X
Luis Gonzaga 21 VI
Luis Obispo 19 VIII
Luis Rey 25 VIII
Macario 28 II, 1950
Macario Alejandrino 2 I
Macario Obispo 10 III
Macedonio 12 IX, 1851
Maclovio 15 XI, 1851
Macrina 14 I, 1851
Magín 19 VIII
Malaquías 3 XI, 1851; 14 I, 1950

Malrubio 21 IV, 1950
Manuel 17 VI
Marcelina 17 VII
Marcelino 2 VI
Marcelino Papa 26 IV
Marcelo 16 I
Marcial 30 VI, 1851
Marciano 2 XI, 1851
Marcos 25 IV
Marcos Papa 7 X, 1851
Mardonio 23 XII
Margarita 10 VI
Margarita 20 VII
Margarita de Cortona 22 II
María Ana 17 IV, 1950
María Cleofas 9 IV
María Magdalena 22 VII
María Magdalena de Pazzis 25 V
María de Cervellón 25 IX, 1950
Mariana de Jesús 17 IV, 1851
Marina 18 VII, 1851
Mario 19 I, 1950
Marta 19 I, 1950
Marta 29 VII
Martín 29 VII, 1950
Martín 11 XI
Martina 30 I
Martiniano 2 I
Mártires de Zaragoza 3 XI
Mateo 21 IX
Matías 24 II
Matilde 14 III
Mauricio 22 IX
Maura 30 XI, 1950
Maurilio 13 IX, 1950
Mauro Abad 15 I, 1851
Mauro Obispo 21 XI, 1851
Maximiano 21 VIII, 1851
Maximiliano 12 X, 1950
Máximo 11 V, 1851
Máximo 25 I, 1851
Máximo Obispo 8 VI
Mayolo 11 V, 1851
Medardo 8 VI, 1851
Melesio Obispo 12 II, 1851
Melesio Obispo 4 XII, 1851
Melito 24 IV
Melitón 1 IV
Melquiades 10 XII
Menas 11 XI, 1950
Metodio 7 VII, 1950
Miguel 29 IX, 1851
Miguel, Aparición 8 V
Miguel de los Santos 5 VII, 1851
Milburga 23 II

Sebastián de Aparicio 25 II,
    1851
Secundino 1 VII, 1851
Segundo 1 VI
Senorina 22 IV
Serafín 12 X, 1851
Serapia 3 IX, 1851
Serapión 14 XI
Sergio 7 X, 1851
Servando 23 X, 1950
Severiano 21 II
Severiano 8 XI, 1950
Severino 11 II, 1851
Severo 8 XI
Severo Obispo 1 II
Sidonio 23 VII
Sidronio 11 VIII
Siete Dolores 17 IX
Silverio 20 VI
Silvestre 31 XII, 1851
Silviano 4 V
Silvino 12 IX, 1851
Simeón 18 II
Simeón Stilita 5 I
Simitrio 26 V
Simón 28 X
Simón de Rojas 28 IX
Simplicio 2 III, 1851
Sinforosa 18 VII, 1950
Sixto 28 III
Sixto 6 IV, 1950
Sofia 1 VIII, 1851
Sofia 18 IX, 1950
Sofia 30 IX, 1950
Sóstenes 28 XI
Sotero 22 IV
Susana 24 V
Susana 11 VIII, 1950
Taide 19 X, 1851

Taurino 11 VIII, 1950
Tecla 23 IX
Telésforo 5 I
Teodomiro 25 VII, 1851
Teodora 1 IV
Teodoro 9 XI
Teodoro 27 XII, 1950
Teodosia 29 V, 1851
Teódulo 17 II
Teofanes 12 III, 1851
Teófilo 8 I, 1851
Teresa 19 V, 1851
Teresa de Jesús 15 X, 1851
Tiburcio 11 VIII
Tiburcio 14 IV, 1851
Tiburcio Martir 9 IX, 1851
Timoteo 22 VIII
Timoteo Diácono 19 XII
Timoteo Obispo 24 I
Tirso 28 I
Tito 4 I, 1851; 6 II, 1950
Todo Santos 1 XI
Tomás 21 XII
Tomás de Aquino 7 III
Tomás Canturiense 29 XII
Tomás de Villanueva 18 IX
Torcuato 15 V, 1851
Toribio Arzobispo 27 IV
Toribio Obispo 16 IV, 1851
Tranquilino 6 VII
Transfiguración del Senor 6
    VIII
Trigio 12 I, 1851
Triunfo de la Santa Cruz 16
    VII, 1851
Urbano 25 V
Urso 30 VII, 1851
Úrsula 21 X
Valente 21 V, 1851

Valentin 14 II
Valeria 28 IV, 1851
Valeriano 14 IV, 1851
Valero 29 I, 1851
Velino 26 XI, 1851
Venancio 18 V, 1851
Vérulo 21 II
Vicente 22 I, 1851
Vicente Ferrer 5 IV
Vicente de Paul 19 VII
Víctor 6 III
Víctor Papa 28 VII, 1851
Victoria 17 XI
Victoria 23 XII
Victoriano 11 XII
Victoriano 23 III
Victorino 8 XI, 1950
Vidal 28 IV
Vilfrido 12 X, 1851
Virginia 21 V, 1851
Visitación 2 VII
Vital 4 XI, 1950
Vito 15 VI
Viviana 2 XII, 1851
Vulfrano 20 III, 1851
Waldo 16 V, 1851
Wenceslao 28 IX
Wilfrido 12 X, 1950
Willehado 8 XI, 1851
Wistano 19 I, 1851
Yucundo Mártir 9 I, 1851
Yucundo Obispo 14 XI,
    1851
Zacarías 5 XI
Zeferino 26 VIII
Zenaida 5 VI, 1851
Zenon 23 VI
Zita 27 IV, 1950
Zózimo 26 XII, 1950

---

* Almanacs for 1904, 1908, and 1912 add the February 29 saints' days of the Segunda Translación del Cuerpo de San Agustín and San Macario Mártir, and the 1912 Almanac additionally moves San Pedro Damiano from February 23 to 29.

COLHUA
(IV.A.20 CR?; XII c. to XVI c. A.D.=Metztitlan – 1m)

The Colhua calendar begins its year on 6 Xul Mayapan. Its glyphs, counting system, and calendric names are Nahuatl. Its calendar round and direction colors are not known. It was identified by Jiménez (1961) in the *Relación Genealógica* of 1543 and named Colhua I. It also occurs in the *Anales de Cuauhtitlan*, *Memorial Breve*, and Chimalpahin.

Because its New Year falls earlier in the year than that of any other calendar in Middle America except Mixe and falls on exactly the same day of the day count as the Mayapan New Year 13 months later, one is tempted to wonder whether the latter date should not be assigned to the Colhua calendar as well. The Colhua calendar, however, calls its first month Mixcoatl. If the equation of that with Quecholli is correct, the placement of the calendar should be where it is located in the preceding heading (see A.D. 1124, 1180, 1308).

COLHUACAN
(IV. J.20 CR?; XII c. to XIV c. A.D.=Tepanec + 1m)

Caso (1971:343) rotundly asserts that there was no calendar that began the year with Toxcatl. Jiménez (1961) says there was and labels it Cuauhtitlan. Kubler and Gibson (1951:48) believe that the *Aubin* codex documents such a calendar, but Caso accuses them of misconstruing the Nahuatl text on page 82 of the codex. Kirchhoff (1950) is pro-Toxcatl.

There seems to be little question but that such a calendar existed, whether or not is is aptly named. It began the year on 6 Muan Mayapan, was named terminally, and otherwise followed the calendrical patterns of Nahuatl (see A.D. 1127, 1152, 1369, 1371, 1372).

CORRELATION

The correlation of the native calendars of Middle America with the Christian calendar may be considered from three somewhat separate points of view: ethnohistorical, archaeological, and astronomical. The present work deals only with the first of these lines of evidence, and although the ethnohistorical data seem to me decisive, it is relevant to survey the alternative proposals that have been generated by other considerations.

It has become customary to identify different correlations by the Julian day number thought to be equivalent to the normal base date of the Mayan (-Olmec) Long Count, 4 Ahau 8 Cumku; such Julian day numbers are identified as "Ahau equations" or "correlation constants." The principal proposals cover a span of more than 800 years. They are grouped next, according to the approximate Long Count dates they would assign to the katun 11 Ahau of the Spanish Conquest.

13.2.0.0.0
394,483 (Bowditch 1910)
438,906 (Willson 1924)
449,817 (Bunge 1940)
482,699 (Smiley 1960a)

482,914 (Smiley 1960a)
487,410 (Owen 1975)
      12.9.0.0.0
489,138 (Makemson 1946)
489,383 (Spinden 1930; 1957)
489,384 (Spinden 1924)
489,484 (Ludendorff 1930–37)
492,622 (Teeple 1926)
497,879 (Dinsmoor; see Satterthwaite 1965:630)
500,210 (Smiley 1960b)
507,994 (Hochleitner 1974)
508,362 (Hochleitner 1974)
525,698 (Hochleitner 1974)
553,279 (Kelley 1976)
577,264 (Hochleitner 1972)
578,585 (Hochleitner 1970)
      11.16.0.0.0
583,919 (Suchtelen 1957)
584,280 (Goodman 1905)
584,281 (Martínez 1926)
584,283 (Thompson 1950; Nowotny 1958)
584,284 (Beyer 1937)
584,285 (Thompson 1935)
584,286 (Lounsbury 1978:815)
584,314 (Calderón 1982)
585,789 (Cook 1973)
588,466 (Mukerji 1936)
588,626 (Pogo 1937)
594,250 (Schove 1976)
609,417 (Hochleitner 1974)
615,824 (Schove 1977)
626,660 (Kaucher 1980)
626,927 (Kreichgauer 1927, 1932)
      11.3.0.0.0
660,205 (Hochleitner 1974)
663,310 (Kelley 1983)
674,265 (Hochleitner 1972)
677,723 (Schulz 1955)
679,108 (Escalona 1943)
697,183 (Vaillant 1935; Wauchope 1947)
698,164 (Dittrich 1936)
      10.10.0.0.0
774,078 (Weitzel 1947)
774,080 (Vollemaere 1984)
774,083 (Vaillant 1935)

The ethnohistorical data collected in the present volume support the Thompson 1950 correlation constant of 584,283. It may be noted that among other proposals, only the Spinden (1930) and Bowditch (1910) correlation constants differ from Thompson by exact multiples of the day count.

CUICATEC
(III. C.19 CR?; XVI c. A.D.=Zapotec)

The Cuicatec calendar is known only from Colonial picture manuscripts, three of which come from the Cuicatec area (*Fernández Leal, Porfirio Díaz,* and *Dehesa*). The first two are native in style and content and the third has a ritual-calendrical character (Glass 1975b). The *Porfirio Díaz* has Type III New Years' days and Type II name days (Toscano 1943:129) and presumably uses the Zapotec calendar. The *Dehesa* uses the Tilantongo calendar, with Type III name days. It also contains a table of year correlations using Aztec dates, which carries a notation of the day of the week acting as "year bearer." More surprisingly, it presents a correlational date tying the Tilantongo and Olmec years together. It seems most likely that the "native" Cuicatec calendar was Zapotec, but there was obviously awareness of other calendars.

A very few of the possible Cuicatec day names can be documented from Mechling (1912):

a. —	f. —	k. —	p. —
b. Yinie (wind)	g. —	l. —	q. Taha (earth)
c. —	h. —	m. —	r. —
d. —	i. Nune (water)	n. —	s. Cubi (rain)
e. —	j. —	o. —	t. —

See A.D. 1522.

CUICUILCO
(I. C.19 CR 1 Lord 1549; VII c. to VI c. B.C.)

The evidence for a "Cuicuilco" calendar is a single artifact listed in the date list as belonging to 679 B.C. and read as an Olmec date, 2 Lord. Although the figure can be read as an Olmec glyph, the accompanying numerals cannot. Olmec 2 would be written as two digits with the fingernails showing. This 2 is written as two dots. Because of its archaeological provenience, this is probably the earliest date known. If it is read as a date in Calendar A (Figure 8), it would correspond to 691 B.C. I believe it should be so read, and this justifies the provisional identification of Calendar A as Cuicuilco (see 679 B.C.).

CUITLAHUAC
(IV.H.20 CR?; XII c. A.D.=Teotitlan + 1m)

The Cuitlahuac calendar began its year on 6 Mac Mayapan. It used Nahuatl glyphs, counting, and nomenclature. Its calendar round and directional colors are unknown. Jiménez (1961:141) refers it to the *Anales de Cuauhtitlan*; Kirchhoff (1950:127) relates it to an unlocatable reference in Tezozomoc cited by Veytia; Caso (1967) finds it in a number of other sources (see Nahuatl). Davies (1977) adds Ixtlilxochitl to the list of sources using it.

Whatever its significance may have been in Nahuatl, the Cuitlahuac calendar is the Postclassic version of the Classic Huastec calendar (q.v.) (see A.D. 1127, 1369, 1370, 1371, 1372).

CUITLATEC
(=Tenango?)

The Cuitlatec calendar is unknown. The language itself is unclassified and poorly documented. From notes by Hendrichs (1939) and Weitlaner (1939), the following day names may be considered possible:

a. Téjpu (alligator)	f. —	k. —	p. Zó'o (owl)
b. Ijte'zlá (wind)	g. —	l. Zlidaká (grass)	q. —
c. Zlá'a (house)	h. Puọ̈mọ̈ (rabbit)	m. Píjti (cane)	r. —
d. Uedúgo, Ku (lizard)	i. Ụma (water)	n. —	s. Baxą̣'i (rain)
e. Ujchi (serpent)	j. Ku'i (dog)	o. Kájka (eagle)	t. Tújtu (flower)

The calendar was presumably based on the day or sun (*tamáli, hya*), month or moon (*tujli'i*), and year (*bá'xa*).

Cuitlatec may stand here as representative of the languages of the region of central Guerrero, about which even less is known: Apanec, "Chontal," Chumbia, Cuyumatec, Itzuco, "Mazatec," Pantec, Tepetixtec, Texame, Tezcatec, Tlacotepehua-Tupuztec, Tlatzihuiztec, Tolimec, and Tuxtec. This region could have been an area in which the Tenango calendar survived, at least for a time.

DAYS

The universality of the veintena is a fundamental feature of the calendar of Anahuac. Everywhere sacred, the 20 days have been preserved in oral and written form with remarkable tenacity and conservatism. At the same time, the concepts of the days and *a fortiori* their linguistic and graphic manifestations have undergone some historical diversification, paralleling the diversificaiton of the calendar itself and the religious system underlying it.

On the evidence, the meanings of the day names have changed less than their graphic representations, and the day signs in turn have been more stable than the linguistic forms used to name them.

Over half the days (b. Wind; e. Serpent; f. Death; g. Deer; i. Water; k. Monkey; n. Jaguar; o. Eagle; q. Quake; r. Flint; and s. Rain) seem to have substantially the same meanings in all calendars from Olmec times onward. The approximate history of the conceptualization of the other 9 days is as follows:

a. Olmec *Sun, Day* preserved in Zapotec, but all other systems changed to *Alligator*.

c. Olmec *Night*? preserved in Tikal, but all other systems changed to *House*.

d. Olmec *Hard*? became Zapotec *Black*; all other systems probably *Iguana*.

h. Olmec *Star*? preserved in Zapotec, Tikal, Teotihuacan (and perhaps Huastec and Totonac); all other systems changed to *Rabbit*.

j. Olmec *Foot* preserved in Tikal and Teotihuacan; all other systems changed to Zapotec *Dog*.

l. Olmec *Jaw* became Tikal *Tooth*; all non-Mayan systems follow Zapotec *Sun* until Tilantongo (and Totonac) *Grass*.

m. Olmec *Cane* preserved in Tikal, Yucuñudahui, Toltec, and Tilantongo but changed to *Twist* in Zapotec, Teotihuacan, Tarascan, Huastec, and Totonac.

p. Olmec *Owl*? changed to Tikal *Howler Monkey* and Tilantongo (-Huastec) *Buzzard*.

t. Olmec *Lord* preserved in Tikal, Teotihuacan, and possibly Toltec; all other systems have Zapotec *Flower*.

A simple genetic tree will not account for all of these developments: some of the calendars are syncretistic. Zapotec, for example, wrote day c as *House* but called it *Night*, and day r as *Flint* but called it *Cold*. Totonac got its glyphs for *Dog, Star*, and *Twist* from Zapotec, its glyph for *Wind* from Teotihuacan and its glyph for *Grass* from Tilantongo. Some syncretisms are linguistic rather than graphic: the Tarascan name for a was borrowed from Totonac and Mixe-Zoque, and the latter also may have provided the Huastec, Otomi, Ocuiltec, and Mazahua terms for h. However, Huastec got its name for a from Nahuatl.

Because of their widespread occurrence, the written day names are the best indicators we have of the history of Middle American writing systems. It is still a very partial history.

The Olmec writing system appears to have been used primarily by the speakers of a proto-Zoquean language in Morelos, southern Puebla, central and southern Veracruz, Oaxaca, Chiapas, and southern Guatemala in Preclassic times. What may be a late and altered form of it puts in an elusive appearance in Peten sites of the Classic. It was probably replaced by the Tilantongo script in the Postclassic.

The Zapotec script is found only in central Oaxaca in Preclassic and Classic times, when it seems to have been used primarily by proto-Zapotecans. It was also replaced by the Tilantongo glyph system in the Postclassic.

The Mayan writing system was used by Late Preclassic and Classic proto-Yucatecans and proto-Cholans in Yucatan, Quintana Roo, Belize, Peten, Tabasco, northeastern Chiapas, and northern Guatemala and Honduras. It was probably the script of the Kekchi and other highland Mayas of Guatemala and Chiapas by the Late Classic, and it continued in use in the lowlands throughout the Postclassic.

The Teotihuacan script is found in Central Mexico and in southern Guatemala beginning in the Late Preclassic and continuing into the Classic period. It may have been used primarily by speakers of a proto-Otomanguean language, at least in its northern range, but if the language of Teotihuacan is problematic, that of Kaminaljuyu is even more so. During the Late Classic, a series of local scripts replaced that of Teotihuacan:

Ñuiñe in the Mixteca Baja, where it antecedents may have been Zapotec rather than Teotihuacan writing and where the language was Mixtec.

Xochicalco in southern Mexico state, Morelos, and perhaps Guerrero, probably used by Mixtec, and perhaps by "Cuitlatec" speakers.

Cholula in Puebla and Tlaxcala, probably used by Nahuatl speakers.

Toltec in northern Mexico state, Hidalgo, and perhaps Queretaro and Guanajuato, probably used by Nahuatl speakers.

Tarascan in Nayarit, Jalisco, and Michoacan, presumably used by Tarascan speakers.

Huastec in northern Veracruz and eastern San Luis Potosi, used by Huastec speakers.

Totonac in central Veracruz, used by Totonac speakers.

In all of these territories, the Classic writing systems may have been replaced in the Postclassic by that of Tilantongo.

The Tilantongo script seems to have been the Postclassic writing system throughout Central, Western and Southern Mexico, and certainly it was so in Mixtec, Nahuatl, Otomi, Tlapanec, Chocho, and Zapotec. Some language groups may have developed their own divergent scripts during this period, to judge from the day names of at least one Cuicatec codex (*Dehesa*). It is uncertain what writing system was employed by the highland Maya of Chiapas and Guatemala in the Postclassic period, but certainly the Tilantongo script was known there, whether it was the exclusive form of writing or not.

The following figures summarize the day glyphs and day names of the scripts and languages of Anahuac. Details of provenience will be found under the relevant listings of the index.

Fig. 15a. THE DAY GLYPHS IN THIRTEEN SCRIPTS

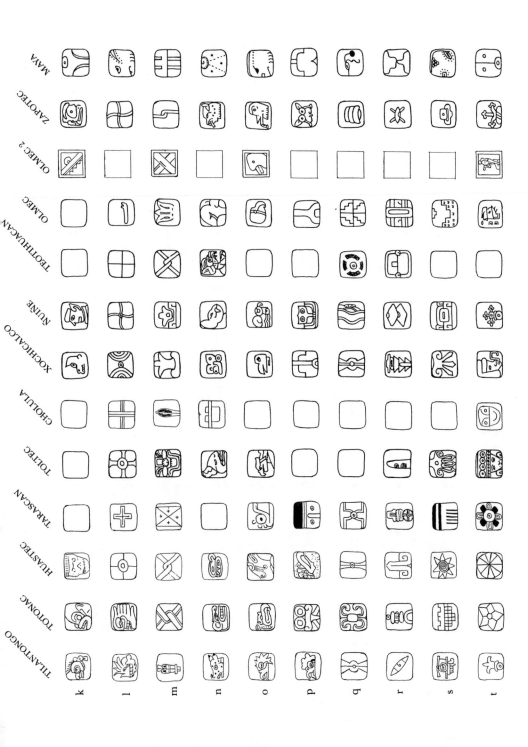

Figure 15b. THE DAY NAMES IN 53 LANGUAGES

DAY NAMES

Language	a.	b.	c.	d.	e.	f.	g.	h.	i.	j.
Yucatec	Imix	Ik	Akbal	Kan	Chicchan	Cimi	Manik	Lamat	Muluc	Oc
Chontal	Imox		Votan	Ache, Can	Chacchan	Tox	Chimay, Ceh	Lamat	Mulu'	Oc
Chol	Imux	Ik	Ac'ab	Canan	Chacchan	Tox	Cuc	Lambat	Moloc	Tz'i
Kekchi	Esem	Ik'	Aq'am'				Mayik			Tzfi
Pokom			Akbal	Qat	Qan	Quimmi	Kieh	Qanil	Toh	Tz'i'
Cakchiq.	Imox	Iq	P'acp'al	K'at	Kan	Camey	Queh	Ganel	Toh	Tz'ih
Tzutuhil	I'mox	Iq'	Aqbal	K'aat	Kan	Kye'mel	Kyéh	Qa'nel	Toh	Tz'i'
Quiche	Imox	Iq	Aq'äb'		Kan	Queme	Queh	Qanil		Chii
Uspantec			Akbal				Kyèeh			
Ixil	Imux	Iq	Akbal	Kach	Can	Kamel	Che	Kanil	Cho	Tz'i
Aguacatec		Ik	Akbal	C'ach	Kan	Camey	Chej	K'anil	Choj	Tciik
Mam	Imix	Ik	Watan	Kets	Abak	Kimex	Tce	K'nel	Tcoj	Elab
Kanhobal	Imux	Ik'	Watan	Kana'	Ab'a	Tox	Cheh	Lambat	Mulu'	Elak
Jacaltec	Imux	Iq	Woton	Kana'	'Ab'ak	Tox	Che	K'anil	Mulu'	'Elab'
Chuh	Himox	Ik'	Akual	K'ana	Chan	Tox	Keh	Lanib'at	Mulu'	Tz'i'
Tojolabal	Ain	Ik	Akal	Osou	Chan	Xemenejich	Cheh	K'anal	Ch'aa	Sul
Chicomuc.	En	Yoh, Yigh	Votan	Chanan	Abagh	Tox, Tog	Vitím	Coi	Yexjá	Elab, Elah
Tzeltal	Mox	Ik'	Akal	Othoow	Tsan	Tsem	Moxic	Lambat	Mulu	Akan
Huastec	Cipac	Iho	Ango, Chi				Its'mal	Koy	Aab	Ombi, Laco
Chiapanec	Ico	Triñ	Ango	Ukú	Olo	Gagame		Oko	Imbu	Ambe
Mangue		Nena	Midu	A'gü	Ule	Angaru		Uku	Iyu	Runya
Subtiaba	Piu	Lawa	Lajuĩ	Lamollo'	Apu	Lamaya	Sihi	Sidu	Laja'	Galtsiqui
Tequistl.	Lepalc'o'		Guguá		Laynofal		Lawala	Tonomma	Iya	Chumuá
Tlapanec			Irrauta	Áluba				Chuagí		Alu
Cacaopera		Uin	Taughe	Yupue	Yárra	Kúla	Yán		Li	Shuy
Jicaque	Uen	Soror	Taughe	Mertsiókaan	Salala	Tepe	Aguingui	Mong	Set	Xuy
Lenca	Huayo	Soor	Tz'uma	Xuvayo	Sala	Xila	Auinge	Mon	Uax	Chuso
Xinca	Vepi	Tan	Tec, Tzu	Huu'n	Ambui, Púki	Teró	Túma	Lur	Uy	Tuy
Zoque	Hukpii	Sava	How	Toqui	Tzan	Cacuy	Mñea	Coya	Na	Ho'o
Mixe	Uśpi'n	Xa'a, Xap	Tsuje	Slúcuc	Tsaa'añ	Uh	Koy	Naan	Ni'in	Šu'ni
Sayula	Uxpi	Jamu	Akxtak'a		Tsana'y	Oqui'c	Jaytsu'	Coya	Neje	Chichi
Totonac	Ne	Uun	Queela		Luuhua	Liinfin	Huugu'i	Tampanámac	Ch'uuchut	Ico
Isthmus	Chi	Bi	Ngata	Yace	Nda		Jiña	Ana		Tella
Zapotec	Cua'ña	Pee	Nu	Cuatsi	Pélla	Quiqueni	Chîna	Peela	Niça	Xne'
Chatino		Cui'i			Cuañ	Ncujui	Cuiñá		Hitya	
Chinantec		Tci			Tcè		Kwaaning	Lökwü	Mui	
Cuicatec		Yinie							Nune	
Chocho			Chin		U'u Yee		Uu Xkaahndü	Caha	Da	Uu Niinä
Ixcatec			Njaa		Yo	Saahwąamuértü	Cuaa, Idzu	Uu Xiatsé	Ndaa	Hua
Mixtec	Quevi	Tsiindyü	Cuau, Mau	Quu	Xucua'	Mahua	Xutaj[5]	Xay, Sayu	Tuta, Duta	Xuhue[3]
Triqui	Cuchucuan	Chi	Hue[3]e	Xiracaj[3]-Nne[35]	Luechuchaa	Gahui[5]	Sonducho	Xato[3]	Nne[34]	Luechee
Amusgo	Ke Tsú	Nane[5]	Huachioo, Waá		Ke Tsuwii	Tsoó	Ke Tsuwii	Tiusu	Da (teyó)	Ke Tsué
Nahuatl	Cipactli	Simanćuhe	Calli	Cuetzpallin	Coatl	Miquiztli	Mazatl	Tochtli	Atl	Itzcuintli
Nicarao		Ndié	Calli	Gazpallin	Goate	Miquiztli	Maçat	Toste	At	Izquindi
Pipil	Cipat	Ehecatl	Calli	Quetzalli	Coatl		Mazatl	Toxtli	Atl	Ytzcuintli
Metzitl.	Cipactli	Hecat	Calli	Xilotl		Tzontecomatl	Macatl	Tochtli	Atl	Izcuin
Otomi	Tetechi-Hucauh	Ehecatl	Ngü	Botaga	Chini	Yayay Atu	Phanixantoehoe	Qhua	Dehe	Yoo
Matlatz.	Toqhuay	Ecatl	Bani	Xichari	Ye	Rini	Pari	Chon	Thahui	Tzini
Ocuiltec	Beori	Dahi	Baani	Korga	Xiskalaha	Tei	Paári	Cuá	Taui	Tosini
Mazatec		Ithaatin	Hyu			Coviu			Tä	Nya
Mazahua	Hyas'ü	Tjaö	Ngümü			U'u			Ndehe	Yo
Tarascan	Uxpi	Ndajma, Tarhiyata	Kuahta	Uahtzáki	Akuitze	Uárhini			Itsi	Ufchu
Cuitlatec	Tëjpu	Ijie'zlä	Zlá'a	Uedígo Ku	Ujchi				Ümä	Ku'i

DAY NAMES

	k. Chuen	l. Eb	m. Ben / Bin	n. Ix	o. Men	p. Cib	q. Caban	r. Etz'nab	s. Cauac	t. Ahau
Yucatec	Chuen	Eb	Ben	Ix	Men	Cib	Caban	Etz'nab	Cauac	Ahau
Chontal		Eh	Bin	Hix, Balum	Men	Chibin	Chaban	Chaab, Chaa	Chauac	
Chol	Batz', Chueen	Eb	Bin	Ix	Tz'ic	Chab, Chib	Cabnal		Chac	Ahau
Kekchi	Batz', Chuen	E	Ah	Hix	Tz'ikin	Ajmajk			Cauuck, Cac	Ajuau
Pokom		Eg	Aj	W'ahlam	Tz'ikin	Ahmak	Noh		Kojjok	Ajqual
Cakchiq.	Batz	Ey	Aj	Ix, Balam	Tz'ikin	Ah'mak	Noh	Tihax	Caok	Hunahpu
Tzutuhil	Pats'	Ey	Ah	I'x	Tz'ikin	Ahmac	Noh	T'ihax	Ka'woq	Ah'pup
Quiche	Batz'	'Ee	Ah	Ix, Balam	Tz'ikin	Ahmáak	Noj	Tihax	Caoc	Hunahpu
Uspantec	B'aatz'	E	Aj	I'x	Tz'ikin	A'mak	Noj			Hunahpu
Ixil	Batz	E'	Aj	I'x	Tz'ikin	Ajmak	Noj	Tihax	Kauak	AjPu'
Aguacatec	Batz	E	Aj	Ix	Tz'ichin	Ajmak	Kisk'ap'	Chij	Cyok	Hunajpu
Mam	Batz	Eyuup	Ben	Ix	Tsikin	Chabin	Noh	Tcij	Tciok	Ahau
Kanhobal	Bats	Ewup	Ah	Hix	Tsikin	Chap'in	Kiskab'	Chinax	K'aq	'Ahaw
Jacaltec	Bats	Eyub'	B'e'en	Hi'ix	Tsikin	Chab'in		Chinax	Kaq	Ahau
Chuh	Ba'a'tz	Eh	Ah	Tsutúichó	Ts'ikin		Chauán	Chinax	Chawok	Vichil
Tojolabal	B'atz'	Kaman	Sin	Hix	Tsiquin		Chix, Chige			Aghual
Chicomuc.		Enoh, Enob	Been				Hom	Tuhil Ixlabon	Chauuk	Huits
Tzeltal	Batz	Ts'an	Pakab	Ix, Pathum	Tsitsin	Chabin		Chinax	Choki	Eme
Huastec	Uthu'	Ihi	Ama	Umbú	Aguagua	Tot		K'amal-Tuhub	Cabogh, Cahog	Ele
Chiapanec	Ambi	Jji	Ure	Endi	Moonkoyo	Ahau			Muxú	Di
Manque	Ambi	Sinyu						Ugo	Amari	Liba'
Subtiaba	Lutu	Dodosgona'ma	Albeba		Diui	Angú		Esee	Unde	
Tequistl.	Galmigú			Galjaguar	Galtijuli	Galaguilacene	Galfounal	Labic	Lagwi	
Tlapanec										
Cacaopera		Niní-ca	Naná	Namá	Guásirri	Kúsma	Lumshi	Súmu	Irra	Báx-ka
Jicaque	Yaru	Juyú	Mishi	Bua	Chiciopon	Cus		Ishalal	Mol	Shuna
Lenca		Ne	Mixi	Lepa	Sira	Cúti	Huyicnár	Ixalal		Zhuna
Xinca	Poxo, Iru	Nórui	Alma	Uilay			Mics		K'unu	Túlo
Zoque	Tzáui	Tetz	Socuy	Tziquin Cang	Hon	Pa'a	Uhx	Tza	Hoy	Yúmi
Mixe	Haiiñ	T'i'ts	Keṕ	Kaa	Huiik	Ju'ca'n	Usu	Tahp	Mfy	Hug'ñ
Sayula	Jamu	Tutsu	Payi	Caja'u	Ju'cu				Paju	
Totonac	Muuxni	Tuhuáan	K'aatiit	Misin	Monkxmi	Pichaahua	Tachiqui	Chhuix	S'een	X'ánat
Isthmus	Migu			Jee	Cuxú	Ciáa				
Zapotec	Pilloo	Piy	Quij	Pèche	Quilloo	Naa	Tixóo	Gopa	Gappe	Lào
Chatino		Cuicha	Liiya	Cuichi	Cocu	Quiñi			Tyoo	Quee
Chinantec										
Cuicatec										
Chocho										
Ixcatec	Uu Ndúčé	Xiikà	Xo	Uu Xàandü	Uu Yàhaa	Uu Xaatshi	Taha	The	Cubi	Tshuu
Mixtec	Ñuu	Cuañe	Thii	Vidzu	Sa[3]	Cuii	Cho	Xuu Ndà	Chiu	Uaço
Triqui	Guruhui[3]	Coj[3]o	Huyo Yo'[3]	Staj[3]u	Xa'u[34]	Xata'[3]u	Nii'njee	Si, Cuxi	Tyuuxtii	Yaj[3]a
Amusgo		Sondaa	Ndohó	Luichiayaa	Ke Tchii	Staé	Qhi Yun[2] Yo'ol		Co, Guman[5], Cobataa	Ndáhá
Nahuatl	Ozomatli	Ndé	Acatl	Ocelot	Quauhtli	Cozcaquauhtli	Ollin	Tecpatl	Natuwa	Xochitl
Nicarao	Oçumatle	Malinalli	Ecat	Oçelot	Goate	Tecolotl	Olin	Tecpatl	Quianuti	Sochit
Pipil	Ozumatli	Malinal	Acatl	Teyollocuani	Quauhtli	Teotl Itonal	Teepilanahuatl	Tecpatl	Quiavit	Xochitl
Metztitl.	Ozoma	Malinalli	Acatl	Ocelot	Cuixtli	Thecha	Nahui Ollin		Ayutl	Omexochitonal
Otomi	Tzepa	Itlan	Xithi	Tzani	Xeni	Yabin	Quitzhey	Yaxi	Quisahutl	Doeni
Matlatz.	Tzonyabi	Chaxttey	Thihui	Xotzini	Ichini		Thaniri	Odon	Yeh	Ettuni
Ocuiltec		Tzinbi	Tathui	Tochini					Yalbi	Tâni
Mazatec	Kúni	Sibí		Xra	Jatse	Kye	Mbi'i	Ndosiwi	Cumaabi	Naxó
Mazahua		Iyu	Isimba		Chabéy'i				Tsin	Ndáhnä
Tarascan	Ozóma	Pjiny'o	Pijtu	Puki	Uakúsi	Tukúru	Yúrhiri	Tzhinápu	Y'ebe, Mánikua	Tsitsiki
Cuitlatec		Zlidaká			Kájka	Žó'o			Baxç'i	Tújtu

Figure 15c. CONCEPTS OF THE DAYS IN THE 24 BEST-DOCUMENTED CALENDARS

	a.	b.	c.	d.	e.	f.	g.	h.	i.	j.
Yucatec	alligator	wind	night	iguana	serpent	death	deer	rabbit	rain	foot
Chontal	alligator	wind	grain	iguana	serpent	death	quetzal	rabbit	rain	foot
Chol		wind	night		serpent	death	deer	rabbit	rain	dog
Kekchi										
Pokom										
Cakchiq.	alligator	wind	night	net	serpent	death	deer	rabbit	rain	dog
Tzutuhil	alligator	wind	night	net	serpent	death	deer	rabbit	rain	dog
Quiche	alligator	wind	night	lizard	serpent	death	deer	rabbit	storm	dog
Uspantec										
Ixil	alligator	wind	night	net	serpent	death	deer	rabbit	rain	dog
Aguacatec		wind	night		serpent	death	deer	rabbit	pay	dog
Mam	alligator	wind	night	net	serpent	death	deer	rabbit	water	dog
Kanhobal	alligator	wind	grain	net	soot	death	deer	rabbit	rain	dog
Jacaltec	alligator	wind	grain	iguana	soot	death	deer	rabbit	rain	dog
Chuh	alligator	wind	grain	iguana	soot	death	deer	rabbit	rain	dog
Tojolabal										
Chicomuc.										
Tzeltal	alligator	wind	house	iguana	soot	death	deer	rabbit	rain	dog
Huastec										
Chiapanec										
Mangue										
Subtiaba										
Tequistl.	light	wind	house	lizard	serpent	death	deer	greet	water	dog
Tlapanec										
Cacaopera										
Jicaque										
Lenca	sun	wind			serpent	death		rabbit	water	
Xinca										
Zoque										
Mixe	root	wind	palm	hard	serpent	world	rabbit	deer	river	vine
Sayula										
Totonac										
Isthmus										
Zapotec	day	wind	night	black	serpent	head	deer	rabbit	water	dog
Chatino										
Chinantec										
Cuicatec										
Chocho										
Ixcatec										
Mixtec	alligator	wind	house	iguana	serpent	death	deer	rabbit	water	dog
Triqui										
Amusgo										
Nahuatl	alligator	wind	house	iguana	serpent	death	deer	rabbit	water	dog
Nicarao	alligator	wind	house	iguana	serpent	death	deer	rabbit	water	dog
Pipil	alligator	wind	house	quetzal	serpent	death	deer	rabbit	water	dog
Metztitl.		wind	house	cornbud	serpent	skull	deer	rabbit	water	dog
Otomi	fish	air	house	lizard	serpent	dead	deer	rabbit	water	dog
Matlatz.	alligator	wind	house	iguana	serpent	death	deer	rabbit	water	dog
Ocuiltec										
Mazatec										
Mazahua										
Tarascan										
Cuitlatec										

	k.	l.	m.	n.	o.	p.	q.	r.	s.	t.
Yucatec	monkey	tooth	cane	jaguar	eagle	owl	quake	flint	storm	lord
Chontal	howler				bird	spider	quake	obsidian	storm	
Chol	monkey	tooth	cane	jaguar	bird	owl	quake	flint	storm	lord
Kekchi		tooth	cane	jaguar	bird	owl		flint	storm	lord
Pokom	monkey	tooth	cane		bird	owl	incense	flint	storm	
Cakchiq.	monkey	tooth	cane	jaguar	bird	owl	incense	flint	storm	hunter
Tzutuhil	monkey	tooth	cane	jaguar	bird	owl	incense	flint	rain	hunter
Quiche									storm	hunter
Uspantec									rain	
Ixil	monkey	tooth	cane	jaguar	bird	owl	incense	flint	storm	hunter
Aguacatec	monkey	tooth	cane	jaguar	quail	owl	incense	flint	storm	hunter
Mam	monkey	tooth	cane	jaguar	bird	owl	incense	flint	storm	hunter
Kanhobal	monkey	tooth	cane	jaguar	bird	spider	quake	flint	fire	lord
Jacaltec	monkey	tooth	cane	jaguar	bird	spider	incense	flint	fire	lord
Chuh	howler	tooth	cane		bird	spider	quake	flint	storm	lord
Tojolabal										
Chicomuc.										
Tzeltal	howler	tooth	cane	jaguar	bird	spider	quake	flint	storm	lord
Huastec										
Chiapanec										
Mangue										
Subtiaba										
Tequistl.										
Tlapanec	monkey	twist	cane	jaguar	eagle	buzzard	sun	stone	rain	flower
Cacaopera										
Jicaque										
Lenca										
Xinca				jaguar						
Zoque										
Mixe	ashes	tooth	cane	jaguar	tobacco	edge	quake	soot	grass	eye
Sayula										
Totonac										
Isthmus										
Zapotec	monkey	sun	cane	jaguar	eagle	crow	quake	cold	cloud	
Chatino										
Chinantec				jaguar						flower
Cuicatec										
Chocho										
Ixcatec										
Mixtec	monkey	grass	cane	jaguar	eagle	buzzard	quake	cold	rain	flower
Triqui										flower
Amusgo										flower
Nahuatl	monkey	grass	cane	jaguar	eagle	buzzard	quake	flint	rain	flower
Nicarao	monkey	grass	cane	jaguar	eagle	owl	quake	flint	rain	flower
Pipil	monkey	grass	cane	heart	eagle	god	prince	flint	turtle	flower
Metztitl.	monkey	tooth	cane	jaguar	cobweb	god	quake	nails	rain	flower
Otomi	monkey	grass	cane	beast	peer	buzzard	quake	flint	rain	flower
Matlatz.	monkey	grass	cane	jaguar	eagle		quake			flower
Ocuiltec										
Mazatec										
Mazahua										
Tarascan										
Cuitlatec										

GREGORIAN

The Gregorian calendar was promulgated by Pope Gregory XIII on 4 X 1582 Julian so that the following day was 15 X 1582 Gregorian. This was made official at various dates elsewhere, so that the first day of the Gregorian calendar was

Rome	15 X 1582	(Recinos 1950:161)
Spain	14 V 1583	(Caso 1967:99)
Mexico	15 X 1583	(Ibid.)
Antigua	29 I 1584	(Recinos 1950:161)
Sololá	2 II 1584	(Ibid.)

The Cakchiquel adopted the Gregorian calendar as soon as it was announced in Sololá. The Tzotzil and Tzeltal not only adopted it, but froze their calendars to it and stopped counting leap years from their respective New Years' days of 1584–85. Post-1584 dates in Quiche are given their Gregorian equivalents in native sources. The Yucatecans did not acknowledge Gregorian dating until the nineteenth century, and a number of other groups continued to use frozen Julian dates as reference points for correlation purposes (see A.D. 1548, 1616).

An anachronistic retroactive application of Gregorian dating is sometimes a chronological problem as well (see A.D. 1557, 1579).

The relation between the Julian and Gregorian calendars is as follows (Schram 1908):

Beginning 1 III 1700 B.C.	Gregorian + 14 days = Julian	
1500	+ 13	
1400	+ 12	
1300	+ 11	
1100	+ 10	
1000	+ 9	
900	+ 8	
700	+ 7	
600	+ 6	
500	+ 5	
300	+ 4	
200	+ 3	
100	+ 2	
A.D. 100	+ 1	
200	0	
300	− 1	
500	− 2	
600	− 3	
700	− 4	
900	− 5	
1000	− 6	

1100	− 7
1300	− 8
1400	− 9
1500	− 10
1700	− 11
1800	− 12
1900	− 13

GUITIUPA
(III.G.1 CR?; XVI c. to XX c. A.D.=Tzotzil + 3m, f1584)

The Ascensión Guitiupa Tzotzil calendar begins the year on 5 Ceh Mayapan (see *1584 and Tzotzil).

HUASTEC
(III.M.0 CR?; II c. to XVII c. A.D. = Tarascan − 1m Classic or
Teotitlan + 1m Postclassic)

Although it is only fragmentarily documented, the Postclassic Huastec calendar almost certainly used Type IV New Years' days and began 6 Mac Mayapan, counting the day count from 1 to 13 and the days of the month from 1 to 20, and naming the years terminally. The start of its calendar round and its direction colors have not been established. This is the calendar best designated as Cuitlahuac, shared by Huastec and Nahuatl (q.v.).

The Tablet of Building 4 at Tajin (De la Torre and Pérez 1976:68) documents the use of bar and dot numerals and possibly day names as early as the second century A.D. It seems most likely that, at that date, Tajin was using the Teotihuacan calendar.

At Tajin, however, there are also two dates of the Classic period indicating the use of Type III year bearers, and this is the calendar I have designated as Huastec. The Postclassic Cuitlahuac calendar is simply the terminally dated form of the same count. It is the Classic calendar that is designated in the preceding heading. The calendar is not documented after the seventeenth century, and the modern Huastecs retain only disjointed beliefs that may once have related to its gods and rituals. There is no identifiable trace of the calendar they may once have shared with the other Mayan peoples some 2,500 years ago. The day glyphs are largely cognate with those of Teotihuacan and Totonac.

The units of the Huastec calendar are the day or sun (*k'ichaa*), month or moon (*its'*), day count (?), year (*tamub*), and calendar round (?). The Huastec year sign appears to have been similar to that of Yucuñudahui and identifies Type III days:

Tajin: Anno 4 Cane
(De la Torre and Pérez
1976:36, Figs. 4–5)

Aguacate: Anno 2 Cane
(De la Fuente and Gutiérrez
1980:ccxxxiii b)

(It may be noted that the Aguacate cartouche appears on the left shoulder of an old man whose right shoulder contains an identical cartouche containing the day sign Serpent but no coefficient. If this were a correlational date with the Olmec calendar—the only one in which Serpent is a year bearer—the expected sign would be 3 Dog.)

A set of probable Huastec day glyphs follows. Most of them are almost certainly day glyphs because they occur with numeral coefficients between 1 and 13. Their positions are less secure, though those that are visually cognate with other glyphic systems are probably correctly placed (c, d, s, and t are dubious). Their sources are identified later.

a. 2 Alligator. Ajalpan (De la Fuente and Gutiérrez 1980:clix a).
   2 Alligator. Chicontepec de Ahuateno (Ibid.:cccxviii a,b).
   Alligator. Tampico (Ibid.:cclxxxv).
b. 2 Wind. Tajin (De la Torre and Pérez 1976:40, Figs. 9–10).
c. 3 House? Tajin (Ibid.:56, Figs. 31–32).
d. Iguana? Tajin (Ibid.:66, Figs. 43–44).
e. Serpent. Aguacate (De la Fuente and Gutiérrez 1980:ccxxxiii a).
f. 4 Death. Tajin (De la Torre and Pérez 1976:45, Figs. 15–16).
g. 2 Deer. Tajin (Ibid.:44, Figs. 15–16).
   Deer. Unknown (De la Fuente and Gutiérrez 1980:ccclviii).
h. 2 Rabbit. Tajin (De la Torre and Pérez 1976:56, Figs. 25–26).
i. Water. Tajin (Ibid.:56, Figs. 31–32).
j. 1 Dog. Tajin (Ibid.:44, Figs. 15–16).
k. Monkey? Tamuin (De la Fuente and Gutiérrez 1980:ccxcii c).
l. 7 Grass. Rio Panuco (Ibid.:cciv b).
m. Anno 2 Cane. Aguacate (Ibid.:ccxxxiii b).
   Anno 4 Cane. Tajin (De la Torre and Pérez 1976:37, Figs. 4–5).
n. 3 Jaguar. Hacienda Tobalo (De la Fuente and Gutiérrez 1980:cxvii a).
   3? Jaguar. Tajin (De la Torre and Pérez 1976:68, Fig. 46).
o. 4 Eagle. Ozuluama (De la Fuente and Gutiérrez 1980:ccclx).
p. 1 Buzzard. Unknown (Ibid.:ccclvii).
   Buzzard. Unknown (Ibid.:ccclvi).
   3 Buzzard. Tajin (De la Torre and Pérez 1976:41, Figs. 9–10).
q. 2 Quake. Tampacayal (De la Fuente and Gutiérrez 1980:cccvi).
r. 1 Flint. Texupesco (Ibid.:cccxxiv a).
   2 Flint. San Luis Potosi? (Ibid.:cccxxiv a).
s. 2 Rain? Tampico? (Ibid.:ccclxiv).
t. 4 Flower? N. Veracruz (Ibid.:ccclxiv).

From Larsen (1955) and B. Edmonson (1983), the following names for the days may be regarded as possible, though modern Huastecs do not recognize them as such, and no historical references to them are known:

a. Cipac (alligator)	f. Tsem (death)	k. 'Uthu' (monkey)	p. T'ot (buzzard)
b. 'Ik' (wind)	g. Its'amal (deer)	l. Ts'ah (vine)	q. Hom (incense)
c. 'Akal (night)	h. Koy (rabbit)	m. Rakab (cane)	r. K'amal T'uhub (flint)
d. 'Othoow (lizard)	i. Aab (rain)	n. Ix, Pathum (jaguar)	s. Muxi' (storm)
e. Tsan (snake)	j. Akan (foot)	o. Ts'itsin (bird)	t. Huits (flower)

Some of the possible day names suggest borrowing from Nahuatl (a, k) and Mixe (h).

A codex in the Bibliothéque Nationale in Paris (Mss. Mexicains No. 65-71, fol. 94v.–102v.) annotates the depiction of the Mexican monthly feasts with the month names of Otomi and Huastec (Lehmann 1920:2:880):

H. –	M. Amab (plant)	R. (An Baxi)	E. Quisa (sun?)
I. Ahit (count)	N. –	A. (Tzimaxygui)	F. –
J. –	O. Cooch (fat?)	B. (Damaxygui)	G. –
K. Tuy (wax?)	P. Chuc Tzeb	C. Chanub (?)	
	(?)	(An Tzhoni)	
L. Pitich (?)	Q. (Itzhoni)	D. (An Tzyni)	X. –

Possible Huastec glosses (B. Edmonson, personal communication) are indicated. Otomi names in parentheses are those of the Otomi months: Q. An Tangotu; R. An Baxi; A. An Ttzenboxegui; B. An Tamaxegui; C. An Tzhoni; D: An Thaxmo. The glyphs for the Huastec months are not known.

A marginal note in the same source places 1 Ahit on 21 III, which matches it to the Otomi month I., An Tatzhoni (see Otomi, below), identifying the date as Julian and unfrozen and probably belonging to the years 1616–19. If any of these assumptions is wrong, there is no way of placing the Huastec calendar in relation to others. Month I. is the fourth month of the Otomi year, but this European date for Ahit indicates that the Huastec year began 2 months after that of Otomi and "solves" the Huastec year as congruent with Cuitlahuac (see A.D. 1616).

HUAVE
(= Mixe?)

The Huave calendar is undocumented. Possible day names are to be found in Stairs and Stairs (1981).

a. Jüm (alligator)	f. Andeow (death)	k. Echweac, Moching (monkey)	p. Potuit (buzzard)
b. Iünd (wind)	g. Xicuuw (deer)	l. Olüiqueran (tooth)	q. Ateam (quake)
c. Nit (palm)	h. Coy (rabbit)	m. Olam (cane)	r. Lixleaban (flint)
d. Ix (iguana)	i. Ijchiür (rain)	n. Lüw (jaguar)	s. Soex (grass)
e. Ndiüc (serpent)	j. Pet (dog)	o. Mojngol (eagle)	t. Mbaj (flower)

The units of the calendar were the day or sun (*nüt*), month or moon (*caaw*), day count (*?ateow*), and calendar round (?).

The first three Huave numerals are inflected for different classes of nouns and have specific forms for counting days and years. The numerals are:

	Rectangular	Day	Year
1	Nop	Noic	Nomb
2	Ijpüw	Ic	Iüm
3	Arojpúw	Er	Aroomb
4	Apiquiw	-	-
5	Acoquiaw	-	-
6	Anaíw	-	-
7	Ayaíw	-	-
8	Ojpeacüw	-	-
9	Ojquiyéj	-	-
10	Gajpowüw	-	-
11	Gajpanoic	-	-
12	Gajpic	-	-
13	Gajpar	-	-

The numeration is reminiscent of the sacred numeral systems of Mixe and Mixtec, and the day name for Rabbit (Coy) is a borrowing from Mixe, perhaps tending to confirm my speculation on geographic grounds that Huave may have been on the Mixe calendar with Type III year bearers.

ISTACOSTOC
(III.H.1 CR?; XVI c. to XX C. A.D. = Guitiupa + 1m, f1584)

First documented from San Andrés Istacostoc (San Andrés Larraínzar) by E. Pineda (1845), this Tzotzil calendar displaces the beginning of the year to what had been the fifth Tzotzil month (Mayan Mac, or month H.).

It seems possible that the change was motivated by a desire to move Ch'aik'in outside of the season of Carnival: in any case, it had that effect. Shrove Tuesday can fall from 2 II to 9 III. (It fell on the latter date in 1943.) In the Istacostoc calendar 1 Ch'aik'in falls on 12 III. The Istacostoc year actually begins on different dates in different years (between 16 III and 21 III), seemingly because there is no authoritative way of establishing the Mayan date.

In recent times, the Christian calendar has been fixed among the Tzotzils by the *Calendario Galván*, but most Tzotzils cannot read it, and there is a consequent lag to the adoption of leap-year corrections in relation to the better known feasts. These are generally held on the correct Gregorian dates, but awareness of the corresponding Mayan dates is a function of individual memory, corrected or not for recent leap years, depending upon fairly haphazard consultation with more or less authoritative others. The fixed (and frozen) date of 1 Tz'un and the variable date of

Carnival (with its somewhat elastic relationship to Ch'aik'in) are major factors in remembering the Tzotzil date (Whelan 1967).

San Andrés Istacostoc began the year on 1 Mok 21 III 1845 Gregorian (E. Pineda 1845), 16 III 1932 Gregorian (Becerra 1933), and 18 III 1948 (Berlin 1951). In the latter year, it was joined by Santa Marta Yolotepec, San Miguel Mitontic, and Santa Catarina Pantelhó (Ibid.). On the other hand, it is recorded as reverting temporarily to 1 Muk'tasak on Mardi Gras, 25 II 1941 (Schulz 1942).

ISTHMUS
(= Zapotec?)

The following are possible day names in Isthmus Zapotec (Cruz 1935:112–15):

a. Ñe (alligator)	f. —	k. Migu (monkey)	p. Cuxú (owl)
b. Bi (wind)	g. Jiña (deer)	l. —	q. —
c. —	h. Ana (rabbit)	m. —	r. —
d. —	i. —	n. Jee (jaguar)	s. —
e. Nda (serpent)	j. Ico (dog)	o. Ciáa (eagle)	t. —

Nothing more is known of its calendar.

IXCATEC
(= Tilantongo?)

The units of the Ixcatec calendar are the day ($t^{y}h\!\!\!\!\!\!\;i$), moon or month (*nduusà*), day count (?), year (*xhngà*), and calendar round (?).

A number of possible day names are reported by Fernández (1961):

a. —	f. Sąahwąamuértú (death)	k. 'Uu Ndŭcé (monkey)	p. 'Uu Xaatshì (vulture)
b. Tsiindyù (wind)	g. 'U Xkaahndù (deer)	l. 'Naa'ñuu (tooth)	q. Nii'njee (earth)
c. Njaa (house)	h. 'Uu X' átsé (rabbit)	m. Thįi (reed)	r. Xu Ndà (flint)
d. —	i. Ndaa (water)	n. 'Uu Xàandú (jaguar)	s. Tyuuxtii (rain)
e. 'Uu Yèe (serpent)	j. 'Uu Niinà (dog)	o. 'Uu Yáhaa (eagle)	t. Tshuu (flower)

It seems likely that Ixcatec used the Tilantongo calendar, but we have no relevant data for establishing the point.

IXIL

(II.L.1 CR 1 Che 1506; XX c. A.D. = Quiche)

The Ixil year begins on 4 Kayab Mayapan. The day count is counted from 1 to 13 and the days of the month from 1 to 20. Its senior year bearer was 1 Deer and its calendar round began on 1 Che in 1506. Its directional colors are unknown. The day names are given by Lincoln (1942:107):

a. Imux (alligator)	f. Kamel (death)	k. Batz (monkey)	p. A'mak (owl)
b. I'q (wind)	g. Che (deer)	l. E (tooth)	q. Noj (incense)
c. Akbal (night)	h. Kanil (rabbit)	m. Aj (cane)	r. Tihax (flint)
d. Kach (net)	i. Cho (rain)	n. I'x (jaguar)	s. Kauak (storm)
e. Kan (serpent)	j. Chii (dog)	o. Tzikin (bird)	t. Hunahpu (hunter)

Lincoln (1942:117–18) reports six lists of Ixil month names, only two of them (numbers 4 and 5 from the same Chajul informant) ostensibly complete. Altogether, thirty-six different names are listed. L., Q., A., B., C., H., and J. are cognate with month names in other systems. Seventeen (including four from the cognate list) can be shown to be mutually exclusive. By placing the cognates first and then making a best fit of the rank orders of the mutually exclusive names, I get the following "solution:"

	(1)	(2)	(3)	(4)	(5)	(6)
L. Cajab Ki	—	(5a)	—	(2a)	—	8 (9b)
M. Pac Tzi	—	—	—	15	12	10
N. Koj Ki	(5c)	6	—	4	—	—
O. Tal Cho	6	7	4	5	—	11
P. Nim Cho	7	8	5	6	—	12
Q. Chotz Cho	8	9	(6d)	8	7	(13e)
R. A Ki	—	(10f)	—	7	13	(14g)
A. Tz'ikin Ki	—	—	—	(9h)	(8h)	5
B. Yax Ki	—	—	7	(3i)	—	(2j)
C. Mol Che (Masat)	2	14	1	1	—	3
D. Tzil Ki	—	(1k)	9	17	17	(41)
E. Petzetz Ki	—	11	(8m)	10	15	—
F. Xukul Ki	—	(2n)	—	11	16	—
G. Yowal	—	12	—	12	9	—
H. Chente Mac	1	3	—	14	11	6
I. Nol Ki	3	13	2	16	14	—
J. Muen Chin	4	4	—	13	10	1
K. Xet Ki	—	—	3	(18o)	(18o)	(7p)

Letters and rank orders given in parentheses indicate the following possible synonyms:

L. (a) Och Ki,   (b) Onchil
N. (c) Chochol   —
Q. (d) Mech Ki,   (e) Tzijep
R. (f) Avax Ki,   (g) Mam A Ki
A. (h) Kucham   —
B. (i) Mekaj,   (j) Mu
E. (k) Tzu Ki,   (l) Tzanakbal
F. (m) Hui Ki   —
G. (n) Tzunun Ki   —
K. (o) Zoj Ki,   (p) Lajab Ki

The nineteenth month (X.) is called the five days (O Ki).

The units of the Ixil calendar are the day (*ki*), month names (*toj amak, tachbal amak*), year (*'ab*), day count (?) and calendar round (*ualyab*).

Although its day names are identical to those of Quiche, the Ixil month names are quite different. Seven of them are cognate with those of Kanhobal: Q., A., B., C., H., J., and X (see A.D. 1940, 1967).

IXIMCHE
(XV c. to XVII c. A.D.)

The unique Cakchiquel calendar employed only in the dating of events in the *Annals of the Cakchiquels* from 1493 to 1604 has its base date in the Tukuche Revolution of Iximche on 11 Ah 2 Ru Kab Tamuzuz Cakchiquel (11 Ben 15 Muan Mayapan, 20 V 1493 Julian). This day (11.13.12.15.13) is treated as the end of a cycle of 400 days (*huná*) or 0.0.0.0 in purely vigesimal notation, and its anniversary recurrences are noted by the 20-day month (ik'), by the *huná* and by the *may* (= 20 *huná*), all of which always begin on Ix and end on Ah; the coefficient of the day ending the *huna* diminishes by 3 each *huná* (11 Ah, 8 Ah, 5 Ah, 2 Ah, 12 Ah, etc.). The month endings diminish by 6 (7, 1, 8, 2, 9, 3, 10, 4, 11, 5, 12, 6, 13) and the coefficients of *may* endings by 8 (11, 3, 8, 13, etc.). The passage of 1 *may* (1.0.0.0 Iximche) is the equivalent of 1.2.4.0 Tikal. Thus

	Iximche			Mayapan
0.0.0.0	11 Ah	=	11.13.12.15.13	11 Ben 15 Muan
1.0.0.0	3 Ah	=	11.14.15.1.13	3 Ben 5 Kankin
2.0.0.0	8 Ah	=	11.15.17.5.13	8 Ben 5 15 Ceh
3.0.0.0	13 Ah	=	11.16.19.9.13	13 Ben 5 Zac
4.0.0.0	5 Ah	=	11.18.1.13.13	5 Ben 15 Ch'en
5.0.0.0	10 Ah	=	11.19.3.17.13	10 Ben 5 Mol
5.6.0.0	10 Ah	=	11.19.9.2.13	10 Ben 20 Ch'en

The last date given is the latest one known in the Iximche calendar.

The numerology dictates that 9 huná = 10 tun, 13 huná = 20 chol q'ih, and 73 huná = 80 haab. An even huná will thus fall on every tenth tun from its starting date (5.13., 15.13., 5.13, etc.). It will fall on 15 Muan every 80 years. And whereas the name day of the tun ending repeats after 18 tzol kin (13 tuns), the name day of the huná repeats after 20 chol g'ih (13 huná).

Some confusion or potential confusion is introduced by the fact that the Cakchiquel use *huná* to refer to both the 400-day and the 365-day year, the latter being also called 'a' (see A.D. 1493).

IZAPA
(II.B.19 CR?; I c. A.D. = Cuicuilco + 1d)

The only readable calendrical inscription from Izapa is a tantalizing date of 7 Death on Miscellaneous Monument 60. The day sign is enclosed in a double cartouche over a trilobate glyph usually read as a day marker. The numeral is bar and dot, and the style of writing is strikingly similar to that of Kaminaljuyu (see 147 B.C.). It is possible that a year date is intended: perhaps that is the import of the double cartouche, as may be true at Kaminaljuyu as well. If so, it indicates the presence of the only known calendar with Type I year bearers.

The inscription probably belongs to the first century A.D., thus placing the date in the year 47.

The mythological significance of 7 Death, one of the Quiche lords of Xibalba, rouses the additional speculation that this may have been the beginning of the Izapan calendar round—1 year later than that of the Olmec and conceivably rejecting the year 1 because of Teotihuacan influence. Could the trilobate glyph be New Fire, as it appears to have been in Kaminaljuyu? (See A.D. 47.)

JACALTEC
(III.L.1 CR 1 K'anil 1518; XX c. A.D. = Kanhobal)

The Jacaltec year begins on 5 Kayab Mayapan. It counts the day count from 1 to 13 and the days of the month from 1 to 20 and begins its calendar round with 1 K'anil (1518). Its directional colors are unknown. The units of the calendar are the day (*tsaiik*), moon or month (*cahau*), day count (?), year (*habil*), and calendar round (?). The day begins and ends at sunset and is named completively (LaFarge and Beyers 1931:172). The year bearers (*iquum habil*) are Type III.

The day names are given by LaFarge and Beyers (1931:160–61):

a. Imux (alligator)	f. Tox (death)	k. Bats (monkey)	p. Chap'in (spider)
b. Iq (wind)	g. Che (deer)	l. Ewup (tooth)	q. Noh (incense)
c. Watañ	h. K'anil	m. Ah (cane)	r. Chinax (flint)
d. Kana'	i. Mulu' (rain)	n. Hix (jaguar)	s. Kaq (fire)
e. Ab'a (soot)	j. Elak (dog)	o. Tsikin (bird)	t. Ahau (lord)

The day names include both the howler (Bats) and spider monkeys (Chap'in). The month names are not reported in Jacaltec. The calendar is congruent with Kanhobal (q.v.) (see A.D. 1927).

JICAQUE
(III.B.0 CR?; XVI c. A.D. = Kaminaljuyu + 1d)

The Jicaque year began on 5 Yaxkin Mayapan. It probably counted the day count from 1 to 13 and the days of the month from 0 to 19. It was probably named initially because Kaminaljuyu, the Mixe, and the Lenca all used initial year bearers. There is no suggestion that it had the Long Count, despite its proximity to Copan, but neither did Kaminaljuyu, from which it presumably derived. Its month names and directional colors are unknown.

The units of the Jicaque calendar were the day (*guaguc, anasetiau, puntée, shagua*), moon or month (*muy*), and year (*chiquin, ioalar*). Possible day names from the vocabularies compiled by Lehmann (1920) are:

a. —	f. Tepe (dead)	k. Yaru (monkey)	p. Cus (buzzard)
b. Soror (wind)	g. Aguingui (deer)	l. Juyú (grass)	q. Lumshi (quake)
c. Taughe (night)	h. Mong (rabbit)	m. Mishi (cane)	r. Ishalal (flint)
d. Yupue (iguana)	i. Set (water)	n. Bua (jaguar)	s. Mol (cloud)
e. Salala (serpent)	j. Shuy (dog)	o. Chiciopon (eagle)	t. Shuna (flower)

The "day names" for b, c, e, g, h, j, k, p, q, r, and t are highly similar to those of Lenca (q.v.). A correlational date of 1530 establishes the New Year of "Hibueras and Honduras" at III.B.0. The reference is almost certainly to Jicaque because the date is nowhere near that of any Mayan calendar (see A.D. 1530).

JULIAN

The Julian calendar differed from the Gregorian one in having a completely regular cycle of leap years every 4 years.

The Julian year forms a cycle of its own with the day count. The days appear as year bearers for 4 years at a time but must skip a day for leap year, thus moving to a new year-bearer set. The first of these 4 years may be considered as a kind of leap-year bearer. The count will return to the same year-bearer set every 20 years, and to the same leap-year bearer every 80, albeit with a different coefficient. These 80-year coefficients of the leap-year bearers diminish by fours: 13, 9, 5, 1, 10, 6, 2, 11, 7, 3, 12, 8, 4, repeating only after 13 x 80, or 1,040 years. Yucatecan sources manifest some awareness of these numerological features of the Julian calendar, as do some of the other native calendars.

The Julian system remained in use among a number of Indian groups after the institution of the Gregorian calendar in 1584, particularly those who had frozen their correlations to it (see A.D. 1531, *1548, 1596, 1606, *1617, *1629, *1632).

KAMINALJUYU
(II.B.0 CR?; III c. B.C. to VII c. A.D.?; = Olmec + 1d − t − 360*)

The calendar of Kaminaljuyu is attested only by a single threeway correlational date at 147 B.C., equating the years 8 Wind Teotihuacan, 1 Lord Olmec, and 7 Wind Kaminaljuyu. The inscription makes it clear that the writing system of Kaminaljuyu was that of Teotihuacan. Other considerations suggest that the Kaminaljuyu calendar was that of the Zoquean-speaking Xinca (q.v.), about whose calendrics we know very little. It is clear that Kaminaljuyu was familiar with Olmec writing as well.

The Kaminaljuyu calendar is derived directly from Olmec, as is indicated in the heading, by advancing 1 day and abandoning terminal naming and the Long Count. The Teotihuacan calendar is derived from that of Kaminaljuyu by the transformation − 105d + 14*. Kaminaljuyu is also the source of the Jicaque (− Lenca?) calendar, which in turn gave rise to that of the Mixe. The suggestion is strong that Olmec-Kaminaljuyu-Jicaque-Lenca-Xinca-Mixe constitute a mainly Zoquean cultural, linguistic, and calendrical complex.

It may seem that these are far-reaching conclusions to derive from a single unsupported reading of one inscription, but it may be noted that the mathematical odds against the chance occurrence of so tight a calendrical pattern, although impossible to calculate, are certainly very high (see 147 B.C.).

KANHOBAL
(III.L.1) CR?; XX c. A.D. = Quiche + 1d)

The Kanhobal year begins on 5 Kayab Mayapan, counting the days of the day count from 1 to 13 and those of the month from 1 to 20. Its calendar round may have begun with that of Jacaltec, but is undocumented, as are its directional colors.

The year bearers (*iquum habil*) are Type III. The units of the calendar are the day (*k'uu*), moon or month (*cahau*), day count (?), year (*habil*), and calendar round (?). The day names are given by LaFarge (1947:164) from Santa Eulalia:

a. Imux (alligator)	f. Tox (death)	k. Bats (monkey)	p. Chabin (spider)
b. Ik' (wind)	g. Cheh (deer)	l. Eyuup (tooth)	q. Kixk'ap (quake)
c. Watan	h. Lambat	m. Ben (cane)	r. Chinax (flint)
d. Kana'	i. Mulu' (rain)	n. Ix (jaguar)	s. K'aq (fire)
e. Abak (soot)	j. Elab (dog)	o. Tsikin (bird)	t. Ahau (lord)

The day names include both the howler (Bats) and spider monkeys (Chabin).

Termer (1930:391) gives a list of Chuh month names from Santa Eulalia. These are listed in the present work under Chuh (q.v.). LaFarge's informants disagreed about the names of the months and their order. His "suggested arrangement," taking Termer's list into account is as follows (LaFarge 1947:168):

L. Wex	Q. Kanal	D. Mol	I. Onéu
M. Saqmai	R. Yaxul	E. —	J. Sivil
N. Nabich	A. Yaxakil	F. —	K. Tap
O. Moo	B. Watsikin	G. —	
P. Bak'	C. Cuxem	H. Mak	X. Oyeb K'u

LaFarge's reconstruction omits 4 months named Sihom with color prefixes (Ibid.:169). My interpretation rests on placing these and other months (Q., R., A., B., C., D., E., F., G., H., I.) where they would go in cognate systems (and modifying the orthography slightly):

L. Wex (time)	Q. Xuhem	D. K'eq Sihom (black flower)	I. Oneu
M. Saqmay (white 20)	R. Kanal (yellow)	E. Yax Sihom (green flower)	J. Sivil (vapor)
N. Nab Ich (1-moon)	A. Wa Tsikin (1-bird)	F. Zaq Sihom (white flower)	K. Tap (crab)
O. Moo (parrot)	B. Yaxakil (green time)	G. K'aq Sihom (red flower)	
P. Bak' (bone)	C. Mol (gather)	H. Mak (cover)	X. Oyeb Ku (5 days)

The Kanhobal calendar is shared by the Jacaltec and Chuh and perhaps by Tojolabal and Chicomuceltec as well (see A.D. 1932).

KEKCHI
(II.N.0 CR?; XVII c. A.D. = Tikal, f1548)

The Kekchi year began on 4 Uayeb Mayapan or 0 Pop Tikal, counting the days of the day count from 1 to 13 and those of the months from 0 to 19, naming the year initially, and presumably beginning the calendar round when Tikal did, on 1 Quake, though that is not documented. Its directional colors are unknown. The units of the calendar were the day (*cutan*), trecena (*oxlajuel?*), month or moon (*po*), day count (*ajlabal cutan*), year (*chihab*), cycles of 4 (*gkoban*) and 20 years (*may*), and calendar round or "century" (*k'e*). The day names can be reconstructed from Haeserijn (1979), Sedat (1955) and the *Lanquin Calendar* (Anon. 1931, Thompson 1932a) as follows:

a. Esem	f. Camik (death)	k. Batz', Chueen (monkey)	p. Chab, Chib (owl)
b. Ik' (wind)	g. Mayik (deer)	l. E (tooth)	q. —
c. Acab' (night)	h. —	m. Ah (cane)	r. Itznab, Chaa (flint)
d. —	i. Moloc (rain)	n. Hix (jaguar)	s. Cauuck, Cac (storm, fire)
e. Chacchan (serpent)	j. Tz'i (dog)	o. Tz'ic (bird)	t. Ajauu (lord)

The Kekchi month names, apparently much mixed with Chol, are recorded in the *Lanquin Calendar*:

N. Pop (mat)	A. Chichin (birds)	F. Zac (white)	K. Ahquicou
O. Icat (1-cat)	B. Ianguca	G. Chac (red)	L. Ccanazi
P. Chacc'at (2-cat)	C. Mol (gather)	H. Chantemac (1-cover)	M. Olh
Q. —	D. Zihora	I. Uniu	
R. Cazeu	E. Yax (green)	J. Muhan (owl)	X. Mahi y Ccaba (nameless)

Actually month N. Pop is documented by Haeserijn (1979), along with alternative names for B. Raxkim (green time), D. Cheen (well), I. Kan (yellow), and perhaps O. Gkatoc, R. Yulic, and X. Gkimuch (?). Haeserijn also provides additional month names that I am unable to place: Rakol, Gkalec, and Gkoloc. The *Calendar* is clear in placing the five nameless days (*holob cutan mahi y ccaba*) after Olh and beginning them with the frozen Julian date 9 VII 1548 (see A.D. *1548).

LACHIGUIRI
(=Zapotec?)

The Lachiguiri or Yautepec Zapotec are calendrically unknown. Presumptively they used the Zapotec calendar.

LACHIXOLA
(III.F.19 CR?; XX c. A.D. = Chinantec + 2m, f1548)

The Lachixola calendar is identical with the Lalana Chinantec calendar (q.v.) except for beginning the year 2 months later.

LALANA
(see Chinantec)

LEAP YEAR

The aboriginal calendars of Middle America did not use a leap year. The seasonal and agricultural implications of their month names gradually slipped away from the periods of the solar year they may originally have designated, and, in the best documented case, the Tikal calendar did nothing about it during its long life span. Astronomical corrections were known and were written down, but the calendar itself was not changed.

Adjustments may have been made in some pre-Conquest calendars by moving the date of New Year 20 days at a time. Such a "correction" could have been applied as frequently as every 80 years, but there is no evidence that such changes were in fact made, and at least nineteen calendars can be shown to have gone substantially longer without correction. Calendar changes of 1 day also occurred, but, on the evidence, they were far too rare to have served the purpose of leap year.

The European introduction of leap year did not cause any change in this respect in most of the native calendars. A number of them adapted to it in one of two ways: (1) by adopting a "frozen" date in the European calendar as a convenient way of calculating other European dates, or (2) by accepting a "frozen" date in the European calendar as the true date of their own New Year and calculating other native dates from it. In the first case, the true Christian date of the native new year falls back 1 day each leap year in comparison with the "frozen" date. The Mayapan calendar followed the first track; the Tzotzil calendar followed the second.

The Mayapan calendar adopted the frozen date of 16 VII 1548 Julian, and calculated other Christian dates with that as the New Year throughout the life of the Mayapan (and Valladolid) calendar. Europeans inquiring about today's native date, however, were given the correct "unfrozen" calendar round, and the Yucatecan calendar round remained unaltered in its relation to other unfrozen native calendars.

For reasons that are not altogether clear, the Guitiupa Tzotzil froze their New Year to 3 III 1584 Gregorian, rather than to the expectable 26 II. Some manipulation of the 5 days of Chaikin seems to have been involved. Both native and European correlational dates for this calendar approximate 3 III, but the Tzotzil day count has been disrupted and lost.

A number of other calendars adopted the Mayapan *modus vivendi* at the same time (after 29 II 1548 Julian): Chiapanec, Chinantec, Pokom, Aztec, and Texcoco. Mazatec may have done so later when its New Year came to rest on January 1 (see A.D. *1548, *1616) and Mixe earlier (see 1531).

The introduction of the Gregorian calendar, which reached the Indians in 1584, may have been the impetus to the freezing of the Guitiupa, Istacostoc, Mitontic, and Cancuc calendars after the 1584 leap year (see A.D. *1584).

Neither of the types of frozen calendars involved the assimilation of leap year into the native system. The Guitiupa calendar, for example, maintains its frozen date by simply ignoring February 29, and that seems to have been the general practice. Anything else, of course, would show up as an interruption of the universal day count, and that simply didn't happen. The circumstantial account to the contrary by Las Navas (see A.D. 1553) must be discounted, as the Tlaxcalan calendar has not been shown to reflect the insertion of leap years in its later dates. Similar allegations about other calendars have been made frequently (see, e.g., Mazatec), but no such claim has ever been substantiated through demonstration of an actual 1-day slippage in some aboriginal cycle with reference to European dates.

The "freezing" of dating correlations was a calculational and mnemonic convenience to the Indians for coping with the European leap year, but although it produced some confusion, it was not allowed to alter the day count, perhaps not even among the Tzotzil, whose day count it destroyed. There was no native leap year in Middle America at any time.

LENCA
(= *Jicaque?*)

The units of the Lenca calendar were the day or sun (*caxi*), moon (*lets'a*), and year (*púlan*). From the vocabularies compiled by Lehmann (1920), the following day names could have existed:

a. Caxi (sun)	f. Xila (dead)	k. Yaru (monkey)	p. Cus (buzzard)
b. Soor (wind)	g. Auinge (deer)	l. Ne (teeth)	q. Lumxi (quake)
c. Tau (night, house)	h. Mon (rabbit)	m. Mixi (cane)	r. Ixalal (flint)
d. Merts'oíkan (iguana)	i. Uax (water)	n. Lepa (jaguar)	s. Cuy (rain)
e. Sala (serpent)	j. Xuy (dog)	o. Sira (bird)	t. Zhuna (flower)

That such day names did exist in Lenca is strongly suggested by a number of the place names of Honduras. In the following compilation, the day names are followed by the numbers from 1 to 13 and coupled with place names that appear to correspond. References are to page numbers in L (Lehmann 1920:1) or M (Membreño 1901).

1. Caxi ita	(Sun 1)	—	(L674)
Uax ita	(Water 1)	Guarita	—
2. Sala pa	(Serpent 2)	Zazalapa	(M116)
3. Mon lagua	(Rabbit 3)	Monleguo	(L721)
4. Uax eria	(Water 4)	Guajirí	(M35)
Lepa eria	(Jaguar 4)	Lepaera	—
5. Lepa say	(Jaguar 5)	Lepasale	(M57)
6. Uax hui	(Water 6)	Gualjuí	(M37)
7. Soor huisca	(Wind 7)	Sorosca	(M94)
8. Mon tepca	(Rabbit 8)	Monteca	(M67)
9. Cuy calapa	(Rain 9)	Cuyuculapa	(M24)
10. isis	—	—	—
11. isis-l-ita	—	—	—
12. isis-la-pa	—	—	—
13. Xila isis-lagua	(Death 13)	Salisgualagua	(M92)

Monleguo is the old name for Pueblo Viejo.

Although Membreño interprets a number of these place names as Nahuatl, I believe the case for a Lenca calendar round looks very solid. This would strengthen the case for comparable calendars in Xinca and Jicaque as well. A particularly telling point is Lehmann's listing of Caxi ita as "1 Sonne" without, apparently, interpreting it as a day name, which it almost certainly is. The preponderance of Type IV days in place names may give a hint about the Lenca year bearers, but the similarity of its day names leads me to place Lenca on the Jicaque calendar.

LOGUECHE
(= Zapotec?)

The Logueche or Miahuatlan Zapotec are calendrically unknown. Presumptively they used the Zapotec calendar.

LONG COUNT

Probably in *355 B.C. (q.v.), the Olmec instituted what has come to be called the Long Count, a calendar based upon the 360-day tun rather than the 365-day year. What made it "long" was that they customarily counted and recorded periods of 20 tuns (katuns) and 400 tuns (baktuns). (For some astronomical purposes, they calculated in even larger vigesimal multiples of tuns.) The documented use of the Long Count by the Olmec was from the first century B.C. to the thirteenth century

A.D. The Yucatecan and Cholan Maya also used it from the third century A.D. until the seventeenth.

In addition to counting katuns and baktuns by 20s, they also counted them by 13. Thirteen katuns (260 tuns) made up the may, or katun cycle. Thirteen baktuns made up the baktun cycle, usually called the Mayan era. Olmec and Mayan dates in the Long Count specify the baktun (from 0 to 12), the katun (from 0 to 19), the tun (from 0 to 19), the uinal (from 0 to 17), and the day (from 0 to 19). The Olmec did the job sparsely with a string of numbers, such as 10.19.18.14.5 (see A.D. 1223), often adding the day-count date. Such an addition is partially redundant because the last number of a Long Count date already tells the name of the day. In the date cited, the 5 identifies the day as Serpent, or Chicchan. Most Mayan Long Count dates add a full calendar round date, giving the date in the day count and the year count, for example, 1 Eb 0 Yaxkin (see A.D. 320). Both Olmec and Maya Long Count dates are commonly introduced with an Initial Series glyph to indicate that such a date follows (see A.D. 162).

Among the Maya, dating by tuns came to have greater importance than dating by years, and during the Classic it was customary to erect monuments primarily focussed upon period endings: the *hotun* (5 tun), *lahuntun* (10 tun), *holahuntun* (15 tun) and katun (20 tun) intervals were particularly emphasized. All such dates end in zero, which is to say that all tuns end on the day Ahau (Lord). The Maya were thoroughly aware of that and were familiar with the sequence of coefficients borne by that day in the various cycles:

Tuns: 13, 9, 5, 1, 10, 6, 2, 11, 7, 3, 12, 8, 4 Ahau
Katuns: 6, 4, 2, 13, 11, 9, 7, 5, 3, 1, 12, 10, 8 Ahau
Baktuns: 4, 3, 2, 1, 13, 12, 11, 10, 9, 8, 7, 6, 5 Ahau

The starting point of the Maya (-Olmec) era was the ending of the last baktun of the preceding cycle, 13.0.0.0.0, which became 0.0.0.0.0 4 Ahau 8 Cumku. Fixing this date in European time is the essential problem of the Mayan correlation.

The latest Long Count dates that can be tied to the archaeology of the Classic period come to an end at 10.5.0.0.0 (see chap. 2) with 10 Ahau. After that, the katun count ("short count") was kept, and it is well established that the founding of Merida in 1542 took place in a katun 11 Ahau. The shortest interval between the two katun endings would place the founding of Merida in the Long Count at 10.10.0.0.0. The subsequent occurrences of 11 Ahau are at 11.3.0.0.0, 11.16.0.0.0, 12.9.0.0.0, and 13.1.0.0.0, and every 260 tuns thereafter.

Neither history nor archaeology nor astronomy has been able to choose decisively among these alternatives, though professional opinion tends increasingly to support the 11.16.0.0.0 date (see A.D. 1539). Although there are persuasive astronomical arguments for some of them, I believe that other correlations based upon intervals other than *exactly* 260 tuns from the 584,283 correlation can be dismissed as ethnohistorically untenable. It is curious that no such alternative has been proposed: there is no 490,683 or 677,883 correlation (see Correlation).

The invention of the Long Count by the Olmec was simplicity itself. They already had a calendar-round system, terminally dated, and ending every fourth year on the day Lord. On some year ending on, say 1 Lord, they simply had to decide to keep count of tun endings as well as years. Hence all tuns end on Lord. Simple extrapolation in vigesimal counting would produce the baktun. Almost certainly, however, the motive for pursuing such a count was astrological and astronomical, and the real achievement of the Long Count was the production of complex, sophisticated, and astonishingly accurate astronomical records. The monumental recording of dynastic history was in a sense a by-product of this achievement, but the two things go hand in hand because dynasties owed their mandates to the control of time.

LOXICHA
(= Zapotec? + 9*)

In the southern Zapotec area identified with the separate language of Loxicha or Pochutla, Carrasco (1951) reported the discovery of a unique day count based on 9 named days and the counting of 20 trecenas to form the 260-day cycle. Further information is given by Weitlaner (1956), establishing that the Loxicha calendar was in use at Candelaria (Carrasco), Magdalena and Santa Lucia (Weitlaner), and San Bartolo and San Agustín Loxichá (both authors). In all cases, it is a ritual calendar for the timing of ceremonies and is part of the esoteric knowledge of professional diviners. Although Santa Lucia is the most conservative of these villages, San Agustín is the best reported.

The nine day names are the names of gods. They are reported from San Agustín (Weitlaner 1956) and (incompletely and out of order) from San Bartolo (Carrasco 1951). Caso (1965b:945) provides the translations:

	San Agustín	San Bartolo	Translation
1.	Mdi	Wndi	Lightning
2.	Ndozin	Ndozin	Death Messenger
3.	Ndo'yet	Mudugyet	Death
4.	Beydo	—	Wind
5.	Dubdo	—	Maize
6.	Kedo	Wnkido	Judge
7.	Ndan	Wndan	Creator
8.	Mxe	—	Evil
9.	Mbaz	Ombwadz	Earth
13.	(Widzin)	(Wlizin)	

The term *Widzin* (*Wlizin*) is a qualifier added to the name of the thirteenth day of each trecena.

The day count is divided into five named periods of 52 days each; each one is called a day or time (*ze*):

	San Agustín	San Bartolo
1.	Ze Gon	Ze
2.	Ze Blazgač	Ze Yačil
3.	Ze Yate Tan	Ze Yatiu Dan
4.	Ze We	Ze Wi
5.	Ze Blagay	Ze Yalu

Each of the *ze* names applies both to the 52-year period and to the first of its four trecenas. The other three trecenas bear names beginning with Sgab (škau, škab, šab) "rite of . . . ":

2. Sgab Lodios (odlios; '?of god')
3. Sgab Gabil (gobil, dola; '?')
4. Sgab Lyu ('earth')

Informants deny that the numbers combine with the nine god names to name the days, and the day names are not used for naming children nor for prognostication (except that 7 Mdan is considered a bad day for the birth of a child). Informants also deny that this count has anything to do with the Christian year or an agricultural or 18-month calendar.

In order to make the 9 days come out even with the 260-day count, the first day of Ze Gon "uses up" two gods, and is counted as Mdi-Ndozin. Thus the structure of the whole calendar is: $8 + (9 \times 28) = 5 \times (4 \times 13) = 260$. The various counts are apparently made separately but they may be combined here for analytical and heuristic purposes. The count runs:

```
 1 Ze Gon  1 Mdi-Ndozin
 2 Ze Gon  2 Ndo'yet
 3 Ze Gon  3 Beydo, and so on, to:
 8 Ze Gon  8 Mbaz
 9 Ze Gon  9 Mdi
10 Ze Gon 10 Ndozin
11 Ze Gon 11 Ndo'yet
12 Ze Gon 12 Beydo
13 Ze Gon 13 Dubdo Widzin
(14 Ze Gon) 1 Sgab Lodios 1 Kedo
(15 Ze Gon) 2 Sgab Lodios 2 Mdan, and so forth
```

There is no reported mechanism for coordinating the count among diviners, and different villages keep different counts. Thus Carrasco (1951) reports that 12 Sgab Gabil fell on 10 IX 1949 Gregorian in San Agustín, whereas 10 Sgab Lodios fell on

12 IX 1949 (2 days later) in Candelaria. We do not know whether the first date, for example, is the 38th, 90th, 142nd, 194th, or 246th day in the Loxicha day count, so we cannot correlate it with the Christian year or with the universal day count. If we choose the latest of these possibilities, we get the following solutions:

1949 10 IX G 12.16.15.14.7 5 Manik 5 Mol = 38 Ze Blagay 12 Sgab Gabil 12 Beydo San Agustín Loxichá

1949 12 IX G 12.16.15.14.9 7 Muluc 7 Mol = 23 Ze Blagay 10 Sgab Lodios 10 Mdi Candelaria Loxichá

The Candelaria correlation links its 231st day to the 189th day of the universal day count. The San Agustín correlation links its 246th day to the 187th day of the universal day count. Thus the San Agustín day count begins on the 203rd day of the universal one, 8 House, whereas the Candelaria day count begins 15 days later, on 10 Flint. But this phrasing of the relationship may be off by some multiple of 52 days in either case or in both.

MAM
(II.L.1 CR 1 Noj 1532; XX c. A.D. = Quiche)

The Mam year begins on 4 Kayab Mayapan. It counts the days of the day count from 1 to 13 and those of the month from 1 to 20. Its senior year bearer is Quake and its calendar round and other features are congruent with those of the Quiche. Its directional colors are unknown.

The units of the calendar are the sun or day (*'ij*), month of 20 days (*wen en 'ij*), moon or month (*xau*), calendar round (?), tun (*guaxakláj xau*), year (*'ab 'ij, haab*), and calendar round (?). The New Year ceremony is called Xoj K'au.

The Mam day names are reported by Oakes from Todos Santos Cuchumatanes (1951:142, 248–52):

a. Imix (alligator)	f. Kimex (death)	k. Batz (monkey)	p. Ajmak (owl)
b. Ik (wind)	g. T'ce (deer)	l. E (tooth)	q. Noj (incense)
c. Akbal (night)	h. K'nel (rabbit)	m. Aj (cane)	r. Tcij (flint)
d. Kets (net)	i. T'coj (water)	n. Ix (jaguar)	s. Tciok (storm)
e. Kan (serpent)	j. T'ciik (dog)	o. Tsikin (bird)	t. Hunajpu (hunter)

The month names are not reported (see A.D. 1946).

MANGUE
(?)

The units of the Mangue calendar may have been the day or sun (*mbu*), month or moon (*yu*), and year (?). From vocabularies in Lehmann (1920), the following are possible day names:

a. —	f. Gagame (dead)	k. Ambi (monkey)	p. —
b. Tiú (wind)	g. Umbongame (deer)	l. Iji (teeth)	q. —
c. Angu (night)	h. Uku (rabbit)	m. Ure (cane)	r. Ugo (flint)
d. Ukú (iguana)	i. Imbu (water)	n. Umbú (jaguar)	s. —
e. Ule (serpent)	j. Ambe (dog)	o. Moonkoyo (eagle)	t. Ele (flower)

From a collection of place names in Lehmann (1920) it is possible to find examples of what appear to be day count place names with coefficients from 1 to 10 using half of the putative day names, although the numerals had to be reconstructed by comparing Mangue and Chiapanec. Lehmann (1920 2:802) also cites Oviedo (1535 3:111; 4:96, 98) as documenting two chiefs with the names Jaguar (N-ambu-e) and Dog (N-ambi). It seems certain that the Mangue had the day count and the calendar round, but it is not possible to place their new year.

Mangue Place Names

Number	Day	Day Count	Place Name
1. tique	iji	(1 Tooth)	Ma-*tiqui-icí*
2. o	umbu	(2 Jaguar)	Mom-*o*-t-*ombo*
	imbu	(2 Water)	M-*o*-n-*imbo*
3. ui	tiu	(3 Wind)	Ma-*bi-tra*
			Ma-*bi-ti*
			Ma-*vi-tia*
	ure	(3 Cane)	Nam-*oy-ure*
			Nam-*uy-ure*
4. aha	umbu	(4 Jaguar)	N-*aga*-r-*ando*
5. ao	ure	(5 Cane)	Y-*u*-l-*ure*
6. amba	ele	(6 Flower)	M-*ombo-ri*-ma
			N-*amba-ri*-na
7. endi	ele	(7 Flower)	L-*indi-ri*
			N-*endi-ri*
8. aho	umbu	(8 Jaguar)	M-*aho*-me-t-*ombo*
9. eli	umbu	(9 Jaguar)	D-*iri-omo*
			B-*eri-ombo*
	angu	(9 Night)	D-*iri-ange*-n

Number	Day	Day Count	Place Name
			D-*iri-aje*-n
	ambe	(9 Dog)	D-*iri-amba*
	ambi	(9 Monkey)	D-*iri-ambe*
10. enda	ele	(10 Flower)	L-*ende-ri*
			N-*ende-ri*
	ure	(10 Cane)	N-*anda-y-ure*

MANI

The *Book of Chilam Balam of Mani* generally employs the Mayapan and Valladolid calendars and, in fact, always does so in the segments of the text that parallel passages of the *Chumayel* and *Tizimin*. It contains, however, three complete counts of the year from January to December, giving a correlation of an aberrant Mayan almanac to European dates frozen to a date of July 16 for 1 Pop, and using Type I year bearers (Craine and Reindorp 1979:20–38, 39–49, 144–54). All three almanacs relate to the end of a year 10 Imix and the beginning of a year 11 Cimi, and all three are primarily concerned with the divinatory and ritual significance of the day count, and particularly of the burner cycle.

Although the manuscript carries a note dating the copying of the second almanac to May 12, 1755, the fact that the correlations are frozen Julian dates of 1548 makes the dating of composition impossible for any of them. It is tempting to suppose that the original compositon belongs to a year in which 10 Oc actually fell on January 1 (and 11 Cimi on July 16), preferably in the first half of the eighteenth century. There was no such year. The last occurrence of 10 Oc on January 1 Gregorian before 1755 was in A.D. 754. The corresponding Julian date would be A.D. 1500. Such dates do not repeat for 1,040 years. The type I Mani calendar is thus a chimera.

MATLATZINCA
(IV.I.20 CR?; XVI c. A.D. = Tepanec)

The Matlatzinca (Pirinda) calendar begins the year on 6 Kankin Mayapan, counting the days of the day count from 1 to 13 and those of the months of the year from 1 to 20, and naming the year terminally. Its directional colors and the beginning of its calendar round are not known. The *Calendario de 1553* gives the Matlatzinca day names, which have been published by Caso (1946:4) and are quoted here as corrected by Barlow (1951:70), who had access to the Peabody Museum photostat of the original manuscript in the Bibliothèque Nationale in Paris, as Caso did not:

a. Beori (alligator) f. Rini (death) k. Tzonyabi (monkey) p. Yabin (buzzard)
b. Ithaatin (wind) g. Pari (deer) l. Tzinbi (grass) q. Thaniri (quake)
c. Bani (house) h. Chon (rabbit) m. Thihui (cane) r. Odon (flint)
d. Xichari (iguana) i. Thahui (water) n. Xotzini (jaguar) s. Yalbi (rain)
e. Chini (serpent) j. Tzini (dog) o. Ichini (eagle) t. Ettuni (flower)

All names are preceded by *In*. Weitlaner (1939) gives a vocabulary of Ocuiltec in which the following potential day names appear:

a. — f. Tei (death) k. — p. —
b. — g. Paári (deer) l. Sibí (grass) q. —
c. Baani (house) h. Cuá (rabbit) m. Tathui (cane) r. —
d. — i. Taui (water) n. — s. Cumaabi (rain)
e. — j. Tosini (dog) o. Tochini (eagle) t. Tâni (flower)

The eighth day suggests Mixe, but most of the rest are clearly Matlatzinca.

The *Calendario de 1553* also provides the names of most of the Matlatzinca months, given here as published by Barlow (1951:71):

I. Thaçari (1 ?) R. Thaxiqui (2 moss)
J. Dehuni (toasting corn) A. Thechaqui (crane)
K. Theçamoni (?) B. Thechotahui (twins)
L. Thirimehui (1 porridge) C. Teyabihitzin (falling)
M. Thamehui (2 porridge) D. Thaxitohui (grandfather)
N. Iscatholohui (1 dead) E. —
O. Mathitohui (2 dead) F. —
P. (Ytz) Bacha (broom) G. —
Q. Thoxiqui (1 moss) H. —
X. Tasyabiri (?)

All names except P. are preceded by *In*.

The correlational position of the Matlatzinca calendar was worked out by Caso (1946) in a brilliant analysis (see A.D. 1553).

MAYAN

The calendars of all but two of the Mayan languages are at least partially documented. The exceptions are Tojolabal and Chicomuceltec. The rest share an overlapping pattern of cognate day and month names (the latter are considerably more diverse than the former). They differ among themselves about which day begins the month and which month begins the year, and about whether the first day of the month is 0 or 1. All but three of them (Campeche, Palenque, and Huastec) agree in naming the year for its first day.

Four of them (Campeche, Palenque, Tikal, and Mayapan) are distinguished from all other Middle American calendars except Olmec by their use of the 360-day tun. The Iximche calendar of the Cakchiquel makes a similar use of the 400-day huná.

Ten of the Mayan calendars use Type II year bearers (Aguacatec, Mam, Cakchiquel, Pokomam, Pokomchi, Ixil, Quiche, Chol, Kekchi, Tikal). Tlapanec and Chiapanec are the only non-Mayan groups that began the year on these days, which were also used by the calendar of Teotihuacan.

Ten of the Mayan calendars are Type III (Tzotzil, Tzeltal, Guitiupa, Istacostoc, Mitontic, Cancuc, Chuh, Jacaltec, Kanhobal, Palenque).

Three are Type IV (Mayapan, Valladolid, Campeche). The remaining Mayan calendars are only fragmentarily documented (Chontal, Chorti, Tzutuhil, Uspantec).

MAYAPAN
(IV.N.1 CR 1 Kan 1529; XVI c. to XIX c. A.D. = Campeche + i, f1548)

The Mayapan calendar began its year on 1 Pop on July 16, 1548 Julian. It counted the days of the day count from 1 to 13 and those of the month from 1 to 20, and named the year for its first day. Its calendar round began on 1 Pop, July 24, 1529 Julian, and its directional colors were east–red; north–white; west–black; south–yellow; and the center–blue/green. Its day and month names and glyphs were those of the Tikal calendar (q.v.), although the colonial *Paris* and *Madrid* codices do not give the month glyphs.

The units of the calendar are the day or sun (*kin*), moon or month of the year (*u, ik*), month of the tun (*uinal*), day count (*tzol kin*), 360-day stone (*tun*), year (*haab*), 20-tun pile of stones (*katun*), calendar round (*hunab*), 13-katun cycle (*may*), 20 katun bundle of stones (*baktun*), and the Spanish day (*día*), week (*semaná*), month (*mes*), and year (*año*). The novena and trecena were recognized and even deified, but do not appear to have been given unit names except as "the nine gods and the thirteen gods."

In addition to beginning its year 2 days later, the Mayapan calendar differs from that of Tikal in abandoning zero counting, and in counting the katun and the may (but not the uinal, tun, or baktun) initially rather than terminally. It appears to have been inaugurated on 11 Ix 1 Pop 11.15.19.12.14, or 11 Ix 2 Pop Tikal, and seated a new 13 katun cycle (may) on 11 Ahau 7 Uo Mayapan. It was replaced by the Valladolid calendar in 1752 on 3 Cauac 1 Pop Mayapan 12.6.15.11.19.

The Mayapan calendar continued to use Long Count dating, but so sparingly that it provides no precise and explicit correlation of it with Colonial Spanish dates. Dating of tun endings is clear (see 1544) and there are perhaps two direct references to the baktun (see 1480, 1618), but the primary link between Colonial and Long Count dating is the count of the may. This count is complicated by the fact that the Tikal calendar counted katuns terminally, whereas the Mayapan calendar counted them initially. Thus the baktun-ending ceremony for 5 Ahau (Tikal) was actually held 80 days after the beginning of 3 Ahau (Mayapan) in 1618 (q.v.).

The proof that the katun 11 Ahau of the conquest of Yucatan was 11.16.0.0.0 and not some other occurrence of that day in the Long Count cannot be provided by decisive links between Colonial Yucatecan traditions and archaeologically documentable events. The gap of 600 years between the last Classic monuments and the first Colonial manuscripts is simply too great to be filled by genealogy (Yucatan's is notably poor) or oral history (Yucatan's is fragmentary and metaphorical). The proof is rather the known placement of the other units of the Mayapan calendar (see 1539, 1544, 1553, 1596). The argument (Kelley 1983) that there was a discontinuity in the day count between the Tikal and Mayapan calendars is falsified by every date in this volume.

There are no native correlations with the Mayapan calendar. Its isolation is not so surprising in view of the fact that it was inaugurated almost exactly at the moment of the Spanish conquest of Yucatan. It was clearly the direct successor to the Tikal calendar and the immediate predecessor to that of Valladolid (q.v.). Its relation to the Campeche calendar (q.v.) is in doubt because of the uncertainty over how the latter counted the year, but scientific economy makes it extremely likely that it was, in fact, derived from the Campeche calendar rather than directly from that of Tikal, and it is so coded in the heading (see A.D. 1537, 1539, 1541, 1544, 1553, 1559, 1593, 1596).

MAZAHUA
(= Otomi?)

The units of the Mazahua calendar were the day (*paa*), month or moon (*zana*), and year (*kjë'ë, tsjë'ë*). It almost certainly had the day count and calendar round as well. From Kiemele (1975) some of its possible day names are:

a. Hyas'ü (light)	f. U'u (death)	k. —	p. —
b. Ndajma (wind)	g. Pjant'e (deer)	l. Pjiny'o (grass)	q. Mbi'i (quake)
c. Ngümü (house)	h. Kjwa'a (rabbit)	m. —	r. Ndosiwi (flint)
d. Korga (iguana)	i. Ndehe (water)	n. —	s. Y'ebe (rain)
e. Xixkalaha (serpent)	j. Y'o (dog)	o. Chabëy'i (eagle)	t. Ndähnä (flower)

My equation of the Mazahua calendar with Otomi is a guess.

MAZATEC
(IV.E.20 CR?; XX c. A.D. = Aztec, f1616)

The Mazatec calendar begins its year on 6 Yax Mayapan, counting the days of the day count from 1 to 13 and those of the month from 1 to 20 and naming the year terminally. Its directional colors and the beginning of its calendar round are unknown, as are its day names. The units of the calendar are the day (*nǐchjín*), month or moon (*sà*), trecena (*téjan*), day count (?), year (*guno*), and calendar round (?).

Weitlaner and Weitlaner (1946:195) published lists of month names: one from Pozo de Aguila (incomplete), one from San José Independencia, and four from Huautla. The most accurate linguistic recording is one of the latter by Hansen and Pike:

E. Mé (want)	J. Hnó (owl)	O. N'é (yellow)	B. Kį (wood)
F. Nt'aò (wind)	K. Tò (fruit)	P. Khoa (greens)	C. Khá (broken)
G. Xki (count)	L. Méhe (fat)	Q. Nhtò (worm)	D. Ntà (cloud)
H. Khoi	M. Mahti (anger)	R. Nčhe (huaje)	
I. Hį (blood)	N. Ntà	A. Sa (grow)	X. Kìkì Ntàąǫ

Month N. takes the prefix *Sì* and D. the prefix *Kì*; all the others (except X.) take the prefix *Čą̌*. There is some disagreement about the order of P., Q., and R. Four of the six sources agree in making *Nčhe* the thirteenth month, but they are evenly split about which of the remaining 2 months precedes or follows it.

The calendars were collected in 1936, and there was general agreement on beginning the year on 1 I Gregorian. The best guess is that this is a frozen date of *1616, chosen because of the coincidence with the Christian year, making the Mazatec calendar congruent with Aztec.

Schulz (1955:240), doggedly pursuing the conviction that the Middle American calendars used leap years, quotes George M. Cowan as explaining the Oaxaca Mazatec leap year this way:

na[h] kintà	the 20th day of the month Kinta
kintà kah	the intercalated day
jnko kintà'ao[h]	the 1st day of the 5-day month

Since the Mazatec day count appears to have lapsed, there is no way of proving this, and it is thoroughly unlikely.

From Wasson *et al.* (1974:passim), Jamieson (1978), and Lehmann (1920), the following Mazatec day names are possible:

a. —   f. Coviu (dead)   k. Kúni (monkey)   p. Kye (buzzard)
b. Tjaò (wind)   g. Nàsinkihạ   l. Iyu (teeth)   q. Hbàninankí
                          (deer)   (quake)
c. Hyu (night)   h. —   m. Ntèh (cane)   r. Ntòh (stone)
d. Çinkrì (iguana)   i. Tá (water)   n. Xra (jaguar)   s. Tsin (rain)
e. Yè (serpent)   j. Nya (dog)   o. Jàtse (eagle)   t. Naxó (flower)

See A.D. *1616.

MAZATLAN

The Mixe of Mazatlan continue to employ the traditional Mixe day count in a unique trecena calendar, applying the day names to 13 days at a time. A number of different cycles are counted; all of them are called day counts (*xii maay'g*). They may be coded as follows:

T1.	13 days	—	—
T13.	169 days	—	—
T20.	260 days	1 day count	D1.
T28.	364 days	—	—
T40.	520 days	—	—
T130.	1690 days	6 day counts	D6.
T260.	3380 days	13 day counts	D13.

The largest of these cycles (D13) is 3,380 days, or 9.7.0 in calendrical notation. It is held to begin on 1 Flower. Once in each cycle, its trecena of 3 Wind falls on the correct date in the general day count, as follows:

1885 9 XII G	12.13.11.2.2	3 Ik 0 Ceh
1895 12 III G	12.14.0.9.2	3 Ik 15 Pax
1904 13 VI G	12.14.9.16.2	3 Ik 5 Zip
1913 14 IX G	12.14.19.5.2	3 Ik 0 Mol
1922 16 XII G	12.15.8.12.2	3 Ik 15 Ceh
1932 18 III G	12.15.18.1.2	3 Ik 10 Kayab
1941 19 VI G	12.16.7.8.2	3 Ik 0 Zotz'
1950 20 IX G	12.16.16.15.2	3 Ik 15 Mol
1959 22 XII G	12.17.6.4.2	3 Ik 10 Mac
1969 24 III G	12.17.15.11.2	3 Ik 5 Cumku
1978 25 VI G	12.18.5.0.2	3 Ik 15 Zotz'

The Mazatlan calendar also maintains a 13-year cycle (Y13), composing a kind of truncated quarter of a calendar round, in which the years are named as follows:

1970	1 Tahp (Flint)
1971	2 How (House)
1972	3 Naan (Rabbit)
1973	4 Kep̃ (Cane)
1974	5 Tahp (Flint)
1975	6 How (House)
1976	7 Naan (Rabbit)
1977	8 Kep̃ (Cane)
1978	9 Tahp (Flint)
1979	10 How (House)
1980	11 Naan (Rabbit)
1981	12 Kep̃ (Cane)
1982	13 Xap (Wind)

The naming of the thirteenth year precludes the continuation of the day count: it simply starts over with 1 Tahp. However, the naming of the years does imply that for 12 years at a time the days are counted normally, day by day, thus potentially corresponding to the count of the D13 cycle only once per cycle. Such segmental counts are, in any case, interrupted and discontinuous, and it is not clear how the thirteenth year was counted, because 13 Wind is 104 days later than the expected 13 Flint. Any given relationship between this count and the general day count would be good for only 12 years.

   In Mazatlan, then, any particular day may have several different day names, perhaps four at the very least: (1) the true universal day name; (2) the day name in the 13-year (or Y13) count just described; (3) the name of the trecena in the D13 cycle in which the day happens to fall; or (4) a derived D13 day name reached by counting forward from the last beginning of a trecena to the day in question. This plurality of cycles explains some of the aberrant day-name correlations that have been published on the Mixe calendar:

1.	1923 1 I G	4 Tsa'añ (Serpent)	(*Cuadernillo de Camotlan*, Lipp, personal communication)
2.	1941 24 V G	1 Hugɨ'ñ (Flower)	(Carrasco *et al.* 1961)
3.	1946 1 VI G	2 Hukpi (Alligator)	(Carrasco *et al.* 1961)
4.	1951 24 V G	1 Hugɨ'ñ (Flower)	(Carrasco *et al.* 1961)
5.	1954 29 V G	10 Kaa (Jaguar)	(Carrasco *et al.* 1961)
6.	1978 7 X G	1 Ho'o (Dog)	(Lipp 1982:205)

None of these dates matches the universal day count. They are off by 54, 24, 11, 12, 25, and 25 days, respectively. Dates 1, 2, 3, 4, and 6 mark the 4th, 1st, 2nd, 1st, and 131st trecenas of the D13 cycle. Date 5 is counted ahead day by day from the D13 date of the 101st trecena, 13 days earlier. Thus, all of them are in the Mazatlan calendar.

The date of the Y13 New Year in the Mazatlan calendar is frozen to 15 X Gregorian, rather than to 1 XI Julian, a 16-day discrepancy. This suggests an independent derivation of the Y13 cycle from an unbroken Mixe day count direct to the Gregorian calendar in the present century (between 1900 and 1923). It corrected for the 13 missed Julian leap years from 1532 to 1582 and those of 1700, 1800, and 1900, aligning Mazatlan with other calendars frozen to 1584, and further confirming the original placement of the Mixe New Year (see Mixe). It is a great tribute to the modern calendrical expertise of Mazatlan (see A.D. 1978).

METZTITLAN
(IV.B.20 CR?; XVI c. A.D. = Tilantongo – 1m)

The Metztitlan calendar begins its year on 6 Yaxkin Mayapan. Its writing system, counting of the day count and the month, and naming of the years are Nahuatl. Its calendar round and directional colors are unknown. The *Relación de Metztitlán* (Chávez 1923) lists its somewhat divergent day names:

a. Tetechi Hucauh	f. Tzontecomatl (skull)	k. Ozoma (monkey)	p. Teotl Itonal (god day)
b. Ecatl (wind)	g. Macatl (deer)	l. Itlan (tooth)	q. Nahui Ollin (4 quake)
c. Calli (house)	h. Tochtli (rabbit)	m. Acatl (cane)	r. Tecpatl (flint)
d. Xilotl (corn bud)	i. Atl (water)	n. Ocelotl (jaguar)	s. Quisahutl (rain)
e. Coatl (serpent)	j. Izcuin (dog)	o. Cuixtli (cobweb)	t. Omexochitonal (2 flower day)

It is worthy of remark that the names of the year bearers are in the standard Nahuatl nomenclature, whereas most of the rest are not.

The *Relación* also gives the names of the months:

B. Panquetzaliztli (flag raising)
C. Atemoliztli (falling water)
D. Tititl (storm)
E. Xochitoca (flower sowing)
F. Xilomaniztli (corn feast)
G. Tzahio (dog flaying)
H. (Quechuli)
 I. −Huei Tozoztli (2 vigil)
 J. Popochtli (youths)

K. Etzalcualiztli (eating bean soup)
L. Tzinco Hū (1 lords)
M. Huey Tecuylhuitl (2 lords feast)
N. Micca Ylhuitl (1 dead feast)
O. Huey Micca Ylhuitl (2 dead feast)
P. Huechpaniliztli (road sweeping)
Q. Pachtli (1 moss)
R. Huey Pachtli (2 moss)
A. Quechuli (macaw)
X. Nennontemi (full in vain)

The insertion of Quechuli for H. is clearly an error, but the Otomi substitutions are correct for G. Tlacaxipehualiztli and L. Tecuilhuitontli.

The Metztitlan calendar is the northernmost in Nahuatl and shows clear signs of having been influenced by Otomi (q.v.). It is undocumented except for the *Relación* (see 1555).

MITONTIC
(III.I.1 CR?; XIX c. A.D. = Istacostoc + 1m)

The Colonial Tzotzil calendar of Mitontic is derived from that of Istacostoc by a 1-month advance in the New Year's date. It is discussed in relation to the other calendars of Tzotzil (q.v.).

MIXE
(III.A.0 CR 1 Flint 1503;
XVI c. to XX c. A.D. = Olmec + 3d − 1m − 360 + 13*, f1532)*

The Mixe calendar begins its year on 5 Xul Mayapan, counting the days of the day count from 1 to 13 and those of the year count from 0 to 19, and naming the years initially. Its senior year bearer is 1 Flint (1503), and its directional colors are unknown, though its directional winds are south–fire/hot; southeast–gray/purple; north–green; northwest–dry/cold (Lipp 1982:106).

The names of the days are given in the (unpublished) *Cuadernillo de San Lucas Camotlán* of 1927 and have been published by Carrasco *et al.* (1961:163–66), Weitlaner and Johnson 1963:46, 58–59), and Lipp (1982:203–5):

a. Hukpii (root)	f. 'Uh (world)	k. Haiim̃ (ashes)	p. Pa'a (edge)
b. Xa'a, Xap (wind)	g. Koy (rabbit)	l. Tɨɨts (tooth)	q. 'Uhx (quake)
c. How (palm)	h. Naan (deer)	m. Kep̃ (cane)	r. Tahp (soot)
d. Huu'n (hard)	i. Nɨɨn (river)	n. Kaa (jaguar)	s. Mɨy (grass)
e. Tsa'añ (serpent)	j. Ho'o (vine)	o. Huuik̃ (tobacco)	t. Hugɨ'ñ (eye)

Notable features of the day names are the inversion of Deer and Rabbit and the semantic uniqueness of a, c, d, f, j, k, o, p, r, s, and t. The Mixe day names appear to be both archaic and idiosyncratic.

The units of the Mixe calendar (Quintana 1729; Lipp 1983) are the day, sun or trecena (*xɨɨ*), month or moon (*po'o*), day count (*xɨɨ tu'u, xɨɨ maay'g*), year (*ipx po'o* 'twenty months'), and probably the calendar round (?).

Like Mixtec, Mixe has a special set of number names used only for calendrical counting. The secular and sacred numbers are:

	Secular	Sacred
1	tuuk	tuum
2	meck	mac
3	toohk	tuuk
4	maktašk	makc
5	mugoošk	mokš
6	tuhtɨɨk	tuHt
7	weštuuk	kuy
8	tuktuuk	tuugut
9	taštuuk	taaš
10	mahk	maHk
11	mahktu'k	kɨ'ɨn
12	mahkmeck	kɨ'ɨš
13	mahktuuhk	pɨgač

The Mixe month names have been published by Weitlaner and Johnson (1963:50–51). The following list is from Lipp (1982:187–89):

E. Mɨh Kahpu'ut (1-town)
F. Haak Kahpu'ut (2-town)
G. Mɨh Xɨwɨ'ɨ (1-oak grove)
H. Haak Xɨwɨ'ɨ (2-oak grove)
I. Kuyduus (bat)
J. Mɨh Kaa (1-jaguar)
K. Ipts Tɨgɨ'ɨ (in house)
L. Taak Am (wetlands)
M. Nɨɨ Tsɨ'ɨ (water squash)

N. Mɨh Xo'ox (1-viper)
O. Haak Xo'ok (2-viper)
P. Hotsoon (liver)
Q. Aaxo'om (mouth)
R. 'Ap (grandfather)
A. Nɨxamɨ'ɨ (dog days)
B. Mɨh Oo (1-bath)
C. Haak Oo (2-bath)
D. Mɨh Tsaatso'k (1-stone)
X. Mu'ts Tsaatso'k (2-stone)

Eleven correlational dates are available from Cotzocon relating the Mixe day count to the European calendar. Page references are to Weitlaner and Johnson (1963) (WJ). Day-name equivalents are calendrical rather than translational.

1.	1888 27 VI J	5 Hugɨ'ñ (Flower)(200)	11 Death (206)(WJ61)
2.	1941 22 I J	8 Huu'n (Iguana)(204)	11 Deer (167) (WJ61)
3.	1935 4 IX J	5 Tahp (Flint)(18)	5 Flint (18) (WJ61)
4.	1957 23 II J	2 Huuik (Eagle)(15)	11 House (63) (WJ61)
5.	1958 16 VII J	12 Haiiɱ (Monkey)(51)	12 Monkey (51) (WJ59)
6.	1958 1 VIII J	2 Koy (Deer)(67)	2 Deer (67) (WJ59)
7.	1958 1 IX J	7 Tahp (Flint)(98)	7 Flint (98) (WJ59)
8.	1959 17 IV J	6 Huuik (Eagle)(175)	1 Death (66) (WJ59)
9.	1959 20 V J	9 Hugɨ'ñ (Flower)(100)	8 Rain (99) (WJ59)
10.	1959 1 VI J	8 Tɨɨts (Grass)(112)	7 Monkey (111) (WJ59)
11.	1959 1 VII J	13 How (House)(143)	11 Alligator (141)(WJ59)

The source of these correlations was an old notebook in the possession of Ladislao Reyes, a Cotzocon curer. The dates are in the Julian calendar. Dates 3, 5, 6, and 7

are correct as they stand. Dates 9 and 10 are 1 day off and date 11 is 2 days off.
Dates 1, 2, 4, and 8 contain unexplained errors of 6, 37, 48, and 109 days, respectively. Cited in connection with the births of individuals, dates 1, 2, and 4 may in
fact be personal names and not dates at all. It seems clear that the day count was
kept at Cotzocon until 1958, albeit not very accurately, and that European leap
years and calendar changes may have been responsible for some calendrical confusion.

   Additional day-count correlations in the Gregorian calendar are reported
from Guichicovi. Page references are to Weitlaner and Johnson (1963) (WJ) and
Caso (1963) (C).

12.	1961 31 X G	4 Miɨy (Rain)(199)	6 Alligator (201) (WJ47)
13.	1963 13 VIII G	5 Ho'o (Dog)(70)	7 Grass (72) (C71)
14.	1963 3 XI G	9 Ho'o (Dog)(152)	11 Jaguar (154) (WJ48)

The day count appears to be kept at Guichicovi as well, but it seems to have lost 2
days.

   A number of ostensible day-count correlations in the trecena calendar of
Mazatlan (q.v.) relate only very erratically to the general day count and may be
disregarded here.

   We also have a number of year correlations from Mazatlan, by Lipp (1982)
(L), Guichicovi, by Weitlaner and Johnson (1963) (WJ), and perhaps Zacatepec by
Miller (1952) (M).

1.	1957 1 XI	13 Rabbit	11 Rabbit	Zacatepec?	(M)
2.	1959 31 X	1 Rabbit	13 Flint	Guichicovi	(WJ47)
3.	1960 31 X	3 Jaguar	1 House	Guichicovi	(WJ47)
4.	1961 30 X	3 Flint	2 Rabbit	Guichicovi	(WJ47)
5.	1962 30 X	4 House	3 Cane	Guichicovi	(WJ48)
6.	1963 30 X	5 Rabbit	4 Flint	Guichicovi	(WJ48)
7.	1970 15 X	1 Flint	11 Cane	Mazatlan	(L254)
8.	1978 15 X	9 Flint	6 Cane	Mazatlan	(L187)

The 3 Jaguar date in 1960 is an obvious error for 2 Cane. Actually, the Zacatepec
and Guichicovi sources agree with Miller (1956:49) and with Weitlaner and Johnson
(1963:50) in giving a frozen date of November 1 for the New Year: the October
dates from Guichicovi are the dates of the name day of the New Year. Only the
Zacatepec date bears an acceptable relation to the Aztec (and hence to all other)
year names, and I believe it to be correct, confirming the placement of the Mixe
New Year at III.A.0 (see 1531).

   The Guichicovi dates in October are actually the dates taken to be the
name day of the year. When the Mixe calendar was frozen, the priests observed
that its first day fell back into the Uayeb days of the preceding year, 1 day at each

leap year, and that whichever year bearer fell in Uayeb would be the correct name of the following year. After the first 5 leap years, however, it would no longer bear the correct coefficient, so a separate count of year coefficients would have to be kept. Apparently it has been but only in Zacatepec. There is a slippage of 2 years in the year count of Guichicovi and of 1 year in that of Mazatlan. It may be that this system of keeping track of year bearers through the rotation of "name days" in Uayeb is responsible for the otherwise mysterious assertion by Herrera that the Mixe festival of the dead fell on November 2 plus or minus 2 days (see 1531). The aberrant New Year's date of 15 X is in the Mazatlan calendar (q.v.).

Like the Tzotzil, the Mixe froze their calendar to the European one rather than vice versa. It seems almost inevitable that this would destroy the day count, as it did in Tzotzil and seems to be doing in Mixe. Not a single date of all those available to us can be taken at face value as the true day-count date for a given Gregorian day because what is left of the Mixe day count is frozen to Julian time.

The choice of 1 Flint as the beginning of the new trecena calendar of Mazatlan (q.v.) may perhaps indicate the traditional seniority of that day and hence the beginning of the pre-Conquest Mixe calendar round (see A.D. 1531).

MIXTEC

There is general agreement that Mixtec is not, or at least not any longer, a single language, but rather a series of intergrading dialects, considerably differentiated at the extremes of its distribution. It is somewhat arbitrarily divided into three geographic ranges: A. Mixteca Alta; B. Mixteca Baja; and C. Mixteca Costeña. Most of our ethnographic and linguistic information comes from the first two (see, however, Josserand, 1983).

Historically, the Mixtec have used at least two distinctive calendars, Yucuñudahui and Tilantongo (q.v.), and two corresponding writing systems, and Mixtec is a strong candidate for having been the language of Teotihuacan as well.

The linguistic complexity of its history makes it impossible to cite clear canonical forms for the Mixtec day names or its numerals, and its month names remain linguistically undocumented. The problem is complicated by the fact that both day names and numerals had secular and sacred forms, and actual usage even in calendrical contexts appears to have confounded them in some degree, particularly after the Conquest.

This may be illustrated first with the numerals. The first four columns (left) in the next list give the secular forms: (1) in proto-Mixtec from Josserand (1983); (2) in Colonial Alta from Alvarado (1962, originally 1593); (3) in modern Baja from Schultze-Jena (1938); and (4) in modern Alta from Dyk and Stroudt (1965). The last three columns (right) give the apparently sacred usage from glosses in three codices: (5) *Sierra*, (6) *Nativitas*, and (7) *Tulane*. (Lists 2, 3, 5, and 6 are taken from Dahlgren, 1954:370; list 4 is from Moser, 1977:169; the readings from *Tulane* are my own.)

	Secular				Sacred		
	(1)	(2)	(3)	(4)	(5)	(6)	(7)
1		ee ec	i	ɨɨn	gau	ca	co go
2	uwi	vvui	owi	uuu	co	co a	ca gu
3	oni	uni	uni	unii	ga	co	ni xa
4	kɨwɨ'	qmi	kumi'	cuuun		q	
5	o'ǫ xe'ų	hoho	o'o	uhuun	q	q c	
6	iyǫ	iño	iño	iñuu	ñu	nu ñ	no, nooh
7	uxe	usa	uśa	usiaa	xa	sa	
8	one	una	una	unaa	na	na	na
9		ee	í	ɨɨɨn	que	q(ue)	
10	uxi	usi	uśü	uxii	xi	si	ci
11		usi ee	uśü í	uxi iin	xi	si	ci go
12	uxi uwi	usi vvui	uśü owi	uxi uu	ca	ca	cii ca cii gu
13	uxi oni	usi uni	uśü uni	uxi uni		si	ci xa

The problem of the day names may be similarly described. From left to right, the next list presents (1) proto-Mixtec forms from Josserand (1983); (2) those from Caso (1965b:956); (3) Alvarado's secular colonial vocabulary of 1593 from Smith (1973:24–25) and Moser (1977:167–68), with modern additions by Smith (in brackets); (4) the sacred vocabulary from Smith (1973:24–25) and Moser (1977:167–68); (5) Dahlgren's list (1954:369), taken from *Nativitas* and *Sierra* (the latter in brackets); and (6) my readings from *Tulane* with an addition from *Sánchez Solís* from Smith (1973:24) (in brackets).

	Secular and Mixed				Sacred		
	(1)	(2)	(3)	(4)	(5)	(6)	
a.		Quehui	Coo Yechi	Quevui	Q(ue)vi		Alligator
b.	Tati'	Chi	Tachi	Chi	Chi		Wind
c.	We'yi	Huahi Cuau	Huahi	Mau Cuau	Mav [Mao] [Cuav]	Hona	House
d.		Quu	(Ti) Yechi	Quu Q(ue)	Q		Iguana
e.	Koo'	Coo Yo Yucoco	Coo	Yo			Serpent
f.		Moku	Ndehi Ndeye Sihi	Mahu(a)	Mahua		Death
g.	Isu	Cuaa Cuav	Idzu Sacuaa	Cuaa Cuav		Huiçu	Deer
h.		Sayu Xay	Idzo	Sayu Xay	Sayu [Xayu]	Guihri Cuhri	Rabbit
i.	nDute	Duta Cuta	nDuta	Cuta		Duta [Tuta]	Water
j.		Ua Huaa	(Te)ina	Hua Huaa	Va		Dog
k.		Ñuu Ñooy	Codzo	Ñuu Ñooy Ñao	Ñuu	Nyo Niaa	Monkey
l.	Yuku	Cuañe	Yucu	Cuañe Cuuñi	Cuañe	Cacu	Grass
m.	nDoo'	Huiyo	Ndoo'	Huiyo Huiya	[Huiyo] Viyo	Caxũ	Cane
n.		Vidzu	Cuiñe	Vidzu Huidzu	Vidzu		Jaguar
o.		Sayacu Xa	Yaha	Sayacu Sa			Eagle
p.		Cuij	[(Ti)sii] [Tijii]	Cuij	Cuij	Tixi	Buzzard
q.		Qhi	Yotnaa Yonehe	Qhi	Qhi		Quake
r.	Wixį	Si Cuxi	Yuchi	Si Cusi	Uisi [Cuxi]	Cuxi	Cold
s.	Sawi'	Dzahui Co	Dzavui	Co			Rain
t.		Uaco Coo Coy	Ita	Uaco Huaco	Vaco		Flower

The units of the Mixtec calendar are the day (*kiwi'*), month (?), year (*kwiya*), day count (?), and calendar round (*ee dziya, ee dzini, ee toto* 'one crown, garland').

The Yucuñudahui and Tilantongo calendars were not only used during the Classic and Postclassic periods, respectively, throughout the Mixteca, they were widely known in other areas and among other peoples. The former is the calendar of Xochicalco and was also known and used at Teotihuacan. It was ancestral to all the other calendars of the Classic period to the north of Oaxaca. The latter was probably the calendar of the Ixcatec, Triqui, and Amusgo and was adopted by the Chocho and Tlaxcala as well. It was the source of all the Postclassic calendars north of Oaxaca, including Toltec and Aztec, and its writing system became the universal standard of Central and Northern Mexico and had a marked influence on the Postclassic Zapotec and the Maya of Chiapas and Guatemala.

MONTHS

In comparison with the days, the native months of the Middle American calendars are poorly documented. As the accompanying figure (16) shows, we have complete lists of month names for only nine Mayan and seven non-Mayan languages and partial documentation of an additional nine.

Many of the month names are obscure. They are often archaic, and they frequently refer to ritual rather than natural symbols. Furthermore, they are historically unstable, differing significantly between even very closely cognate languages and calendars (e.g., Quiche and Cakchiquel, Tzeltal and Tzotzil, Chol and Yucatec, Kanhobal and Chuh). They are also calendrically unstable. Tititl, for example, occurs as month C. or D.; Izcalli as D. or E.; Tlacaxipehualiztli is G. in Nahuatl and Cakchiquel but F. in Quiche; Canaazi appears to be L. in Chol but Q. in Pokom. There are numerous other examples.

Although the veintenas of the year count do seem to be referred to the agricultural year in various tantalizing ways, they do not constitute an agricultural calendar and probably never did. There is considerable evidence, in fact, that it was the divinatory calendar of the day count that was most closely related to agricultural decision making. The year count was primarily a ritual calendar for general civic ceremonial, and its primary focus was on the New Year rites. There is notable agreement among the otherwise diverse calendars in the conceptualization of the nineteenth month, which was everywhere considered to be a dangerous dead space at the end of the year.

The year was punctuated by other ceremonies tied to the year count and the months, though it is not always easy to differentiate them from those timed by the day count. Some of the concepts behind the year-count ceremonies were certainly widespread and ancient and seem, on the face of it, to have left traces in the naming of the months themselves. It is at least highly suggestive, for example, that month A was called Xul 'End' in Yucatec and month B was called Yax Kin 'Green Time, New Time, First Time.' These are very good descriptions of the positions of these 2 months in the Olmec calendar, which was probably the first calendar the ancestral Yucatecans knew. At least six other Mayan languages have preserved a similar name for month B. In a similar vein, the Totonac month B 'Middle' was the tenth month of the Postclassic Totonac year.

Other concepts embodied in the month names of five or more different languages are summarized in the following list. The number of languages sharing the concept is indicated in parentheses.

A. Bird (11)	F. White (6)	K. Stew (7)	P. —
B. Green (7)	G. Flay (8)	L. Lord (5)	Q. Moss (6)
C. Gather (5)	H. Cover (9)	M. Lord (5)	R. —
D. —	I. —	N. —	
E. Bird (7 +)	J. Owl (6)	O. —	X. Lost (18?)

No one calendar ever used this collection of ritual ideas at any one time; they are simply among the more widely current notions that provided names for the months indicated.

Figure 16b assigns English names to the months that are suggestive of their etymologies in those cases where it seems possible to do so, in order to facilitate comparison.

Figure 16a. NAMES OF THE MONTHS IN 25 LANGUAGES

	A.	B.	C.	D.	E.	F.	G.	H.	I.	J.
	Xul / Chichin	Yaxkin / Yaxkin	Mol / Mol	Ch'en / Ik Zih	Yax Zih / Yax Zih	Zac / Zac Zih	Chac / Chac Zih	Mac / Mac	Kankin / Oncu	Muan / Muan
Yucatec	Xul	Yaxkin	Mol	Ch'en	Yax Zih	Zac	Chac	Mac	Kankin	Muan
Chol	Chichin	Yaxkin	Mol	Ik Zih	Yax Zih	Zac Zih	Chac Zih	Mac	Oncu	Muan
Chontal										
Kekchi	Chichin	Ianguca	Mol	Zihora	Yax	Zac	Chac	Chante Mac	Uniu	Muhan
Pokom	T'zikin Kih	Mox Kih	Tik Cheik	Yax	Zac	Tsi	Tzip	Chante Mac	Uniu	Muan
Cakchiquel	Tz'ikin Q'ih	Cakam	Ibota	Qatik	Izcal	Pa ri Che	Tacaxepual	Nabe Tamuzuz	Ru Cab Tamazuz	Zibisik
Quiche	Tz'ikin Q'ih	K'aqam	Balam	Nabe Zih	U Kab Zih	R Ox Zih	Chee	Tacaxepual	Tz'iba Pop	Zaq
Ixil	Tz'ikin Ki	Yaxki	Mol	Pctzctz Ki	Yax	Hui Zih	Koh Ki	Chante Mac	Och Ki	Muen
Kanhobal	Wa Tsikin	Yaxakil	Mol	K'eq Sihom	Yax Sihom	Zaq Sihom	K'aq Sihom	Mac	Oncu	Sivil
Chuh	Kanal	Yaxkin	Yaxul	Savul	Xujm		Mol	Mak	Oncu	Sivil
Tzeltal	Pom	Yaxkin	Mux	Tz'un	Batz'ul	Zacilab	Ahel Chac	Mac	Olalti	Hulol
Tzotzil	Pom	Yaxkin	Mux	Tz'un	Batz'ul	Zi Zac	Mukta Zac	Mac	Olalti	Ulol
Chiapanec		Mua	Tupiu	Tuhu	Muhu	Turi	Manga	Puri	Cuturi	Cupané
Nimiyua	Haomé	Mahuui	Toho	Mua	Topia	Tumuhu		Cupamé	Puri	Puhuari
Mixe	Nixam'i	Mih Oo	Haak Oo	Mih Tsaatso'k	Mih Kahpu'ut	Haak Kahpu'ut	Mih Xiwi'i	Haak Xiwi'i	Kuyduus	Mih Kaa
Chinantec	Moh	Mii	Nyö	Tau Jaü	Ya	Niu	Luä	Jau	To Jaü	Huh
Mazatec	Sa	Hi	Kha	Kimiä	Me	Ni'ao	Xki	Khoi	Hi'	Hrtö
Totonac										
Huastec		Calcusot								
Otomi	Tzhoni	Thaxme	Chamup	Buc	Thudocni	Quisa	Ttzayoh	Tzhotho	Ahit	Tzbiphi
Matlatzinca	Thechaqui	Thechotahui	Camlehe / Teyabihitzin	Thaxitohui	Xochitoca	Buoentaxi	Tzahio		Tatzhoni / Thaçari	Dehuni
Metztitlan	Quecholli	Panquetzaliztli	Atemoztli	Tititl	Izcalli	Xilomaniztli	Tlacaxipchualiztli	Tezoztontli	Huei Tozoztli	Popochtli
Pipil	Quecholli	Panquetzaliztli	Atemoztli	Tititl	Izcalli	Atlacahualo / Cuahuitleua	Tlacaxipchualiztli	Tezoztontli	Huei Tozoztli	Toxcatl
Nahuatl	Quecholli / Tepeilhuitl / Pilahuanao-tepeilhuitl / Mixcoatl	Panquetzaliztli	Atemoztli / Tititl	Izcalli	Izcalli	Xilomaniztli / Atlcahualo	Coailhuitl / Xochimanaloyo	Xochimanaloyo	Huei Tozoztli	Toxcatl / Tepopochtli
Tarascan	Cáheri Uapánscuaro		Pcuánscuaro	Curíndaro	Tzitacuarénscuaro	Cihuailhuitl / Purecóraque	Xilopehualiztli / Cuingo	Unispérácuaro		

Figure 16b. CONCEPTS OF THE MONTHS IN 23 LANGUAGES

	A.	B.	C.	D.	E.	F.	G.	H.	I.	J.
	End / Bird	Green / Green	Gather / Gather	Well / Black	Green / Green	White / White	Deer / Red	Cover / Cover	Yellow	Owl / Owl
Yucatec	End	Green	Gather	Well	Green	White	Deer	Cover	Yellow	Owl
Chol	Bird	Green	Gather	Black	Green	White	Red	Cover		Owl
Chontal										
Kekchi	Bird		Gather	Green	Green	White	Red	Cover		Owl
Pokom	Bird	Red	Tree	Burn	White	Dog		Cover		Owl
Cakchiquel	Bird	Red	Mat	Flower	Sprout	Tree	Flay	Termite	Termite	Smoke
Quiche	Bird	Green	Jaguar	Flower	Flower	Flower	Tree	Flay	Mat	White
Ixil	Bird	Green	Gather	Black	Green	White	Red	Cover		Owl
Kanhobal	Bird	Green	Gather		Green		Gather	Cover		Vapor
Chuh	Yellow		Green			White	Red	Cover		
Tzeltal	Incense	Green	Mud	Plant	Amaranth	White	White	Cover		Arrive
Tzotzil	Incense	Green	Mud	Plant	Amaranth	White	Tree	Cover	Bat	Arrive
Mixe	Dog	Bath	Nausea	Stone	Town	Town	Count	Tree	Blood	Jaguar
Chinantec	Sow	Fruit	Broken	Five	Burn	Depth	Flay	Deep		Lie
Mazatec	Grow	Wood		Cloud	Want	Wind				Owl
Totonac					Sun					
Huaste		Middle								
Otomi	Flight	Tortilla	Water	Old	Flower	Corn	Flay	Flight	Count	Turkey
Matlatzinca	Crane	Twins	Falling	Grandpa	Flower	Corn	Flay		Smoke	Corn
Metztitlan	Bird	Flag	Water	Storm	Sprout	Corn	Snake	Vigil	Vigil	Youth
Pipil	Macaw	Flag	Water	Storm	Sprout	Tree	Flower	Vigil	Vigil	Youth
Nahuatl	Mountain / Tornado	Flag	Storm	Sprout	Flower	Water / Woman	Corn / Bird	Flower	Vigil	Drought
Tarascan	Cast		Birth		Rebirth	War		Bone		

Figure 16a. NAMES OF THE MONTHS IN 25 LANGUAGES

	K. Pax	L. Kayab	M. Cumku	N. Pop	O. Uo	P. Zip	Q. Zotz'	R. Tzec	X. Uayeb
Yucatec	Pax	Kayab	Cumku	Pop	Uo	Zip	Zotz'	Tzec	Uayeb
Chol	Canaazi	Ohl	Pop	Ik Kat	Chac Kat	Zotz'	Zec	Cazeu	Mahi i kaba
Chontal		Canaazi	Ohl	Pop	Uo	Chac Kat	Kanazi	Kanahal	Mahi i Ccaba
Kekchi	An Kiku	Sac Kohk	Ohl	Kan Halam	I Kat	Kazcu	Nabe Pach	Ru Cab Pach	Tz'api Qih
Pokom	Tam	Nabe Mam	Ru Cab Mam	Liqin Qa	Makux, Pet Kat	Ru Cab Toqiq	U Kab Pach	Ts'ikil Lakam	Tz'api Qih
Cakchiquel	Uchum	Nabe Mam	U Kab Mam	Liqin Ka	Nabe Toqiq	Nabe Pach	Chotz Che	Xct Ki	Q Ki
Quiche	Ch'ab	Tal Cho	Nim Cho	Mech Ki	U Kab Liqin Ka	Zil Ki	Xuhem	Kanal	Oyeb Ku
Ixil	Pac Tzi	Wex	Saq May	Nab Ich	Mu	Bak'	Tam	Hua Tziquin	Oyeb 'In
Kanhobal	Tap	Bex	Sac May	Nab Ich	Moo	Bac	Chan Uinicil	Ox Uinicil	Ch'ai Kin
Chuh	Tap	Uch	Muc Uch	Huc Uinicil	Mo	Ho Uinicil	Yoxibal Uinicil	Xchanibal Uinicil	Ch'ai Kin
Tzeltal	Okan Ahau	Uch	El Ech	Nichil Kin	Uac Uinicil	Xchibal Uinicil			
Tzotzil	Okan Ahau	El Ech	Numbi	Cutamé	Hum Uinicil		Mahua		
Chiapanec	Tamugii	Iatati	Numbi	Muhu	Iaume	Mundju	Catani	Manga	Nbu
Nimiyua	Turi	Yucu	Nii Tsi't	Numbi	Hatati	Hotsoon	Aaxc'om	'Ap	Nbu
Mixe	Ipts Tigi't	Taak Am		Mih Xo'ox	Haak Xo'ox			Riu Kuin	Muts Tsaatso'k
Chinantec	Lhu	Nò	Lò	Kúan	Lou	I	Jeh	Ncho	Ta Nyiu
Mazatec	Tó	Mehe	Mahti	Sintà	N'e	Nhoa	Nhtô		X Kiki Nià'oo
Totonac									
Huastec	Tuy	Pitich	Amab		Cooch	Chuc Tzeb	Tzen Bo Xegui	Tama Xegui	Dupa
Otomi	Guoeni	Tzen Gohmun	Tan Gomuh	Tzen Gotu	Tan Gotu	Baxi	Tho Xiqui	Tha Xiqui	Tasyabiri
Matlatzinca	Theçamoni	Thirimchui	Thamchui	Iscathalohui	Huei Miccaylhuitl	Ytz Bacha	Pachtli	Huei Pachtli	
Metztitlan	Erzaleualiztli	Tzin Cohu	Huei Tccuilhuitl	Miccailhuitl	Xocotl Huetzi	Huechpaniztli	Teotleco	Tepeilhuitl	Nemnontemi
Pipil	Erzaleualiztli	Teculhuitontli	Huei Tccuilhuitl	Tlaxcailhuitontli	Huei Miccailhuitl	Uchpaniztli	Pachtontli	Huei Pachtli	
Nahuatl	Erzaleualiztli	Tccuilhuitontli / Tochilhuitzintli	Huei Tccuilhuitl	Miccailhuitontli / Tlaxochimaco	Xocotl Huetzi	Ochpaniztli / Tenahuitliztli	Teotleco / Ecoztli	Tepeilhuitl	Nemontemi
Tarascan	Máscuto	Uáscata Cónscuaro	Cáheri Cónscuaro	Hanciñáscuaro	Hicuándiro	Sicuíndiro	Charapu Zapi	Uapánscuaro	

Figure 16b. CONCEPTS OF THE MONTHS IN 23 LANGUAGES

	K. Drum	L. Turtle	M. Dark	N. Mat	O. Frog	P. Stag	Q. Bat	R. Skull	X. Specter
Yucatec	Drum	Turtle	Dark	Mat	Frog	Stag	Bat	Skull	Specter
Chol		Turtle	Mat	Mat	Frog	Bat	Bat	Skull	Nameless
Chontal			Lord	Mat	Frog	Moss			Nameless
Kekchi	Crab	Turtle	Lord	Soft	Damp	Moss	Moss	Moss	
Pokom	Reseed	Squash	White	Soft	Soft	Bone	Moss	Shoots	
Cakchiquel	Arrow	Lord	White	Moon		Bone			Extra
Quiche	Crab	Lord	Possum	Moon				Yellow	Extra
Ixil	Crab	Time	Possum		Parrot			Bird	Five
Kanhobal	Lord	Time	Squash		Parrot		4	3	Five
Chuh	Lord	Possum	Mule	7		5	3	4	Specter
Tzeltal	House	Possum	Anger	Possum	6	2			Lost
Tzotzil	Mosquito	Wet		Viper	1		Mouth		Lost
Mixe	Fruit	Stand		Grow	Viper	Liver		Corn	
Chinantec	Wax	Fat			Fall	Heavy	Hot	Huaje	Unseen
Mazatec	Lord	Fat			Yellow	Greens	Worm		
Totonac									
Huastec		Lord	Lord		Dead	Broom	Moss	Moss	
Otomi		Lord	Porridge	Dead	Dead	Broom	Moss	Moss	Dead
Matlatzinca		Lord	Lord	Dead	Fruit	Sweep	Moss	Mountain	
Metztitlan	Bean	Lord	Lord	Flower	Dead	Sweep	Moss	Moss	Vain
Pipil	Bean		Lord	Dead	Fruit	Sweep		Mountain	
Nahuatl	Bean		Lord	Flower					Vain
Tarascan	Stew	Feast	Feast	Feast	Assembly	Bath	Flay	Cast	

A number of calendars gave the same name to two or more sequential months, distinguishing them by numbers or by concepts tantamount to numbers, such as little and great, or sequences of colors.

The use of numbers is salient in Quiche D.E.F., and in Tzotzil O.P.Q.R., or the inverted sequence from 7 to 3 in Tzel⁺al N.O.P.Q.R. The use of colors is illustrated by Chol (-Kanhobal) Black, Green, White, Red (D.E.F.G.), attenuated to D.E. in Pokom, E.F. in Yucatec, E.F.G. in Kekchi, and F.G. in Tzeltal and Tzotzil. Triads occur in Tzotzil L.M.N. and Ixil O.P.Q., but most examples are pairs, such as Bath (B.C.), Town (E.F.) and Tree (G.H.) in Mixe, Damp (O.P.) and Termite (H.I.) in Cakchiquel, Soft (N.O.) in Quiche, N.O. in Chol and O.P. in Kekchi, and R.A. in Tarascan. Four such paired months appear to be widely adopted Nahuatl inventions:

Vigil (H.I.) in Otomi and Metztitlan
Lord (L.M.) in Otomi, Metztitlan, Cakchiquel, Quiche, Tarascan, and
    Tzeltal
Dead (N.O.) in Otomi, Metztitlan, and Matlatzinca
Moss (Q.R.) in Otomi, Metztitlan, Matlatzinca, and Cakchiquel (P.Q. in
    Quiche)

There have been a number of attempts to link the etymologies of the month names to seasonal phenomena as a way of tying particular calendars to the solar year. Stoll (1889:60) tried to time the Cakchiquel months of Termite to the first annual appearance of the insects for which they are ostensibly named. Graulich (1981) dated the genesis of the Nahuatl month names to A.D. 682 by identifying the last time their seasonal implications were in phase with the actual solar year. Bricker (1982) dated the Mayan names to A.D. 550 on a similar basis. I do not believe much credence can be placed in these efforts, though they remain interesting speculations if they can be otherwise validated.

There would seem to be no reason why each month name should not have its own history. The following compilation is a preliminary exploration of those with apparent seasonal implications. It gives the date of the beginning of the month in 1550 and the beginning of the European month that might best correspond to the name in terms of the agricultural and meteorological pattern of Middle America. The century in the left-hand column places how far back we would have to go to match the two dates; each 25-day difference is equivalent to a century. The agricultural etymologies are particularly dubious for this purpose because different areas may have anywhere from one to three harvests a year.

Century	Month	1550	Likely Date
III	Mixe L. Wet	5 VI	1 V
	Cakchiquel K. Reseed	16 V	1 VII
	Chinantec K. Mosquito	16 V	1 VII
	Chinantec Q. Hot	18 XI	1 XII
IV	Pokom L. Squash	5 VI	1 VIII
	Chinantec B. Fruit	17 XI	1 X
	Maya C. Gather	6 XII	1 X
V	Chinantec E. Burn	17 I	1 III
	Huastec E. Sun	17 I	1 XI
VII	Nahuatl K. Bean	16 V	1 VIII
	Chuh G. Gather	26 II	1 X
	Tarascan K. Stew	16 V	1 VIII
VIII	Otomi I. Flight	6 IV	1 XI
	Mazatec A. Grow	28 X	1 VI
	Tarascan A. Cast	28 X	1 IV
IX	Tzotzil E. Amaranth	17 I	1 VII
	Nahuatl J. Drought	26 IV	1 XI
	Mazatec K. Fruit	16 I	1 X
X	Chinantec A. Sow	28 X	1 IV
	Tzotzil C. Mud	6 XII	1 V
	Nahuatl C. Water	7 XII	1 V
	Nahuatl C. Storm	7 XII	1 V
	Nahuatl D. Sprout	28 XII	1 VI
	Nahuatl E. Sprout	17 I	1 VI
	Mazatec F. Wind	6 II	1 VII
	Cakchiquel O. Damp	9 VIII	1 V
XI	Cakchiquel P. Damp	29 VIII	1 V
	Mazatec D. Cloud	28 XII	1 V
	Nahuatl D. Storm	28 XII	1 V
	Otomi H. Flight	10 III	1 XI
	Yucatec I. Yellow	6 IV	1 VIII
	Quiche R. Shoots	9 X	1 VI
	Mazatec R. Huaje	9 X	1 VI
XII	Nahuatl R. Moss	9 X	1 VI
	Quiche A. Bird	14 VII	1 XI
	Quiche O. Soft	9 VIII	1 V
	Mazatec P. Greens	29 VIII	1 VI
	Kanhobal R. Yellow	9 X	1 VIII
XIII	Tzotzil D. Plant	27 XII	1 IV
	Cakchiquel D. Burn	26 XII	1 III
	Nahuatl F. Water	6 II	1 V
	Quiche N. Soft	20 VII	1 V
	Huastec M. Plant	25 VI	1 IV
	Nahuatl Q. Moss	18 IX	1 VII

Century	Month	1550	Likely Date
XIV	Chinantec E. Burn	17 I	1 III
	Otomi J. Smoke	26 IV	1 III
	Quiche J. Fire	26 IV	1 III
	Kanhobal J. Vapor	26 IV	1 III
XV	Quiche P. Moss	29 VIII	1 VII

Although it may be true that each month name has its own history, the concentration of Nahuatl examples in the tenth century is at least suggestive. Because the year comes full cycle after 365 leap years (1460) years, any of these etymologies could potentially belong to the previous cycle. This could place Yucatec I. Yellow, for example, in the fourth century B.C. rather than the eleventh A.D.

    Although these speculations do not date the calendars in question, they can perhaps shed some light on the history of their month names.

NAHUATL

Twelve different calendars were employed in Nahuatl. They all used Type IV New Years' and Type III name days, counted the days of the day count from 1 to 13 and those of the month from 1 to 20, used similar day and month names, and wrote in Tilantongo Mixtec glyphs. They differed in beginning the year with different months and their calendar rounds with different years. Their directional colors may also have been subject to some variation. The day glyphs and their names are given by Caso (1967):

a   f   k   p

b   g   l   q

c   h   m   r

d   i   n   s

e   j   o   t

a. Cipactli (alligator)

b. Ehecatl (wind)

c. Calli (house)

d. Cuetzpallin (iguana)

e. Coatl (serpent)

f. Miquiztli (death)

g. Mazatl (deer)

h. Tochtli (rabbit)

i. Atl (water)

j. Itzcuintli (dog)

k. Ozomatli (monkey)

l. Malinalli (grass)

m. Acatl (cane)

n. Ocelotl (jaguar)

o. Quauhtli (eagle)

p. Cozcaquauhtli (buzzard)

q. Ollin (quake)

r. Tecpatl (flint)

s. Quiahuitl (rain)

t. Xochitl (flower)

All of the Nahuatl calendars appear to have used the same day names except Pipil and Metztitlan (q.v.).

The units of the Nahuatl calendar are the sun or day (*tonalli*), the moon or month (*metztli*), day count (*tonalpohualli*), year (*xihuitl*), and calendar round (*xiuhmolpilli*).

The month names were more variable, some months having as many as four different names (a., b., c., and d. in the next compilation) in different areas. They are summarized from Caso (1967):

	a.	b.	c.	d.
A.	Quecholli (macaw)	Tepeilhuitl (mountain feast)	Mixcoatl (tornado)	Pilahuanaotepeilhuitl
B.	Panquetzaliztli (flag raising)	— —	— —	— —
C.	Atemoztli (falling water)	Tititl (storm)	— —	— —
D.	Tititl (storm)	Izcalli (sprout)	— —	— —
E.	Izcalli (sprout)	Xochilhuitl (flower feast)	Xochitoca (flower sowing)	—
F.	Cuahuitleua (tree rises)	Xilomaniztli (corn feast)	Atlcahualo (water left)	Cihuailhuitl (woman feast)
G.	Tlacaxipehualiztli (man flaying)	Coailhuitl (snake feast)	Xochimanalayo (flower offering)	Xilopehualiztli (corn starting)
H.	Tozoztontli (1-vigil)	Xochimanalayo (flower offering)	— —	— —
I.	Huei Tozoztli (2-vigil)	— —	— —	— —
J.	Toxcatl (drought)	Tepopochtli (youths)	— —	— —
K.	Etzalcualiztli (eating bean soup)	— —	— —	— —

	a.	b.	c.	d.
L.	Tecuilhuitontli (1-lords feast)	Tochilhuitzintli (rabbit feast)	—	—
M.	Huei Tecuilhuitl (2-lords feast)	—	—	—
N.	Miccailhuitontli (1-dead feast)	Tlaxochimaco (give flowers)	—	—
O.	Huei Miccailhuitl (2-dead feast)	Xocotl Huetzi (fruit falls)	—	—
P.	Ochpaniztli (road sweeping)	Tenahuitiliztli	—	—
Q.	Pachtontli (1-moss)	Teotleco (god arrives)	Ecoztli (arrival)	-
R.	Huei Pachtli (2-moss)	Tepeilhuitl (mountain feast)	—	—
X.	Nemontemi (full in vain)	—	—	—

The glyphs for the months are subject to still further variation even when they are called by the same name. The following set is drawn from Caso (1967:36, Fig. 14) and corresponds to the preceding list a. except that the sixth glyph is for Fc. Atlcahualo. Caso does not cite glyphs for the other month names on the preceding list, except that he does have a glyph for Tepeilhuitl (see Figure 17).

The twelve Nahuatl calendars may be listed as follows in the order of the months with which they begin the year:

A. COLHUA. Jiménez (1961) finds this calendar in the *Anales de Cuauhtitlan* and designates it Colhua I. It begins with Ac.

B. METZTITLAN. Caso (1946; 1971) finds this calendar in the *Relación de Metztitlan* and identifies its month names as Ba., Ca., Da., Ec., Fb., G?., H?., Ia., Jb., Ka., L?., Ma., Na., Oa., Pa., Qa., Ra., Aa. The name for H. is erroneously given as Quecholli, and G. and L. are named in Otomi.

C. TLAXCALAN. Caso (1967:Table X) identifies this calendar from Muñoz Camargo, Zapata, Torquemada (2:295), Lorenzana, Clavijero, Veytia (Rueda 5), Vetancurt, and Veytia, giving its month names as Ca., Da., Ea., Fb., Ga., Ha., Ia., Ja., Ka., La., Ma., Na., Oa., Pa., Qa., Ra., Aa., Ba. This is the Tilantongo Mixtec calendar.

D. TEXCOCO. Caso (1967:Table X) identifies this calendar from the *Calendario de Bobán, 1538*, and names its months as corresponding entirely to list a. Kirchhoff (1950) also finds it in Alva Ixtlilxochitl and labels it Chichimec.

E. AZTEC. Caso (1967), Jiménez (1961), and Kirchhoff (1950) agree in identifying this calendar with the Mexica of Tenochtitlan. Caso (1967:Table X) finds it in the *Codex Borbonicus* and the *Tovar Calendar* and gives the month names as Ea., Fa., Ga., Ha., Ia., Ja., Ka., La., Ma., Nb., Ob., Pa., Qb., Rb., Aa., Ba., Ca., and Da. but notes that the *Aubin Tonalamatl* gives Gc. and Del Castillo has Db., Eb., and Fc.

Figure 17. AZTEC MONTH GLYPHS

F. TEPEPULCO. Caso (1967) identifies this calendar with Texcoco, Jiménez (1961) with Cuitlahuac, and Kirchhoff (1950) with Atzcapotzalco, but actually it is the Otomi calendar, widely used in northern Nahuatl territory. Sahagún's *Primeros Memoriales* give the month names as those of list a., but his *Códice Matritense* from Tlatelolco gives Nb., Ob., Ab. *Códice Telleriano* and *Vaticanus A* give Fb. *Códice Magliabecchi, Códice Ixtlilxochitl, Códice Veytia,* and Cervantes de Salazar give Nb., Ob., and Ec. Tezozomoc, *Mexicanus,* and *Mexicayotl* give Fc. *Costumbres de Nueva España* gives Fb., Lb., Mb., Ob. The Otomi *Códice de Huichiapan* and Cristóbal del Castillo give Db., Eb., Fb., and Del Castillo also adds Nb. Torquemada and Vetancurt give Fb., Nb., Ob., Qb., Rb., and Durán (1581) gives Fb. The Tepepulco calendar is also used by Clavijero, Bobán, and the González letter (Caso 1967: Table X).

G. TEOTITLAN. This is the calendar of the Postclassic Totonac, of the eastern Nahua of Nonoalco and the Pipil, and perhaps of the Nicarao as well. Jiménez (1961) identifies it in the *Relación Genealógica* and labels it Colhua II. Caso (1967; 1971) finds it in Motolinía and gives the month names from the *Relación de Teotitlán* as Ga., Ha., Ia., Ja., Ka., La., Ma., Na., Oa., Pa., Qc., Ra., Aa., Ba., Ca., Da., Ea., Fd. Tovar, *Códice Ramírez, Calendario de Tovar, Mexicano, probanza de Sahagún* and *Acosta II* give Qb., Rb. The *Relación de Tecciztlán* gives Hb., Nb., Ob., Qb., Rb. Sigüenza, Gemelli, and Boturini (Rueda 4) give Fc. Olmos, the *Anónimo* of Mendieta, and *Historia de los Mexicanos por sus pinturas* give Fc., Nb., Ob. Ixtlilxochitl gives Gb., and Durán (1579) gives Fb. and Pb.

H. CUITLAHUAC. Kirchhoff (1950) finds this calendar in *Anales de Cuauhtitlán, Anales de Tlatelolco,* Chimalpahin, *Codex Mendoza,* Mendieta, Veytia, Tezozomoc, Sigüenza, and Boturini. Caso (1967:Table X) gives the month name Fa. It is the Huastec calendar of the Postclassic.

I. TEPANEC. This is the Postclassic calendar of the Tarascans and Matlatzincas, used in Nahuatl by Atzcapotzalco and Tlatelolco (Jiménez 1961).

J. COLHUACAN. Jiménez (1961) finds this calendar in the *Anónimo de Tlatelolco* and labels it "Cuauhtitlan?". Kirchhoff (1950) finds it in *Aubin, Mappe de Tepechpam,* and *Vaticanus A.* Kubler and Gibson (1951) find it in the *Tovar Calendar,* which Caso (1967) disputes. I believe it to be the calendar of Postclassic Tula but have reserved the name Toltec for the antecedent Classic period calendar at that site.

K. TEPEXIC. Kirchhoff (1950) identifies this calendar with Tlaxcala or Huexotzinco and finds it in Zapata. Brotherston (1983) finds it in *Codex Vindobonensis Obverse* and identifies it as Tepexic. I believe it is the Postclassic derivative of the Classic calendar of Puebla and Tlaxcala, which I have labeled Cholula.

L. CHALCA. This is the Postclassic calendar derived from the Classic calendars of the Yucuñudahui Mixtec and Xochicalco. Its Nahuatl use is identified by Jiménez (1961) in Chimalpahin and labeled Chalca.

The variability of the names of the months in Nahuatl is somewhat less consequential than it might be because they are rarely cited anyway. Most dates are given in the form "coefficient, year bearer, coefficient, day, coefficient, month," and the month designation is omitted if it is redundant. The Nahuatl dates are written with dot numerals, and year bearers are identified by the year sign (Chapter 1, Figure 7). Most dates refer only to the year.

For 6 native months out of every year, all the Nahuatl calendars are in agreement about the year bearer, differing only in the coefficient. Any event occurring between May 17 and October 29, 1550, for example, would be ascribed to a year Rabbit, but it would be 4 Rabbit Colhua, 11 Rabbit Metztitlan, 5 Rabbit Tlaxcalan, 12 Rabbit Texcoco, 6 Rabbit Aztec, 13 Rabbit Tepepulco, 7 Rabbit Teotitlan, 1 Rabbit Cuitlahuac, 8 Rabbit Tepanec, 2 Rabbit Colhuacan, 9 Rabbit Tepexic, and 3 Rabbit Chalca. Agreement between sources about the dating of particular events definitively correlates their calendars, but unfortunately such agreement is rare. I have illustrated the problems by including in the date list what the sources have to say about the reign of Quetzalcoatl at Tula and the founding of Tenochtitlan.

A good case has been made (Kirchhoff 1950) for dating the founding of Tenochtitlan to 1368–71 (q.v.). But the dates are actually cited in six different Nahuatl calendars:

Calendar	1368	1369	1370	1371
A. Colhua				
B. Metztitlan				
C. Tlaxcalan				
D. Texcoco			1	
E. Aztec		7	8	
F. Tepepulco		1	2	
G. Teotitlan				
H. Cuitlahuac	1	2	3	4
I. Tepanec				
J. Colhuacan	2		4	5
K. Tepexic	9			
L. Chalca				
	Flint	House	Rabbit	Cane

Kirchhoff quite reasonably narrows the choice to 1369–70. If indeed the event were early in a year Rabbit, it could still have been House in some of the later calendars. For example, all three of the dates given by Chimalpahin and those in the *Anales de Cuauhtitlán* (Lehmann 1938) could very well be correct if Tenochtitlan were founded in a predawn ceremony on 1 Izcalli. In any event, the use of the calendars by different sources is of interest:

	1368	1369	1370	1371
D. Texcoco			*Cuauhtitlan* Ixtlilxochitl *Xolotl*	
E. Aztec		Chimalpahin	*Cuauhtitlan*	
F. Tepepulco		Chimalpahin *Tlatelolco*	Clavijero	
H. Cuitlahuac	*Aubin* Ixtlilxochitl *Tlatelolco* *Mendoza*	Chimalpahin Clavijero *Cuauhtitlan* Mendieta *Tlatelolco* Veytia	Tezozomoc	Boturini Veytia
J. Colhuacan	*Aubin*		*Tepechpam* *Vaticanus A*	*Tlatelolco*
K. Tepexic	Zapata			

Chimalpahin gives the same year in three calendars and *Aubin* in two. The *Annals of Tlatelolco* gives four dates in three different calendars and three different years. The *Annals of Cuauhtitlan* gives two different years in three calendars. Clavijero and Ixtlilxochitl both give two different years in two calendars, and Veytia does the same thing in one.

The prominence of the Cuitlahuac calendar in this connection is somewhat startling, as is the fact that we have no Tepanec date for the founding of Tenochtitlan (-Tlatelolco) and that the *Annals of Tlatelolco* uses two other calendars but not the Tepanec one. The question of the provenience of the various Nahuatl calendars is obscured by the eclecticism of their use, particularly in post-Conquest sources.

Although he was initially responsible for the recognition of thirteen different Nahuatl calendars, Kirchhoff (1950; 1955–56) eventually became persuaded that their year bearers were not synchronized because they were on different day counts. The fact that no less than eight of them can be independently correlated with the Christian calendar falsifies this conclusion. As Broda's (1969:36ff.) discussion makes clear, some confusion is due to the dating of festivals. No doubt there was some flexibility in the programming of such activities, as there is among the modern Tzotzil (q.v.), but the underlying unity of all the calendars is patent and rested on the universal sanctity of the unbroken day count.

NETZICHUS
(= *Zapotec?*)

The Netzichus, Rincon, or Villa Alta Zapotec are calendrically unknown. Presumptively they used the Zapotec calendar.

NICARAO
(= *Pipil?*)

Lehmann (1920) gives a list of day names in Nicarao from Oviedo (1535):

a. Çipat (alligator)	f. —	k. Oçumatle (monkey)	p. —
b. Hecat (wind)	g. Maçat (deer)	l. Malinal (grass)	q. Olin (quake)
c. Cali (house)	h. Toste (rabbit)	m. Ecat (cane)	r. —
d. Gazpalin (iguana)	i. At (water)	n. Oçelot (jaguar)	s. Quiavit (rain)
e. Goate (serpent)	j. Izquindi (dog)	o. Goate (eagle)	t. Sochit (flower)

The Pipil day names are different for n., q., and s.; the Nicarao names are only orthographic variants of the standard Nahuatl names. It seems likely that the Nicarao calendar was congruent with that of the Pipil and Teotitlan, but nothing more is known of it.

NIMIYUA
(II.L.1 CR?; XVII c. A.D. = *Chiapanec* + *1m*)

The calendar of the Nimiyua or Tia Suchiapa Chiapanec began on 4 Kayab Mayapan and was thus congruent with the Quiche calendar. It is 1 month later than Chiapanec (q.v.).

OLMEC
(I.B.19 CR 1 Lord 1516;
VII c. B.C. to XVI c. A.D. = *Olmec* + *360** after IV c. B.C.)

The earliest documentable calendar in Middle America was Olmec. It employed bar-and-dot numerals, counted the day count from 1 to 13 and the days of the month from 0 to 19, named the year for its last day, and began its calendar round on 1 Lord (1516). Its New Year fell on 3 Yaxkin Mayapan. Its direction colors are unknown. The Olmec calendar reflects the earliest use of the calendar round (see 679 B.C.) and the earliest use of Long Count dating (see 36 B.C.), and it was the longest lived of the Middle American calendars (see A.D. 1522).

The peoples who invented and employed the Olmec calendar were probably the ancestors of the non-Mayan peoples later living in the region where it was
primarily manifested. These were either Zoquean (Mixe, Zoque, Sayula) or possibly Zoquean (Xinca, Lenca, Huave) or Hokan (Subtiaba, Jicaque, Tequistlatec).
The calendars of these groups are largely unknown; only that of the Mixe can be
considered well documented. From what we know of them and from scattered archaeological evidence, the following day glyphs can be provisionally identified and
possible names assigned to them. The sources of the glyphs are identified below.

a. Sun. Unknown (Coe 1965b:760, Fig. 43f; 762, Fig. 47). Cf. Tequistlatec
   'Light'; Mixe 'Root'.
   10 Sun. Ucanal Stela 4, C2 (Graham 1967:Fig. 81).
b. Anno 10 Wind. Monte Alban (Caso 1965b:933, Fig. 3). See 531 B.C.
   1 Wind. Unknown (Caso 1965a:854, Fig. 11h).
   7 Wind. Unknown (Ibid.:854, Fig. 11g).
   8 Wind. Kaminaljuyu Stela 10 (Miles 1965:254, Fig. 13).

c. House?. Unknown (Coe 1970:748, Fig. 18). Cf. Tequistlatec 'House'; Mixe 'Palm'.

d. Iguana. Unknown (Coe 1965b:749, Fig. 22). Cf. Tequistlatec 'Lizard'; Mixe 'Hard'.

e. Anno 12 Serpent. Tenango (Caso 1967:162, Fig. 19).
   12 Serpent. Jimbal Stela 1:A4 (Jones and Satterthwaite 1982:Fig. 78).
   12 Serpent. Jimbal Stela 2:C9 (Ibid.:Fig. 79).
   Serpent. Chalcatzingo (Nicholson 1971:94, Fig. 10).

f. 13 Death. Jimbal Stela 1:B4 (Jones and Satterthwaite 1982:Fig. 78).
   13 Death. Jimbal Stela 2:D9 (Ibid.:Fig. 79).

g. 1 Deer. Jimbal Stela 1:A5 (Ibid.:Fig. 78).
   1 Deer. Jimbal Stela 2:D10 (Ibid.:Fig. 79).

h. 9 Rabbit. Cerro de las Mesas (Coe 1965a:703, Fig. 21).
   Rabbit. Unknown (Coe 1965b:760, Fig. 43c; 761, Fig. 44; 762, Figs. 46, 47, 48; Caso 1967:163, Fig. 21). Cf. Tequistlatec 'Greeting'.

i. 7 Water. Seibal Stela 13:A1 (Graham 1980:2:159).
   Water. Unknown (Coe 1965b:760, Fig. 43d).

j. 1 Foot. Cerro de las Mesas (Coe 1965a:703, Fig. 21).
   Foot. Monte Alban (Caso 1965b:934, Fig. 7j; 935, Fig. 8k). Cf. Yucatec 'Foot'.

k. 5 Monkey. Seibal, Stela 3.
   7 Monkey, Seibal, Stela 3.

l. 1 Jaw. Simojovel (Coe 1965b:747, Fig. 17; 760, Fig. 43b). Cf. Tequistlatec 'Twist'.

m. 6 Cane. Chiapa de Corzo (Marcus 1976:51, Fig. 6).

n. Jaguar. Unknown (Coe 1965b:749, Fig. 22).

o. 2 Eagle. Ucanal Stela 4:C2 (Graham 1967:Fig. 81).
   Eagle. La Venta (Marcus 1976:48, Fig. 4). Cf. Mixe 'Tobacco'.

p. Tree?. Unknown (Coe 1965b:748, Fig. 18). Cf. Mixe 'Tree'.

q. 8 Quake. Tuxtla (Marcus 1976:52, Fig. 7).
   Quake?. Monte Alban I (Caso 1965b:934, Fig. 5s; 935, Fig. 9j).

r. 6 Flint. Tres Zapotes (Marcus 1976:52, Fig. 7). Cf. Tequistlatec 'Sun'.

s. Rain. Chalcatzingo (Coe 1965b:758, Fig. 41). Cf. Mixe 'Grass'.
   Rain. Monte Alban I (Ibid.:934, Fig. 5s).

t. Anno 1 Lord. Kaminaljuyu Stela 10 (Miles 1965:254, Fig. 13).
   Anno 1 Lord. Simojovel (Caso 1965b:932, Fig. 1b; 933, Fig. 3; 854, Fig. 11c,d; Coe 1965b:747, Fig. 17). Cf. Yucatec, Mixe 'Lord'.
   Anno 11 Lord. Monte Alban III Whittaker 1980:35, Fig. 8). 1 Lord. Cacaxtla (McVicker 1985:86, Fig. 3).
   3 Lord. Unknown (Coe 1965a:749, Fig. 20).
   4 Lord. Unknown (Coe 1965a:748, Fig. 19).
   5 Lord. Monte Alban II (Caso 1965b:935, Fig. 9p).
   10 Lord. Monte Alban II (Caso 1965b:937, Fig. 14).

A number of aberrant day glyphs enclosed in square cartouches are found on Classic period monuments at Ucanal, Jimbal, and Seibal in the Guatemalan Peten (see A.D. 848). They are certainly not Mayan, and bear some resemblances to the scripts of Huastec, Teotihuacan, and Olmec. They include three of the four Olmec year bearers and may refer in at least one case to an Olmec year. I believe they may be Late Classic (and provincial) Olmec. They are listed above and reproduced below.

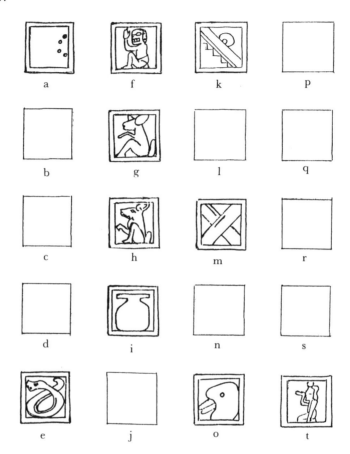

The units of the Olmec calendar are the day, month, day count, tun, year, katun, calendar round, and baktun. A complete calendar-round date was apparently written 'Anno, Yearbearer, Coefficient, Day, Coefficient (Month, Ordinal Number)'. The month phrase may take the alternative form '(Coefficient, Month Name)' and is omitted altogether if it is redundant. Only three of the Olmec month glyphs have been identified; there is no way of providing them with names:

Month B.   Month H.   Month R.

Olmec Initial Series dates appear to have been written 'Initial Series, Month Name, Baktun Coefficient, Katun Coefficient, Tun Coefficient, Uinal Coefficient, Day Coefficient, Day Count Coefficient, Day Name,' thus including a calendar-round date (see 32 B.C.). It seems likely that the Olmec calendar named tuns terminally, as in Maya, but we have no Olmec tun-ending dates. There is no Olmec evidence of novena and trecena enumeration.

Most of the Olmec glyphs contrast clearly with the Zapotec and Mayan glyphs for the same day, so that even a single day glyph may be sufficient to indicate which writing system is being used. In the "Olmec territory" of Chiapas, Tabasco, and southern Veracruz, all of the early dates are Olmec.

At Monte Alban, correlational dates appear (see 563 B.C.), using both Olmec and Zapotec calendars and both writing systems. A Kaminaljuyu inscription records the correlation of the Kaminaljuyu, Olmec and Teohtihuacan calendars (see 147 B.C.). A later Monte Alban date (see A.D. 565) relates the Olmec, Zapotec, and Yucuñudahui calendars and uses all three writing systems, and one correlational date is found as far afield as Tenango relating the Tenango, Olmec, and Zapotec calendars (see A.D. 798). A date at A.D. 861 links the Olmec and Teotihuacan calendars, and one at A.D. 1522 correlates the Olmec and Tilantongo calendars. Such correlational dates are easily recognized, as the Olmec calendar is the only one in Middle America using Type V name days. It is surprising that no satisfactory Olmec–Tikal correlational date has been recognized.

Scientific parsimony forces the construal of the Olmec calendar as using zero and naming the year terminally because the Zapotec calendar, the one most directly derived from it, does so. The interpretation of 1 Lord as the beginning of the Olmec calendar round rests on the interpretation of the (New) Fire glyph as the marker of the calendar round (see Figure 7 and 667 B.C.) (see 679, 667, 531, 36, and 32 B.C.; A.D. 162, 468, 532, 545, 848, 1223, and 1522).

OTOMI
(IV.F.20 CR 2 An Xithi 1539; XI c. to XVII c. A.D. = Aztec + 1m)

The Otomi year begins on 6 Zac Mayapan. It counts the days of the day count from 1 to 13 and those of the month from 1 to 20 and names the years terminally. Its calendar round begins on 2 Cane (An Xithi) in 1539; its directional colors are not known. Its writing system was that of Tilantongo.

The day names are given by Santiago (1632:11) and are reproduced in facsimile by Caso (1967:216):

a. Toqhuay (fish)	f. Yayay atu (dead)	k. Tzepa (monkey)	p. Thecha (god?)
b. Dahi (air)	g. Phanixantoehoe (deer)	l. Chaxttey (grass)	q. Quitzhey (quake)
c. Ngû (house)	h. Qhua (rabbit)	m. Xithi (cane)	r. Yaxi (nails)
d. Botaga (lizard)	i. Dehe (water)	n. Tzani (beast)	s. Yeh (rain)
e. Cqueya (serpent)	j. Yoo (dog)	o. Xeni (peer)	t. Doeni (flower)

All the names are preceded by *An*. The translations given here are Caso's. Soustelle (1937:521–28) interprets Toqhuay as *dok'way* 'stone knife', Xeni as *soni* 'eagle', Chaxttey as *k'ast'ey* 'yellow grass', (An) Thecha(a Oeni) as *xoni* 'turkey', and Yaxi as *doyasi* 'flint'. Carrasco (1950:168–95) reads (An Thecha a) *ceni* 'god, turkey', Tzani as 'biter', Chaxttey as 'green grass,' and Yaxi as 'flint.'

The units of the Otomi calendar are the day (*mapa*) month or moon (*çana, ncayoh*), year (*quenza*), day count (?), and calendar round (?). The *Huichapan Codex* (Santiago 1632:13) also gives the month names:

F. Buoentaxi (corn bud)	O. Tangotu (2-dead)
G. Ttzayoh (dog flaying)	P. Baxi (broom)
H. Tzhotho (1-flight)	Q. Ttzenboxegui (1-moss)
I. Tatzhoni (2-flight)	R. Tamaxegui (2-moss)
J. Tzibiphi (smoke)	A. Tzhoni (flight)
K. Guoeni (turkey meat)	B. Thaxme (tortilla)
L. Ttzengohmuh (1-lords)	C. Candahe (low water)
M. Tangomuh (2-lords)	D. Bue (old)
N. Ttzengotu (1-dead)	E. Thudoeni (flowers)
	X. Dupa (dead days)

All names are preceded by *An*.

Two correlational dates (*1632, *1638) establish the position of the Otomi year. The *Huichapan Codex* suggests that the Otomi simply followed the Aztec calendar round. If so, however, the marginal notes in the codex are probably wrong in equating the year 2 An Xithi Otomi with 2 Acatl Aztec (1507). The Otomi date should correspond to 8 Acatl Aztec (1539) and that must have been the beginning of the Otomi calendar round. This may explain the annotator's error noted by Caso (1967:224) of placing don Martín Cortés in 1 Acatl Aztec (1519) instead of 1 An Xithi Otomi (1531).

The *Huichapan* annals, curiously enough, date events in the Aztec and Teotitlan calendars rather than in that of the Otomi (see A.D. 1443). The Otomi calendar clearly influenced that of Metztitlan (q.v.) and was widely used in Nahuatl as well (see Tepepulco). It may also have influenced Huastec (q.v.). Its later history is unknown.

OXKUTZCAB
(V.N.1 CR?; XVI c. = Mayapan + 1d)

This hypothetical calendar is discussed in Chapter 2, 1544.

PALENQUE
(III.N.20 CR?; VII c. to VIII c. A.D. = Tikal + t − 0)

The Palenque calendar began its year on 5 Uayeb Mayapan, counting the day count from 1 to 13 and the year count from 1 to 20, and naming the year terminally. In all other respects, it was identical to the Tikal calendar. Indeed, the only difference was the single glyph for "seating" (0), which became "ending" (20) in certain inscriptions. Because, however, of the typological importance of terminal and initial counting in Middle American calendrics, the appearance of terminal counting among the Maya, normally initial counters, is a matter of some moment.

The earliest example of Palenque dating is from the katun ending in A.D. 692 (q.v., Chapter 2), somewhat later than the earliest Campeche dates (q.v.), which may also have involved terminal naming. Six other Palenque dates have been found:

9.16.16.6.7	13 Manik 20 Yaxkin (15 VI 767)	Palenque, House C, G1
9.16.17.6.12	1 Eb 20 Yaxkin (14 VI 768)	Yaxchilan, Lintel 9, A1-A3
9.17.5.8.12	9 Eb 20 Yaxkin (12 VI 776)	Naranjo 19, C1-C3
9.17.9.1.7	8 Manik 20 Ceh (29 XII 779)	Piedras Negras, Misc. Sculptured Stone 16, A1-B1
9.17.10.9.17	1 Caban 20 Yaxkin (11 VI 781)	Piedras Negras, Shell, K2
9.18.1.2.12	2 Eb 5 Uayeb (21 XI 791)	Naranjo, Hieroglyphic Stairway, A1

At all of these sites, Palenque dating is a secondary usage; the principal calendar was that of Tikal. Among the Mayan peoples, only the Huastec have actually adopted terminal dating.

PIPIL
(IV.G.20 CR 2 Acatl 1519?; XV c. to XX c. A.D. = Teotitlan)

Pipil day and month names are found in the *Popol Vuh* and *Annals of the Cakchiquels*, and some were actually adopted by the Quiche and Cakchiquel before the Conquest. The Pipil year began on 6 Ceh Mayapan, and the Pipil name for that month (Tlacaxipehualiztli) was adopted by both Mayan groups and actually became the beginning of the Cakchiquel year as well. The Pipil counted the days of the day count from 1 to 13 and those of the month from 1 to 20 and named the year for its 360th day. Its directional colors are unknown. It was calendrically congruent with the calendar of Teotitlan, which is not surprising in view of the Pipil tradition of having come from Nonoalco, presumably in Veracruz.

Presumably written in Aztec glyphs, the names of the days in Pipil are nonetheless different from those of Nahuatl. They are given in the anonymous *Calendario de 1685* as follows:

a. Cipactli (alligator)	f. Miquiztli (death)	k. Ozumatli (monkey)	p. Tecolotl (owl)
b. Ehecatl (wind)	g. Mazatl (deer)	l. Malinalli (grass)	q. Tecpilanahuatl (foreign prince)
c. Calli (house)	h. Toxtli (rabbit)	m. Acatl (cane)	r. Tecpatl (flint)
d. Quetzalli (quetzal)	i. Atl, Quiahuitl (water, rain)	n. Teyollocuani (heart eater)	s. Ayutl (turtle)
e. Cohuatl (serpent)	j. Ytzcuintli (dog)	o. Quauhtli (eagle)	t. Xochitl (flower)

The *Calendario* gives the names of the months as Fc., Ga., Ha., Ia., Ja., Ka., La., Ma., Nb., Ob., Pa., Qb., Rb., Aa., Ba., Ca., Da., Ea. (see Nahuatl). The list is identical to that given by Torquemada (1976:3:422ff.), even to the spelling mistakes. Beginning the list with Atlcahualo (*sic*) would seem to imply that the Pipil calendar was congruent with that of Tepepulco, but it is unlikely that it is in fact a Pipil month list at all, and the more likely placement of the Pipil year is that given in the heading.

POKOM
(II.L.1 CR?; XVIII c. to XX c. A.D. = Quiche, f1548)

The Pokom (Pokomam and Pokomchi) calendar began its year on 4 Kayab Mayapan and was thus congruent with the calendars of Quiche, Ixil, Aguacatec, and Mam. Presumably, it counted the day count from 1 to 13 and the month days from 1 to

20, though this is not directly attested. Its calendar round and direction colors are not documented. From Smith-Stark (1982) and Teletor (1959), the following Pokomam day names are possible:

a. —	f. Quimmi (death)	k. —	p. AjMajk (owl)
b. —	g. Kieh (deer)	l. Eg (tooth)	q. —
c. Aq'am' (night)	h. Q'anil (yellowness)	m. Aj (cane)	r. —
d. —	i. —	n. W'ahlam (jaguar)	s. Kojjok (thunder)
e. —	j. Tz'i (dog)	o. Tz'ikin (bird)	t. Ajgual (lord)

It seems likely that all of the day names were cognate with those of Quiche.

Two lists of Pokomchi month names have been published (Anon., 1931; Narciso 1906). They agree on beginning with Yax:

D. Yax (green)	I. Uniu	N. Kan Halam	A. Tz'ikin Kih (bird time)
E. Sac (white)	J. Muuan	O. Makux, Pet Cat	B. Mox Kih
F. Tsi	K. Tam (crab)	P. Kazeu	C. Tik Cheik (standing trees)
G. Chip, Tzip	L. Sac Cohk (white squash)	Q. Kanazi	
H. Chante Mac (cover)	M. Ohl	R. Kanahal (yellowing)	X. —

In comparison with cognate calendars, and particularly Kekchi, months D. and E. are one month early. F. is omitted from Anon. 1931. H., I., J., M., and A. would then be in the right spots. P. is 2 months early. L. and Q. are transposed. Neither list gives the name for X.

The units of the Pokom calendar are the day (*kih*), moon or month (*poh*), day count (?), year (*'ab, haam'*) and calendar round (?).

Vocabularies of Pokomam (Morán 1720) and Pokomchi (Morán 1725?) provide us with three correlational dates, frozen to 29 II 1548, for three of the months: J., L., and O., making it clear that the 5 days of X. fell at the end of either J. or K., so that the year must have begun with K. or L. The unfrozen date of 1905 (q.v.) confirms this by citing the first day of L. If that was the beginning of the Pokom year, it was congruent with Quiche, and it has been so coded in the heading. The three frozen dates are documented as follows:

J. Muan. el tiempo desde veynte de Abril hasta nueve de Mayo (Morán 1720).

J. Muan: the time from 20 April to 9 May (Pokomam).

L. Canazi. el tiempo q̃ ay desde 4 de junio, hasta los 23 del mismo (Morán 1725?).

L. Canazi: the time there is from June 4 until the 23rd of the same (Pokomchi).

O. Pet cat. el tiempo desde 3 de agosto hasta sus 22 inclusive (Morán 1725?).

O. Pet Cat: the time from August 3 to its 22nd inclusive (Pokomchi).

The 1905 correlational date lands on the right Pokom month of Canazi (Kayab) but 2 days late, thus indicating a late "Narciso calendar" (IV.L.1), named after its discoverer and calendrically equivalent to the calendar of Tenango. The calendrically impossible tradition that this corresponds to the beginning of Yax (Izcalli) would displace the beginning 7 native months back to make it an Aztec calendar with initial dating (IV.E.1), and indeed it seems likely that this tradition can be attributed to Aztec influence on Pokomchi, probably in Colonial times, which may indicate something about the origins of the Baja Verapaz "Pipil" and their calendar. A parallel but pre-Conquest and calendrically distinct (Teotitlan) influence is detectable in the Quiche and Cakchiquel calendars (q.v.).

QUICHE
(II.L.1 CR 1 Nooh 1532; XVIII c. to XX c. A.D. = Tikal – 2m – 0 – 360*)

The Quiche calendar begins its year on 4 Kayab Mayapan, counting the day count from 1 to 13 and the month days from 1 to 20, and naming the year initially. Its senior year bearer is 1 Quake (as in the Tikal calendar) and its calendar round began in 1532. Its direction colors are east–red; north–black; west–white; south–yellow; and perhaps the center–blue/green.

The units of the Quiche calendar are the sun or day (*q'ih*), moon or month (*iq', poval*), day count (*chol q'ih*), year (*'ab*), 400-day year (*may*), and calendar round (*hunab*). A hint in the *Popol Vuh* (line 8158) suggests that they may have known about the *tun* as well.

The day names are given by Ximénez (1720:1:119–20):

a. Imox (alligator)	f. Camey (death)	k. Batz (monkey)	p. Ahmac (owl)
b. Ic (wind)	g. Queh (deer)	l. Ei (tooth)	q. Noh (incense)
c. Acbal (night)	h. Canel (rabbit)	m. Ah (cane)	r. Tihax (flint)
d. Cat (lizard)	i. Toh (rainstorm)	n. Ix, Balam (jaguar)	s. Caoc (rain)
e. Can (serpent)	j. Tzi (dog)	o. Tziquin (bird)	t. Hunahpu (hunter)

A list of days from Santa Rosa Chujuyub is published by Termer (1930:383) with modern "translations." The *Popol Vuh* mentions all of the days except a., d., l., p., and r. and refers to a number of day names from other languages: Pipil (a. Cipac, k. Zimah, l. Q'aam, p. Kaqix), Yucatec (j. Aqan, Icxit, k. Choven, t. Ahav), and Mam (q. Kikap).

The Quiche month names are given in *Chol Poval Ahilabal Q'ih* (The Count of the Cycle and the Numbers of the Days):

L. Nabe Mam (1-lord)	Q. U Cab Pach (2-moss)	D. Nabe Zih (1-flower)	I. Tz'iba Pop (painted mat)
M. U Cab Mam (2-lord)	R. Tz'izil Lacam (shoots)	E. U Cab Zih (2-flower)	J. Kak (fire)
N. Liquin Ca (soft earth)	A. Tz'iquin Q'ih (bird time)	F. R Ox Zih (3-flower)	K. Ch'ab (arrow)
O. U Cab Liquin Ca (2-soft earth)	B. Cakam (red clouds)	G. Chee (trees)	
P. Nabe Pach (1-moss)	C. Balam (jaguar)	H. Tequexepual (flaying)	X. Tz'api Q'ih (extra days)

There is apparent Pipil influence in the naming of L., M., P., Q., and H.

The correlational position of the Quiche calendar has been continuously reaffirmed, and it is the only native calendar that has survived intact to the present time (see A.D. 1722, 1770, 1854, 1930, 1931, 1960, 1977).

SAYULA
(= Mixe?)

Sayula Popoluca is of special calendrical interest as one of the Zoquean languages spoken in historic times in the immediate vicinity of important Olmec archaeological sites. Unfortunately, its calendar has not survived. From the Oluta dialect (Clark 1981) we have the following possible day names:

a. Uśpi'n (alligator)	f. Qqui'c (dead)	k. Jamʉ (ashes)	p. Ju'cu (owl)
b. Jạmu (wind)	g. Jaytsu' (deer)	l. Tʉtsʉ (tooth)	q. Ušʉ (quake)
c. Tsujʉ (night)	h. Coya (rabbit)	m. Pựyi (cane)	r. —
d. Tọqui (iguana)	i. Nʉjʉ (water)	n. Cạja'u (jaguar)	s. Pạjʉ (grass)
e. Tsana'y (serpent)	j. Šu'ni (dog)	o. Ju'ca'n (tobacco)	t. —

The units of the Sayula calendar were the day (*cǫ̌so*), moon or month (*po'a*), day count (?), year (*šivi'i*), and calendar round (?). Nothing more is known of it.

SUBTIABA
(?)

The units of the Subtiaba calendar were the day (*bi*), moon or month (*uku*), and year (*sigu*) and very probably the day count and calendar round as well. From the vocabularies compiled by Lehmann (1920), the following day names are possible:

a. Piu (alligator)	f. Anganu (dead)	k. Lutu (monkey)	p. Angú (buzzard)
b. Nena (wind)	g. Sihi (deer)	l. Sinyu (tooth)	q. —
c. Midu (night)	h. Sídu (rabbit)	m. —	r. Esee (flint)
d. A'g'ú (iguana)	i. Iya (water)	n. Endi (jaguar)	s. Unde (rain)
e. Apu (serpent)	j. Runya (dog)	o. Diui (bird)	t. Di (flower)

TARASCAN
(IV.I.20 CR 1 Auani 1530?;
XVI c. A.D. = Toltec − 1m Classic or Huastec + 1m Postclassic)

The Postclassic Tarascan year probably began on 6 Kankin Mayapan, coinciding with the Matlatzinca and Tepanec New Years. The Tarascans counted the days of the day count from 1 to 13 and those of the month from 1 to 20 and named the year terminally. Their directional colors and the beginning of their calendar round are unknown.

De la Coruña, identified by Baudot (1983) as the author of the *Relación de Michoacán*, refers (1541:143) to the days Deer, Water, Dog, Monkey, Cane, and Knife in Spanish without giving their Tarascan names. From Swadesh (1969), the following day names are possible:

a. Uxpi (alligator)	f. Uárhini (death)	k. Ozóma (monkey)	p. Tukúru (owl)
b. Tarhíyata (wind)	g. Axúni (deer)	l. Uitzákua (grass)	q. Yúrniri (quake)
c. Kuahta (house)	h. Auani (rabbit)	m. Isimba (cane)	r. Tzhinápu (flint)
d. Uahtzáki (lizard)	i. Itsí (water)	n. Puki (jaguar)	s. Mánikua (rain)
e. Akuítze (serpent)	j. Uíchu (dog)	o. Uakúsi (eagle)	t. Tsitsíki (flower)

The name for k is Nahuatl; that for a is Totonac and Sayula. The glyphs for the Tarascan days are documentable only from archaeology:

a. Alligator. Tizapan (Meighan and Foote 1968:197, Pl. 18c).
b. Wind. Tizapan (Ibid.:194, Pl. 15b,c,d; 189, Pl. 10d).
c. House. Tizapan (Ibid.:189, Pl. 10d).
d. Lizard.
e. Serpent.
f. Death.
g. Deer. Tizapan (Ibid.:92, Fig. 13).
h. 1 Rabbit. Tizapan (Ibid.:189, Pl. 10b,c).
   2 Rabbit. Tizapan (Ibid.:189, Pl. 10d).
   3 Rabbit. Tizapan (Ibid.:189, Pl. 10a).
   4 Rabbit. Amapa (Bell 1971:708, Fig. 10d).
   Rabbit. Tizapan (Meighan and Foote 1968:194, Pl. 15c).
i. 3 Water. Tizapan (Ibid.:190, Pl. 11c).
j. Dog. Tizapan (Ibid.:108, Fig. 29).
k. Monkey.

l. 2 Grass. Tizapan (Ibid.:189, Pl. 10d).
    Grass. Tizapan (Ibid.:194, Pl. 15; 195, Pl. 16b; 196, Pl. 17c).
m. Cane. Tizapan (Ibid.:189, Pl. 10c).
n. Jaguar.
o. Eagle. Tizapan (Ibid.:193, Pl. 14a).
p. Owl. Tizapan (Ibid.:189, Pl. 10c).
q. Quake. Tizapan (Ibid.:191, Pl. 12b).
r. Flint. Tizapan (Ibid.:189, Pl. 10a).
s. Rain. Tizapan (Ibid.:197, Pl. 18a).
t. Flower. Tizapan (Ibid.:195, Pl. 16b).

The prominence of Rabbit signs suggests the possibility that Rabbit may have been the senior year bearer.

De la Coruña (1541) mentions most of the Tarascan months. Page references are given in the following listing:

I. −   R. Uapánscuaro (18) (1-casting)
J. −   A. Cáheri Uapánscuaro (10) (2-casting)
K. Máscuto (264) (stew)   B. −
L. Uáscata Cónscuaro (11) (1-feast)   C. Peuánscuaro (nativity)
M. Cáheri Cónscuaro (248) (2-feast)   D. Curíndaro (10)−
N. Hanciñáscuaro (186, 188) (assembly) E. Tzitacuarénscuaro (resurrection)
O. Hicuándiro (191) (bathing)   F. Purecóracua (70, 101, 246) (war)
P. Sicuíndiro (9, 82, 188) (flaying)   G. Cuingo (10, 158, 265) (bird)
Q. Charapu Zapi (10) (little red)   H. Unisperácuaro (161–62) (bone song)
                                        X. −

The Tarascan year has to have begun with one of four months: I., J., G., or H. The relative positions of K., M., R., S., and F. are certain, and their correlational dates are clear (see A.D. 1540, 1541). The order of the others must be considered uncertain. Months C. and E. are given by León (1903:486), quoting Ramirez (1600). Caso's (1967:252) placement of the New Year makes it congruent with Matlatzinca and Tepanec (q.v.), which is plausible *prima facie* on geographic grounds.

The units of the Tarascan calendar are the sun or day (*huryíyata*), moon or month (*kutsí*), day count (?), year (*uéxurini*), and calendar round (?).

The autonomy of its writing system and the independence of its language both argue for the separateness of the Classic Tarascan calendar. Evidence for it is found in archaeological contexts antedating the Tilantongo calendar, and leading to the expectation of an initial dated Classic calendar derived from that of Yucuñudahui. Its most likely placement is that given in the heading (see A.D. 1540, 1541).

TENANGO
(IV.L.0 CR?; VI c. A.D. = Yucuñudahui − 100d)

The Tenango year probably began on 6 Kayab Mayapan, counting the day count from 1 to 13 and the year count from 0 to 19, and naming the year initially. Its calendar round and directional colors are unknown. It cannot be ethnically or linguistically identified, though it seems most likely to have been Otomanguean. It might also have been used by the unclassified Cuitlatec.

From the Tenango Tablets of A.D. 798 (q.v.), the Tenango calendar can be fixed in relation to those of the Zapotec and Olmec. It appears to have used initial dating, suggesting possible relationships with Yucuñudahui, Teotihuacan, and Tlapanec, and it begins the year 100 days before the Yucuñudahui calendar. It probably did use zero dating, though that is uncertain. If it used the 2 to 14 count of days of Teotihuacan and Tlapanec, its starting point was 2 months later. It is recorded in Xochicalco glyphs with Teotihuacan year signs and is unknown apart from one pair of inscriptions, but it seems possible that it could have survived among the undocumented calendars of Guerrero (see Cuitlatec) (see A.D. 798).

TEOTIHUACAN
(II.Q.0 CR?; II c. B.C. to IX c. A.D. = Kaminaljuyu – 105d + 14*)

The Teotihuacan calendar derives from Kaminaljuyu (q.v.). It began its year on 19 Zip Mayapan and used Type II name days, counting the day count from 2 to 14 and the days of the month from 0 to 19. The beginning of its calendar round and its directional colors are unidentified. Its day glyphs and use of bar and dot numerals show both Olmec and Zapotec influence, and the assumption that it used zero counting as they did seems reasonable but is unproved. The initial naming of the year makes it like Maya and unlike Olmec and Zapotec. We do not have a complete calendar round date in the Teotihuacan writing system, and there is no suggestion that it used the Long Count. Although it must have spanned the Classic period, it is sparsely documented only at the beginning and the end of it, in the second century B.C. and the eighth and ninth centuries A.D. A survival of the distinctive 2 to 14 counting of the day count is attested in the Tlapanec calendar of the fifteenth and sixteenth centuries: it is otherwise unique to Teotihuacan.

The use of 2 to 14 counting of the day count is documentable in several places, but it is headlined by a glyph at Teotihuacan itself recording the Yucuñudahui year 13 Flint as 14 Flint Teotihuacan style. It uses the Yucuñudahui year sign but the Teotihuacan day glyph and numeration.

A further confirmation is the special dot sign used at Cacaxtla for the extra unit of the Teotihuacan count (see A.D. 861).

The units of the Teotihuacan calendar were the day, month, day count, year, and calendar round. Its year sign was widely used elsewhere, even as far afield as Honduras and became the standard symbol of the year in all the writing systems of Postclassic Central Mexico. The Yucuñudahui year sign was also used at Teotihuacan.

Half of the day glyphs can be identified. Their provenience follows.

a. Alligator?
b. Anno 2 Wind. Teotihuacan (Caso 1967:149, Fig. 11).
   Anno 2 Wind. Texmilincan, Teotihuacan (Ibid.:161, Fig. 18).
   Anno 5 Wind. Teotihuacan (Ibid.:159, Fig. 14b).
   Anno 7 Wind. Teotihuacan (Ibid.:147, Fig. 6; 159, Fig. 14a).
   Anno 7 Wind. Stela Labrada, Veracruz (Navarrete 1986:13, Figs. 7a,b).
   Anno 7 Wind. Ixtapaluca (Ibid.:104, Fig. 1).
   Anno? 7 Wind. Kaminaljuyu Stela 10 (Miles 1965:254, Fig. 13).
   Anno 12 Wind. Texmilincan, Teotihuacan (Caso 1967:169, Fig. 5a).
   11 Wind. Kaminaljuyu? (Ibid.:156 bis, Pl. 1).

c. 7 Night. Teotihuacan (Ibid.:146, Fig. 3).
d. Iguana?
e. 10 Serpent. Teotihuacan (Ibid.:156 bis, Pl. 1:3).
   Serpent. Teotihuacan (Ibid.:156 bis, Pl. 1:4).
f. 7 Death. Izapa, Miscellaneous Monument 60.
g. Deer. Teotihuacan (Ibid.:156 bis, Pl. 1, Fig. 8d).
h. Anno 8 Rabbit. Teotihuacan (Ibid.:146 bis, Figs. 4–5b).
   Anno 12 Rabbit. Teotihuacan (Ibid.:146 bis, Figs. 5–4a,c).
   Rabbit. Teotihuacan (Ibid.:145, Fig. 2).
i. Water. Unknown (Ibid.:149, Fig. 12).
j. Foot?
k. Monkey. El Castillo Stela 1 (Miles 1965:267, Fig. 18d).
l. Anno 5 Sun. Unknown (Caso 1967:151, Fig. 15b).
   Anno 10 Sun. Teotihuacan (Ibid.:151, Fig. 15b).
   6 Sun. Unknown (Caso 1967:151, Fig. 15a).
   8 Sun. Unknown (Ibid.:151, Fig. 15a).
   10 Sun. Huajuapan (Ibid.:151, Fig. 16).
m. 2 Cane. Unknown (Ibid.:148, Fig. 10d; 157, Fig. 10).
   4 Cane. Unknown (Ibid.:148, Fig. 10c).
   8 Cane. Teotihuacan (Ibid.:156 bis, Pl. 1, Fig. 8b).
   11 Cane. Kaminaljuyu (Miles 1965:263, Fig. 16e).
n. 11 Jaguar. Atetelco, Teotihuacan (Ibid.:166, Fig. 7).
o. Eagle?
p. Owl?
q. Anno 4 Quake. Teotihuacan (Navarrete 1986:20, Fig. 12g).
   Anno 11 Quake. Stela 1, Fracción Mujular, Chiapas (Navarrete 1986:19,
      Fig. 11).
r. Anno 14 Flint. Teotihuacan (Caso 1967:156 bis, Pl. 1, Fig. 1).
s. Rain?
t. Lord?

The citation of Rabbit as a year bearer at Teotihuacan is doubtless owing to the use of the Yucuñudahui calendar. The preponderance of Type II year bearers clearly marks the Teotihuacan usage.

Two correlational dates link the Teotihuacan calendar to Olmec (see 147 B.C., A.D. 861), fairly well spanning its history. Virtually all subsequent dating in Central and Northern Mexico is in provincial Type III calendars modeled after that of Yucuñudahui (q.v.), already in use in the Mixteca in the fifth century (see A.D. 426, and Cholula, Huastec, Tarascan, Toltec, Totonac). The only late survival of 2 to 14 counting was in Tlapanec (q.v.) (see 147 B.C., A.D. 798, 861).

TEOTITLAN
(IV.G.20 CR 2 Acatl 1519; XIV c. to XVI c. A.D. = Otomi + 1m)

The Teotitlan calendar began the year on 6 Ceh Mayapan. Its glyphs, counting, and nomenclature are those of Nahuatl (q.v.). Its directional colors are unknown. It seems to be associated with the eastern Nahua of Puebla, northern Oaxaca, and Veracruz and was congruent with the calendars of the Postclassic Totonac, Pipil, and perhaps Nicarao (see Kirchhoff 1950; Jiménez 1961; Caso 1967).

Dates from 1545, 1549, and 1579 correlate the Teotitlan calendar with the Julian and Gregorian ones independently from other correlations. Although these dates are somewhat less decisive than those for the Tepepulco calendar (q.v.), they amply confirm the placement of the Teotitlan calendar given in the preceding heading (see A.D. 1545, 1549, 1579, 1596).

TEPANEC
(IV.I.20 CR 1 Rabbit?; XI c. to XVI c. A.D. = Cuitlahuac + 1m)

The Tepanec calendar is the Postclassic calendar of the Tarascan and Matlatzinca (q.v.), used in Nahuatl by Atzcapotzalco and Tlatelolco, according to Jiménez (1961). Its year began on 6 Kankin Mayapan. Jiménez refers it to the *Anónimo de Mendieta.*

TEPEPULCO
(IV.F.20 CR 2 Acatl 1539; XII c. to XVII c. A.D. = Otomi)

The Tepepulco year began on 6 Zac Mayapan. This is the Nahuatl version of the Otomi calendar, and its counting, nomenclature, and glyphs are Nahuatl (q.v.). Its calendar round and directional colors are unknown. It is documented in a great many Spanish sources and codices, including Sahagún, none of which is pre-Conquest (see Nahuatl).

The placement of the Tepepulco calendar is firmly established by correlational dates in 1567 and 1612 (q.v.), quite independently of the Aztec-Julian correlations and those of several other intercorrelated Central Mexican calendars (see Otomi, Tarascan, Matlatzinca, Mazatec, Huastec), any one of which would establish the general Christian-day count correlation if the more highly touted Aztec dates did not exist. The Tepepulco correlations are among the most satisfactory of these dates (see A.D. 1298, 1308, 1567, 1612).

TEPEXIC
(IV.K.20 CR?; XI c. to XVI c. A.D. = Colhuacan + 1m)

The Tepexic calendar began its year on 6 Pax Mayapan, counting the day count from 1 to 13 and the year count from 1 to 20, and naming the year terminally. Its calendar round and directional colors are unknown.

It is probable that this is the Postclassic version of the Cholula calendar of the Classic period. It seems quite possible that it was Mixtec. The Nahuatl version has also been tentatively identified with Huexotzinco (Jiménez 1961) (see A.D. 1036, 1048, 1340, 1369, 1424, 1507, 1544, 1596).

TEQUISTLATEC
(= Mixe?)

The units of the Tequistlatec ("Chontal de Oaxaca") calendar are the day (*lidine*), moon or month (*galmuľa*) day count (?), year (*amats'*), and calendar round (?). The day names have been published by Martínez Grácida (1910) and more accessibly by Turner and Turner (1971:362).

a. Lebalc'o' (light)
b. Lawa' (wind)
c. Lajuľ (house)
d. Lamol'o' (lizard)
e. Laynofaľ (serpent)

f. Lamaya (death)
g. Lawala Q'uec (deer)
h. Tonomma (greeting)
i. Laja' (water)
j. Galtsiqui (dog)

k. Galmigú' (monkey)
l. Dodasgona'ma (twist)
m. Albeba (sugarcane)
n. Galjaguár (jaguar)
o. Galtijuli (eagle)

p. Galaguilacene (buzzard)
q. Galfounaľ (sun)
r. Labic (stone)
s. Lagwi (rain)
t. Liba' (flower)

A remarkable feature is the incorporation of *mico* (spider monkey), jaguar, eagle and sugarcane into the day count. Some of the other day names have a suggestive relationship to the Olmec glyphs.

TEXCOCO
(IV.D.20 CR?; XII c. to XVI c. A.D. = Tilantongo + 1m, f1548)

The Texcoco calendar began its year on 6 Ch'en Mayapan. Its writing, counting, and naming systems are those of Nahuatl (q.v.). Its calendar round and direction colors are unknown. It is documented in the *Xolotl, Historia Tolteca-Chichimeca, Anales de Cuauhtitlan* (Kirchhoff 1950:127, 131; Jiménez 1961:139, 141), *Memorial Breve, Crónica Mexicáyotl*, and *Anales de Tlatelolco* (Davies 1977:456–57, 462) (see A.D. 1150, 1170, 1176, 1296, 1339, 1416, 1553).

TIKAL
(II.N.0 CR 1 Caban 1544; I c. to XVIII c. A.D. = Olmec − 1m − 100d)

The Tikal calendar begins its year on 4 Uayeb Mayapan, using bar and dot numerals, counting the day count from 1 to 13 and the days of the months of the year from 0 to 19, and naming the year for its first day. Its calendar round begins (see Chapter 2, 1500) with 1 Quake (1544) and its directional colors are east–red; north–white; west–black; south–yellow; and center–blue/green. Its Long Count begins with 4 Ahau 8 Cumku (Julian day number 584,283: see Correlation). It is first attested in A.D. 292 (q.v.) and was still known in A.D. 1618 when Merida celebrated Baktun 12.0.0.0.0 (Edmonson 1986, Ch. 29). It was finally abandoned in 1752. A complete initial series date is written 'Initial Series (Infix), Coefficient, Baktun, Coefficient, Katun, Coefficient, Tun, Coefficient, Uinal, Coefficient, Kin, Coefficient, Day Name, Lord of the Night, Coefficient, Month Name' (see A.D. 320).

The units of the Tikal calendar are the day (*kin*), moon or month of the year (*ik, u*), month of the stone (*uinal*), day count (*tzol kin*), 360-day stone (*tun*), year (*haab*), night cycle (?), 20-stone pile (*katun*), calendar round (*hunab*), 13-katun cycle (*may*), and 20-katun bundle of stones (*baktun*). For astronomical purposes, higher vigesimal multiples of the baktun are also found. The higher multiples in counting time were the *pictun* (= 20 baktuns), *calabtun* (= 20 pictuns), *kinchiltun* (= 20 calabtuns) and *alautun* (= 20 kinchiltuns). In kinchiltuns, the end of the 13-baktun cycle is 1.11.19.13.0.0.0.0, corresponding to the winter solstice of A.D. 2012.

Except for "baktun," which is an academic coinage, these and other names used in reading the Tikal calendar are customarily those of Colonial Yucatecan Maya, even though most Tikal inscriptions were written in Cholan or in a substantially earlier form of Yucatec.

The following Tikal day glyphs are those of the *Dresden Codex*, the only surviving pre-Columbian Mayan manuscript:

a f k p

b g l q

c h m r

d i n s

e j o t

The names of the Yucatecan days are given by Landa (1966). Many are very archaic, and their meanings are sometimes more inferential and associational than translational:

a. Imix (alligator)

b. Ik (wind)
c. Akbal (night)

d. Kan (iguana)
e. Chicchan (serpent)

f. Cimi (death)

g. Manik (deer)
h. Lamat (rabbit)

i. Muluc (rain)
j. Oc (foot)

k. Chuen (monkey)

l. Eb (tooth)
m. Ben (cane)

n. Ix (jaguar)
o. Men (eagle)

p. Cib (owl)

q. Caban (quake)
r. Etz'nab (flint)

s. Cauac (storm)
t. Ahau (lord)

The glyphs for the months are collected from the monumental inscriptions and illustrated by Thompson (1971):

The Yucatecan month names are also given by Landa (1966):

N. Pop (mat)	A. Xul (end)	F. Zac (white)	K. Pax (drum)
O. Uo (frog)	B. Yaxkin (green time)	G. Ceh (deer)	L. Kayab (turtle)
P. Zip (stag)	C. Mol (gather)	H. Mac (cover)	M. Cumku (dark god)
Q. Zotz' (bat)	D. Ch'en (well)	I. Kankin (yellow time)	
R. Tzec (skull)	E. Yax (green)	J. Muan (owl)	X. Uayeb (specters)

The 9-year cycle of the Lords of the Night is labeled with the glyphic names of the nine gods (*bolon ti ku*):

G1.	G4.	G7.
G2.	G5.	G8.
G3.	G6.	G9.

Their Yucatecan names are found in the *Book of Chilam Balam of Tizimin* (Edmonson 1982:1001–1010):

G1. Hau Nab
   (slice point)
G2. Hutz' Nab
   (split point)
G3. Kuk Nab
   (quetzal point)

G4. Oyal Nicte
   (island flower)
G5. Ninich Cacau
   (wormy cacao)
G6. Chabi Tok
   (digging knife)

G7. Macuilxochitl
   (5 flower)
G8. Hobon y Ol Nicte
   (painted heart
   flower)
G9. Kouol y Ol Nicte
   (pouch heart flower)

The correlation of the Tikal calendar to those of the rest of Middle America is indirect: it depends upon relating Tikal to Mayapan, Mayapan to Julian, and Julian to the other native calendars. The introduction of the Mayapan calendar in 1539 kept the first month of the Tikal calendar, advanced the beginning of the year 2 days, and dropped zero dating of the days of the month.

The correct Julian date of the Mayapan year is given by Landa (see A.D. 1553) as July 16, 1553, but it is a frozen date of one leap year earlier (i.e., before February 29, 1552). That is traditionally proved by the Julian date for the surrender of Cuauhtemoc on August 13, 1521 (Nahuatl 1 Coatl), but any other day-count correlation would in fact prove the same thing. For the cognate Mayan 1 Chicchan to fall on August 13, 1521, the Mayapan year bearer 12 Kan would have to fall on July 15, 1553. This confirms the "modified Thompson 2" correlation (Satterthwaite 1965:628), placing 4 Ahau 8 Cumku on Julian day number 584,283. No other solution is ethnohistorically possible without postulating a break in the continuity and uniformity of the universal Middle American day count, an assumption falsified by every correlational date in the present volume.

The interpretation of Yucatecan dating expressions is contingent upon their careful use of current and completive counting. Thus:

    t u hun pop 'on 1 Pop (completed)' = 2 Pop
    t u hun te pop 'on 1 Pop (current)' = 1 Pop
    t u ca tun 'on the second tun (current)' = tun 2
    t u ca p'iz tun 'on the second (measured) tun' = tun 3

See 1532–44 dates from the *Chronicle of Oxkutzcab* (A.D. 1544). The placement of a calendar round in katun 4 Ahau (1480–1500) identifies the beginnings of the Tikal calendar round as 1 Caban (1491) (see A.D. 1500).

Derived from the Olmec calendar, the Tikal calendar was the primary calendar of the Eastern Maya from the first century on. It was challenged in the seventh century by the Palenque and Campeche calendars (q.v.) and was officially replaced by the calendars of Mayapan (in 1539) and Valladolid (in 1752). The Guatemalan Maya appear to have used the Tikal calendar in early times but eventually inaugurated other Type II calendars (see Pokom, Ixil, Quiche, Mam, Aguacatec, Cakchiquel). Only the Yucatecans, Chol, and Kekchi (q.v.) kept the Tikal calendar into the eighteenth century (see *236 B.C., A.D. *176, 199, 292, 320, 889, 948, 1178, 1236, 1480, 1486, 1500, 1513, 1517, 1532, 1533, 1544, 1618, 1752).

TILANTONGO
(IV.C.20 CR 2 Huiyo 1546; X c. to XVI c. A.D. = Zapotec + 1d – 0)

According to Caso (1965b:940), a new Mixtec calendar was substituted for that of Yucuñudahui (q.v.) by King 5 Quevi "Tlachitonatiuh" of Tilantongo in A.D. 973. It begins the year on 6 Mol Mayapan, counting the days of the day count from 1 to 13 and those of the month from 1 to 20 and naming the year terminally. Its calendar round began on 2 Cane on 7 Ix 6 Mol Mayapan (1546). Its direction colors are unknown.

The Tilantongo Mixtec appear to have originated the writing system subsequently identified with the Aztecs (see Nahuatl), and the Tilantongo calendar is continuously documented in genealogical picture manuscripts from its inception into the sixteenth century. Caso (1977:2:8) identifies twenty-eight of these as Mixtec (24 Alta: A.; 1 Baja: B.; and 3 Costa: C.). Glass (1975b) classifies four Mixtec A. and two Mixtec C. manuscripts as pre-Conquest (A.: *Bodley, Aubin 20, Zouche-Nuttall, Vienna*; C.: *Becker I, Colombino*), and two of these (*Aubin 20* and *Vienna Obverse*) are calendrical in character. Three Colonial Mixtec A. manuscripts (*Becker II, Selden, Zacatepec 25*) are native in style and content.

Although the day glyphs are Aztec, the day names are those of Mixtec (q.v.). The month names are unknown, but the two known month glyphs are also at least similar to Aztec (see A.D. 1555). Correlational dates relate the Tilantongo calendar to Tepexic (1036, 1048, 1339), Zapotec (1445), Aztec and Texcoco (1416, 1500), and the Christian calendar (1520, 1544, 1549, 1551, 1589). The Tilantongo calendar was the dominant calendar throughout the Mixteca and beyond the Mixtec language area into the Chocho and possibly Cuicatec (q.v.) and Nahuatl regions (see Tlaxcalan). Veytia (1973:58) calls it the Toltec calendar, which can neither be proved nor disproved on present evidence, but which I believe to be unlikely (see A.D. 973, 1036, 1048, 1339, 1416, 1445, 1500, 1520, 1544, 1549, 1551, 1555, 1589).

TLAPANEC
(II.J.0 CR 2 Grass 1507; XV c. to XVI c. A.D. = Teotihuacan – 7m)

The Tlapanec calendar began the year on 4 Muan Mayapan, counting the days of the day count from 2 to 14 and those of the months of the year from 0 to 19, and naming the year initially. Its calendar round began on 2 Grass (1507).

The units of the calendar were the day (*bihi*) moon (*guic*), day count (?), year (?), and calendar round (?). A few possible day names are reported in Lehmann's (1920) vocabularies:

a. –	f. –	k. –	p. –
b. –(wind)	g. –(deer)	l. –(grass)	q. –(quake)
c. Guguá (house)	h. Chuagí (rabbit)	m. –	r. –
d. –	i. Iya (water)	n. –	s. –
e. –	j. Chumuá (dog)	o. Yucuu (bird)	t. –

The Tlapanec writing system was Tilantongo, and there are four known Tlapanec picture manuscripts (*Humboldt Fragment I, Lienzo de Tlapa, Azoyu I,* and *Azoyu II*). The position of the calendar is given by its correlation with Aztec (see A.D. 1487, 1522).

TLAXCALAN
(IV.C.20 CR?; XVI c. to XVII c. A.D.= Tilantongo)

The Tlaxcalan calendar is the Nahuatl version of the Tilantongo Mixtec calendar. Veytia (1973:58) identifies both with the Toltec, but the scanty calendrical glyphs of Tula do not permit settling the point.

The Tlaxcalan year begins on 6 Mol Mayapan. Its counting and terminology are those of Nahuatl, and its calendar round and direction colors are unknown. Its month names follow list a. except for Fb. (see Nahuatl). It is documented by Muñoz Camargo, Zapata, Torquemada, Lorenzana, Clavijero, Vetancourt, and Veytia, and perhaps by the Borgia group of codices.

TOJOLABAL
(= Kanhobal?)

The units of the Tojolabal calendar are the day or sun (*k'ak'u*), month or moon (*'ixau*), day count (?), year (*hab', hab'il*), and calendar round (?). A number of its possible day names can be found in Brinton (1893) and Furbee-Losee (1976):

a. Aín (alligator)	f. —	k. B'ats' (monkey)	p. —
b. 'Ik' (wind)	g. Cheh (deer)	l. 'Eh (tooth)	q. —
c. Ak'ual (night)	h. K'anal (star)	m. 'Ah (cane)	r. —
d. —	i. Ch'aa (water)	n. —	s. Chauuk (storm)
e. Chan (serpent)	j. Tz'í (dog)	o. —	t. Ahau (lord)

It seems likely that it was congruent with the Kanhobal calendar.

TOLTEC
(III.0.0 CR 1 Tecpatl 1536?; = Cholula – 1m Classic or Tepanec + 1m Postclassic)

The limited calendrical materials from Tula include use of the Teotihuacan year sign, both dot-and-bar and dot numerals, and perhaps 13 recognizable day signs, 8 of them with coefficients. There are no calendar round or Long Count dates. The glyphs are similar to those of Aztec, Yucuñudahui, Teotihuacan, and Xochicalco but identical to none of them.

The unique glyph I have read as Fist is rather arbitrarily assigned as day c., but there is no way of placing it securely. With this qualification, the possible day signs of Tula are:

a. 5 Alligator, Tula (Caso 1941:92, Fig. 9).
b. 2 Wind. Tula Building B (Acosta 1956:64, Pl. 15).
c. 9 Fist. Tula (Nicholson 1971:108, Fig. 26).
d. Iguana?
e. Serpent. Tula (Acosta 1956:100, Pl. 40).
f. Death. Tula Skull Altar (Dutton 1956:236, Pl. 18d).
g. Deer?
h. 2 Rabbit. Tula Building B (Acosta 1956:68, Pl. 18).
i. Water?
j. Foot?
k. Monkey?
l. Grass. Tula (Acosta 1956:96, Fig. 16).
m. 1 Cane. Cerro de la Malinche (Navarrete and Crespo 1971:12, Fig. 6c).
   4 Cane. Cerro de la Malinche (Ibid.:Fig. 61).
   Cane. Tula (Dutton 1956:206, Pl. 4g; possible year sign).
n. Jaguar. Tula (Ibid.:206, Pl. 4g).
o. Eagle. Tula (Ibid.:210, Pl. 5).
p. Owl?
q. Quake?

r. 6 Flint. Tula (Navarrete and Crespo 1971:12, Fig. 5).
    8 Flint. Cerro de la Malinche (Ibid., Fig. 6h).
s. 1 Rain. Tula (Caso 1967:176, Fig. 12c).
    3 Rain. Tula (Acosta 1956:66, Fig. 10:1).
t. Anno 2 Lord. Tula (Dutton 1956:234, Pl. 17d; cf. Caso 1967:236, Pl. 18a
    and Acosta 1957:154, Pl. 28).

From the use of two different numeral systems, there would appear to have
been two calendars represented at Tula. One, using dot numerals and the possible
year bearers Cane, Flint, and Rabbit may be the Tilantongo calendar attributed to
the Toltec by Veytia (1973:58). I consider it more likely, however, that this is the
Postclassic Colhuacan calendar. The other, using bar-and-dot numerals with the
days Alligator and Fist, probably represents the Classic Toltec calendar, placed as
in the preceding heading, though this cannot be documented. The apparent occurrence of a year sign with the day Lord would appear to imply an Olmec calendar,
but this is a questionable reading.

TOTONAC
(III.L.0 CR?; XVI c. A.D. = Huastec – 1m Classic, Otomi + 1m Postclassic)

The Totonac calendar probably began the year on 6 Ceh Mayapan, counting the
days of the day count from 1 to 13 and those of the month from 1 to 20, and
naming the year terminally. Its calendar round and directional colors are unknown.
From Aschman (1973) the following day names are possible:

a. Uxpi (alligator)	f. Liiníin (death)	k. Muuxni (monkey)	p. Monkxnú (owl)
b. Uun (wind)	g. Huuqu'i (deer)	l. T'uhuáan (grass)	q. Tachiquí (quake)
c. Akxtak'a (house)	h. Tampanámac (rabbit)	m. K'áatiit (cane)	r. Chíhuix (stone)
d. Slúcuc (lizard)	i. Ch'uuchut (water)	n. Misin (jaguar)	s. S'een (rain)
e. Luunua (serpent)	j. Chichí (dog)	o. Pichaahua (eagle)	t. Xánat (flower)

The first day name is the same as the Tarascan and Sayula name. From calendrical
names of persons and gods mentioned by Torquemada, Ixtlilxochitl, and the
Relaciones Geográficas (Broda 1969:86), days b, h, k, m, q, s, and t are attested in
Totonac even though they are only named in those sources in Nahuatl (see also
Melgarejo [1966:35–40]).

The units of the Totonac calendar are the sun or day (*ch'ichiní*), moon or
month (*papá*), day count (?), year (*c'aata*), and calendar round (?).

From Olmos (1912), we have the dates of the beginnings of two Totonac months, indicating that they had Type IV New Years, and that they began the year sometime between B. Yaxkin (Panquetzaliztli) and G. Ceh (Tlacaxipehualiztli). Thus the Totonac calendar could be IV.B. (Metztitlan), IV.C. (Tlaxcalan), IV.D. (Texcoco), IV.E. (Aztec), IV.F. (Otomi), or IV.G. (Teotitlan) (see A.D. 1538, 1539). Influence of any of these on Totonac calendrical ideas would be geographically plausible. I am inclined to guess that its year began on the latest date (IV.G.), which would place it 1 month earlier than Huastec (IV.H.), and make it congruent with the adjacent Nahuatl calendar of Teotitlan. The Totonac name for Yaxkin was Calcusot (cf. *c'aalacx'itaat* 'in the middle'). This would be the tenth month if I have placed the calendar correctly. The other month names are unknown. A number of possible day glyphs can be recovered from the archaeology of Central Veracruz:

a. Alligator. "Tres Picos" (García 1971:534, Fig. 29b).
b. Wind. Nopiloa (Nicholson *et al.* 1971:65, No. 60).
   Wind. Río Blanco (Ibid.:35, Fig. 2b).
   Wind. "Tres Picos" (García 1971:534, Fig. 29c).
c. House. Dicha Tuerta (Nicholson *et al.* 1971:68, No. 73).

d. Iguana. Veracruz (Ibid.:opposite p. 13).
e. 2 Serpent. Río Blanco (Ibid.:36, Fig. 5).
f. Death. "Tres Picos II" (García 1971:540, Fig. 35b).
g. Deer.
h. Rabbit. San Marcos (Nicholson *et al.* 1971:71, No. 78; 68, No. 69).
i. 4 Water. "Tres Picos" (García 1971:534, Fig. 29b).
j. 3 Dog. "Tres Picos" (Ibid.:534, Fig. 29b).
k. Monkey. San Marcos (Nicholson *et al.* 1971:63, No. 56).
l. Grass. Río Blanco (Ibid.:36, Fig. 5).
m. Cane. Nopiloa (Ibid.:67, Nos. 66, 67, 68, 69, 74; 64, No. 58).
    Cane. "Remojadas" (Ibid.:47, Nos. 15, 17; 45, Nos. 10, 13).
    Cane. Los Cerros (García 1971:531, Fig. 24b).
n. Jaguar. Río Blanco (Nicholson *et al.* 1971:35, Fig. 2b).
o. Eagle. Río Blanco (Ibid.:35, Fig. 2b).
p. 2 Owl. Los Cerros (García 1971:531, Fig. 24b).
q. Quake. Los Cerros (Ibid.:531, Fig. 24b).
r. 1 Flint. "Tres Picos II" (Ibid.:540, Fig. 35b).
s. 4 Rain. Río Blanco (Nicholson *et al.* 1971:36, Fig. 5).
t. Flower. Veracruz (Ibid.:frontispiece) (see A.D. 1538, 1539).

TRIQUI
(= Tilantongo?)

The units of the Triqui calendar were the sun or day ($g\ddot{u}i^3$), moon or month ($ahui^{34}$), day count ($nayo'^4$), year ($yo^{13}o$), and calendar round (?).
   The following are possible day names from Good (1979):

a. —	f. Gahui[15] (death)	k. Guruhui[3] (monkey)	p. Xa'u[54] (owl)
b. Nane[5] (wind)	g. Xutaj[5] (deer)	l. Coj[3]o (grass)	q. Yun[2] Yo'o[1] (quake)
c. Hue'[3]e (house)	h. Xato[3] (rabbit)	m. Yo[34] (cane)	r. —
d. Xiracaj[3] Nne[35] (iguana)	i. Nne[34] (water)	n. Taj[3]u (jaguar)	s. Guman[5] (rain)
e. Xucua' (snake)	j. Xuhue[3] (dog)	o. Xata[3] (eagle)	t. Yaj[3]a (flower)

Nothing more is known of Triqui calendrics.

TZELTAL
(*III.E.1 CR?; XVIII c. to XX c. A.D. = Kanhobal – 7m, f1548, f1584*)

The Tzeltal year began on 5 Yax Mayapan. The calendar counted the days of the day count from 1 to 13 and those of the month from 1 to 20. The beginning of its calendar round and its directional colors are unknown. The day names are reported by Núñez de la Vega (1702), quoted and attributed by Clavijero (1964:181–82, 292);

a. (Mox (alligator)

b. Yoh (Yigh) (wind)

c. Votan (house)

d. Ghanan (Canan) (iguana)

e. Abagh (soot)

f. Tox (Tog) (death)

g. Moxic (deer)

h. Lambat (rabbit)

i. Mulu (rain)

j. Elah (Elab) (dog)

k. Batz (howler)

l. Enoh (Enob) (tooth)

m. Been (cane)

n. Hix (jaguar)

o. Tziquin (bird)

p. Chabin (spider)

q. Chix (Chige) (quake)

r. Chinax (flint)

s. Cabogh (Cahog) (storm)

t. Aghual (lord)

Alternate spellings by V. Pineda (1888:180–82) are in parentheses.
The month names are given by V. Pineda (1887:177):

E. Batzul (1-amaranth)

F. Saquiljá (white water)

G. Agelchac (dawn red)

H. Mac (cover)

I. Olatí, Alaltí

J. Tzun (plant)

K. Julol, Julal (arrival)

L. Hoquen-hajab (5 day lord)

M. Yal Uch (1-possum)

N. Muc Uch (2-possum)

O. Juc-vinquil (7-score)

P. Guac-vinquil (6-score)

Q. Jo-vinquil (5-score)

R. Chan-vinquil (4-score)

A. Osh-vinquil (3-score)

B. Mush (mud)

C. Yash-quin (green time)

D. Pom (incense)

X. –

An alternative list is given by the *Calendario Sna Holobil* (1979):

K. Jok'en Ajaw (5 day lord)

L. Ch'in J'uch (1-possum)

M. Muk' J'uch (2-possum)

N. Juk Winkil (7-score)

O. Wak Winkil (6-score)

P. Jo' Winkil (5-score)

Q. Chan Winkil (4-score)

R. Ox Winkil (3-score)

A. Pom (incense)

B. Yax K'in (green time)

C. Mux (mud)

D. Tz'un (plant)

E. Batzul (1-amaranth)

F. Sakil Ja' (white water)

G. Ajil Ch'ak (dawn red)

H. Mak (cover)

I. Olal ti'

J. Jul ol (arrival)

X. Chay K'in (lost days)

Months K., L., M., A., B., C., D., E., F., H., I., J., and X. on the second list
are cognate with those of Tzotzil.

There are four Tzeltal correlational dates, no two of them based on the
same calendrical premises, but all confirming the beginning of the Tzeltal year on
17 I 1585 G or 16 I 1549 J.

> 1888 15 IX G (V. Pineda 1888). Counting this as 1 Batz'ul places 1 Oken Ahau
> on 17 I G (120 days later).
> 1917 6 V J (Schulz 1953). Counting this as a Julian date for 1 Oken Ahau
> places 1 Batz'ul on 17 I G (120 days earlier).
> 1941 30 IV J (Schulz 1942, 1953). This date is calculated from the preceding
> one by subtracting the intervening leap-year days.
> 1978 20 XII G (*Calendario Sna Holobil* [1979]). Counting this as 1 Batz'ul places
> 1 Oken Ahau on 17 I G 1585.

The correct solution is that indicated by the second and third dates: it was 1 Batz'ul
that started the year on 17 I 1585 G or 16 I 1549 J, as is demonstrated by the
positions of the cognate Tzotzil months. These dates also demonstrate the correctness of the 1979 list of month names: the 1887 list places 1 Batz'ul and 1 Oken
Ahau too far apart to fit the previously mentioned calculations and corresponds to
the Tzotzil list only for the first 5 months.

The 1917 date indicates the Colonial invention of a new calendar, identified
in 1941 with Cancuc and hence so named here and classified as III.J.1. It began
the year on 1 Oken Ahau (5 Muan Mayapan), and apparently the Tzeltal kept its
correlational date on Julian time. On the evidence, this would appear to be the only
native correlational date on the list. The others were interpretations by Europeans,
who did not get any of them exactly right. The Cancuc calendar is documented
from Oxchuc and Tenejapa as well as Cancuc itself (see A.D. *1548, *1584).

TZOTZIL
(III.D.1 CR?; XVII c. to XX c. A.D. = Tzeltal − 1m, f1548, f1584)

The Tzotzil year begins on 5 Ch'en Mayapan. The Tzotzil presumably counted the
day count from 1 to 13 and the days of the month from 1 to 20. The beginning of
their calendar round is unknown, as are the names of the days, which were very
likely the same as the Tzeltal names.

The units of the calendar are the day (*'oxil*), moon or month (*'u*), day count
(*otol k'ak'al*), year (*habil*), and next year (*hunab*). The latter expression was probably
once the name of the calendar round. These designations are from Laughlin (1975).
The directions are associated with colors by Holland (1963:92–94): east–white;
north–white; west–black, south–red, (? center)–green. It seems possible that yellow
may once have been east.

The Tzotzil month names are given by E. Pineda (1845):

D. Tzun (plant)
I. Olalti
N. Muc Uch (3-possum)
A. Pom (incense)

E. Batzul (1-amaranth)
J. Ulol (arrival)
O. Hum Uinicil (1-score)
B. Yaxkin (green time)

F. Zi Zac (1-white)
K. Okin Ahau (5 day lord)
P. Xchibal Uinicil (2-score)
C. Mux (mud)

G. Mukta Zac (2-white)
L. Ala Uch (1-possum)
Q. Yoxchibal Uinicil (3-score)

H. Mac (cover)
M. Elech (2-possum)
R. Xchanibal Uinicil (4-score)
X. Chai Kin (lost days)

Months B., F., and H. are calendrically congruent and linguistically cognate with Yucatec.

There are 22 correlational dates, 16 of which agree in placing the beginning of the Tzotzil year on 28 XII 1584 G or 27 XII 1548 J.

a. 1688   28 XII (Rodaz 1688, cited by Pereyra 1723). 1 Tzun. Guitiupa.
b. 1845   3 III (E. Pineda 1845). Counting this as 1 Mukta Zac places 1 Tzun on 28 XII, 65 days earlier. Yolotepec, Tanjobel.
c. 1845   28 II (E. Pineda 1845). Counting this as 1 Mukta Zac places 1 Tzun on 25 XII, 65 days earlier. Chenalho, Mitontic.
d. 1845   21 III (E. Pineda 1845). Counting this as 1 Mac places 1 Tzun on 26 XII, 85 days earlier. Istacostoc.
e. 1845   11 IX (E. Pineda 1845). Counting this as 1 Tzun places 1 Olalti on 25 XII, 105 days later. From (?).
f. 1917   3 III (Schulz 1953). Counting this as 1 Mukta Zac places 1 Tzun on 28 XII, 65 days earlier. From (?).
g. 1932   23 II (Becerra 1933). Counting this as 1 Mukta Zac places 1 Tzun on 25 XII, 60 days earlier. Chenalho, Mitontic.
h. 1932   26 II (Becerra 1933). Counting this as 1 Mukta Zac places 1 Tzun on 28 XII, 60 days earlier. Yolotepec.
i. 1932   16 III (Becerra 1933). Counting this as 1 Mac places 1 Tzun on 26 XII, 85 days earlier. Istacostoc, Chenalho.
j. 1941   25 II (Schulz 1942). Counting this as 1 Mukta Zac places 1 Tzun on 27 XII, 60 days earlier. Istacostoc, Chenalho, Yolotepec.
k. 1941   2 III (Schulz 1942:9). Counting this as 1 Mukta Zac places 1 Tzun on 27 XII, 65 days earlier. Chenalho, Istacostoc, Yolotepec, Tanjobel, Chamula, Aldama, Utrilla.
l. 1941   3 III (Schulz 1942). Counting this as 1 Mukta Zac places 1 Tzun on 28 XII, 65 days earlier. Chenalho.
m. 1941   8 V (Schulz 1942). Counting this as 7 Ulol places 1 Tzun on 28 XII, 131 days earlier. Mitontic.

n. 1941   1 XI (Schulz 1942). Counting this as 5 Pom places 1 Tzun on 28 XII, 305 days earlier. Todos Santos, Chenalho.

o. 1944   2 III (Guiteras-Holmes 1961). Counting this as 1 Mukta Zac places 1 Tzun on 27 XII, 65 days earlier. Chenalho.

p. 1948   18 II (Berlin 1967). Counting this as a Julian date for 1 Mukta Zac places 1 Tzun on 27 XII, 65 days earlier. Chenalho.

q. 1948   18 III (Berlin 1967). Counting this as 1 Mac places 1 Tzun on 28 XII, 80 days earlier. Istacostoc, Yolotepec, Mitontic, Pantelho.

r. 1955   2 III (Guiteras-Holmes 1961). Counting this as 1 Mukta Zac places 1 Tzun as 27 XII, 65 days earlier. Chenalho.

s. 1958   2 III (Pozas 1959). Counting this as 1 Mukta Zac places 1 Tzun on 27 XII, 65 days earlier. Chamula.

t. 1966   2 III (Whelan 1967). Counting this as 1 Mukta Zac places 1 Tzun on 27 XII, 65 days earlier. Chamula.

u. 1968   27 II (Bricker 1968). Counting this as 1 Mukta Zac places 1 Tzun on 29 XII, 60 days earlier. Chamula.

v. 1969   2 III (Gossen 1974). Counting this as 1 Mukta Zac places 1 Tzun on 27 XII, 65 days earlier. Chamula.

These calculations reveal that the Tzotzil calendar was frozen to the Julian one after *29 II 1548, placing the beginning of the year on 27 XII Julian (dates j, k, o, p, r, s, t, v). This dating tradition is reflected in five dates from Chenalho, four from Chamula, two each from Yolotepec and Istacostoc, and one each from Aldama, Utrilla, and Tanjobel.

Another tradition maintains a frozen Gregorian New Year date of *29 II 1584, placing the beginning of the year on 28 XII Gregorian (dates a, b, f, h, l, m, n, q). This is reflected in four dates from Mitontic, one each from Guitiupa, Chalchihuitan, Yolotepec, and Chenalho, and one without provenience.

It will be noted that there is a degree of uncertainty about whether or not to calculate the 5 days of Chay Kin as part of the interval between 1 Tzun and the later date. It appears that dates g, h, j, q, and u do not. This difference may be traditional, depending on whether the date was fixed before or after the New Year was moved from 1 Tzun to some other date. The five dates just cited are arguably relics of the earlier calculation. They come from Chenalho, Mitontic, Yolotepec, Istacostoc, and Chamula.

There remain six problematic dates: c and g, which are 2 days, d and i, which are 1 day too early for the Julian correlation, and u, which is 1 day late for the Gregorian one. Date e may be aimed at the Gregorian correlation but is off by the 12 days of the Julian-Gregorian correction.

The correlational dates cited make it clear that there are at least three different Tzotzil calendars. The classical one began on 1 Tzun (5 Ch'en Mayapan). The second, probably antedating 1584, is first documented at Guitiupa in 1688 and is therefore so named here. It began the year on 1 Mukta Zac (5 Ceh Mayapan). The third is first documented from Istacostoc in 1845 and is therefore so named. It began the year on 1 Mac (5 Mac Mayapan). A fourth calendar is suggested by the aberrant date e. Its year would begin on 1 Olalti (5 Kankin Mayapan), and it may be called Mitontic.

The 15 dates actually cited for 1 Mukta Zac Guitiupa (b, c, f, g, h, j, k, l, o, p, r, s, t, u, v) vary between 18 II and 3 III. In some cases, there is some consistency by village: four of the five dates from Chamula are 2 III (k, s, t, v)— but one (u) is 27 II. On the other hand, the Chenalho dates range from 18 II to 2 III within one village (c, g, j, k, o, p, r). Since 1940, the dating of the Guitiupa year appears to have become more closely associated with the date of Carnival, presumably as a consequence of a shift in ritual ideology, particularly at Chamula. The following compilation illustrates the point:

Year	Date	Shrove Tuesday	Shrove Tuesday to 1 Mukta Zac
1688	a	7 III	0
1845	b	4 II	27
	c	"	24
1917	f	20 II	11
1932	g	9 II	14
	h	"	17
1941	j	25 II	0
	k	"	5
	1	"	6
1944	o	22 II	8
1948	p	10 II	8
1955	r	22 II	8
1958	s	18 II	12
1966	t	22 II	8
1968	u	27 II	0
1969	v	18 II	12

The dates for the Istacostoc New Year (d, i, q) are not involved with Carnival and range from 16 III to 21 III. The expected frozen Gregorian date would be 23 III.

The corresponding New Year's date in the Mitontic calendar is 12 IV (see A.D. *1548, *1584).

TZUTUHIL
(II.L.1 CR 1 Kan 1532; XX c. A.D. = Quiche?)

The day names of the Tzutuhil calendar from San Pedro la Laguna are reported by Rosales (1939:763):

a. I'mox (alligator)	f. Kye'mel (death)	k. P'ats' (monkey)	p. Ah'mak (owl)
b. Iq' (wind)	g. Kyéh (deer)	l. Ey (tooth)	q. Noh (incense)
c. P'aq'p'al (night)	h. Q'a'nel (rabbit)	m. Ah (cane)	r. Ti'hax (flint)
d. K'at (net)	i. Toh (rain)	n. I'x (jaguar)	s. Ka'woq (storm)
e. Kan (serpent)	j. Ts'i' (dog)	o. Tz'i'kin (bird)	t. Ah'pup' (hunter)

The month names have not been reported, but Rosales confirms the calendrical position of the day count (see A.D. 1939) and reports (1939:763) that "Kan es el más superior a todos los demás porque es del cerro y volcanes, también le llaman el día ALCALDE." The strong probability appears to be that the Tzutuhil calendar was identical to that of the Quiche, even though Rosales's note would seem to imply Type V year bearers (see A.D. 1939).

USPANTEC
(=Quiche?)

The units of the Uspantec calendar are the day (*q'i·x*), month (*i·k'*), daycount (?), year (*'a·b'*) and calendar round (*xuna·b'*). Possible day names noted by Kaufman (n.d.) are:

a. —	f. —	k. B'â·č) (monkey)	p. Axmà·k (owl)
b. —	g. KYè·x (deer)	l. 'È· (tooth)	q. —
c. 'Aq'áb' (night)	h. —	m. —	r. —
d. K'á·t (net)	i. —	n. —	s. —
e. —	j. C'i'	o. C'ikín	t. —

It seems likely that the Uspantec calendar is that of the Quiche.

VALLADOLID
(IV.N.1 CR 1 Kan 1529; XVIII c. to XIX c. A.D. = Mayapan + 24*)

The Valladolid calendar was instituted at Valladolid on 3 Cauac 1 Pop Mayapan on 6 VI 1752 Julian (q.v.). It retained all the features of the Mayapan calendar (q.v.) except that it altered the beginning date and the time span of the katun and the time span (but not the beginning date) of the may.

Katun 4 Ahau Mayapan began in 1737 and was due to end in 1757. In 1752, the day 4 Ahau followed directly upon the first day of the year, 3 Cauac. By converting the katun from a period of 20 tuns (of 360 days) to 24 haabs (of 365 days), it would work out that the katun sequence could be maintained but that each katun could be initiated on its name day on the second day of the year. The new 24-year katun was therefore seated on 4 Ahau 2 Pop Mayapan on 7 VI 1752 Julian, to end with the inauguration of 2 Ahau in 1776.

The new katun altered but did not disrupt the may, which was still counted from the beginning of the Mayapan calendar in 11 Ahau. A certain confusion resulted, however, from the retroactive Julian dating of the new katuns, which were sometimes dated backward at 24-year intervals from 1752. Documents using the Valladolid calendar therefore begin the may at 11 Ahau but date that to 1512 instead of to 1539. The new may was to run for 24 new katuns instead of 13 old ones — hence from A.D. 1512 to 2088. The correspondences of the beginnings of the katuns from the fall of Mayapan in the Mayapan and Valladolid calendars are as follows:

Katun	Mayapan	Valladolid
8 Ahau	1441	1392
6 Ahau	1461	1416
4 Ahau	1480	1440
2 Ahau	1500	1464
13 Ahau	1520	1488
11 Ahau	1539	1512
9 Ahau	1559	1536
7 Ahau	1579	1560
5 Ahau	1598	1584
3 Ahau	1618	1608
1 Ahau	1638	1632
12 Ahau	1658	1656
10 Ahau	1677	1680
8 Ahau	1697	1704
6 Ahau	1717	1728
4 Ahau	1737	1752

Thus the *Chumayel* anachronistically dates the beginning of 8 Ahau as 151 years before 1543 (line 2722), puts Juan de Montejo's landing at Ecab in 1526 in 11 Ahau (line 2676), dates Montejo the Nephew's landing at Campeche to 1513 instead of 1540 (line 2705) and the "coming of Christianity" to 1519 instead of 1546 (line 1519).

The Valladolid calendar begins with 12.6.15.12.0 4 Ahau 2 Pop, 7 VI 1752 Julian. In a sense this represents the end of the Long Count for the Mayas who adopted it because there is no point in continuing to count the tuns if the katuns are not based upon them. There is a degree of confusion of tuns with haabs throughout the Colonial period in any case, and after this date it would appear that only haabs were counted. Awareness of the tun continued (see *Tizimin* line 5446), but it was not used (see A.D. 1752).

XINCA
(= *Kaminaljuyu?*)

The units of the Xinca calendar were the day or sun (*pári*), month or moon (*ahua, mola*), year (*ayapa*), and probably the day count and calendar round, though the latter periods are not documented. From the vocabularies collected by Lehmann (1920), the following are possible day names:

a. Huayo (alligator)	f. Teró (dead)	k. Poxo, Iru (monkey)	p. Cúti (buzzard)
b. Tan (wind)	g. Túma (deer)	l. Núrui (grass)	q. Huyicnár (quake)
c. Tz'uma (night)	h. Lur (rabbit)	m. Aima (corn)	r. —
d. Xuvayo (lizard)	i. Uy (water)	n. Uilay (jaguar)	s. K'unu (clouds)
e. Ambui, Púki (serpent)	j. Chuso (dog)	o. —	t. Túlo (flower)

From its relationship to Lenca (q.v.), Xinca almost certainly had calendar-round dating. The possibilities of Mayan, Nahuatl, or other influences upon its calendar make it difficult even to speculate about its calendrical location, but it seems likely that it was at least originally on the calendar of Kaminaljuyu.

XOCHICALCO
(*III.Q.0 CR?; VIII c. A.D. = Yucuñudahui*)

The site of Xochicalco used the Yucuñudahui Mixtec calendar, though it was also aware of that of Teotihuacan, and its writing system manifests rather more Teotihuacan influence than does the Ñuiñe of the Mixteca Baja. The Ñuiñe calendrical glyphs are listed with the Yucuñudahui calendar (q.v.). Those of Xochicalco are given here with their provenience.

a. 8 Alligator. Xochicalco (Caso 1967:167, Fig. 20).
b. Anno 12 Wind. Texmilincan (Ibid.:169, Fig. 5a).
    7 Wind. Xochicalco (Ibid.:169, Fig. 4b; 184, Pl. 1).
    7 Wind. Piedra Labrada, Gro. (Ibid.:169, Fig. 4a).
c. Anno 1 House. Xochicalco (Ibid.:180, Fig. 17g).
    Anno 4 House. Xochicalco (Ibid.:180, Fig. 17d).
    Anno 9 House. Tenango (Ibid.:180, Fig. 17h).
    11 House. Xochicalco (Ibid.:167, Fig. 2b,c).
d. Iguana?
e. 7 Serpent. Xochicalco (Ibid.:167, Fig. 2d).
f. 5 Death. Xochicalco (Ibid.:167, Fig. 2e).
g. Deer. Unknown (Ibid.:180, Fig. 16).
h. Anno 4 Rabbit. Xochicalco (Ibid.:180, Fig. 17b,c).
    6 Rabbit. Unknown (Ibid.:180, Fig. 16).
    7 Rabbit. Xochicalco (Ibid.:167, Fig. 2).
i. Anno 2 Water. Tenango (Ibid.:162, Fig. 19). Tenango calendar.
j. 1 Foot. Xochicalco (Ibid.:173, Fig. 9b).

3 Foot. Xochicalco (Ibid.:173, Fig. 9d).
6 Foot. Unknown (Ibid.:166, Fig. 1).
9 Foot. Xochicalco (Ibid.:173, Fig. 9e).
13 Foot. Xochicalco (Ibid.:173, Fig. 9c).
k. 11 Monkey. Xochicalco (Ibid.:167, Fig. 2g).
13 Monkey. Xochicalco (Ibid.:173, Fig. 2h).
l. 5 Sun. Xochicalco (Ibid.:172, Fig. 8d).
6 Sun. Xochicalco (Ibid.:172, Fig. 8a,b).
7 Sun. Xochicalco (Ibid.:172, Fig. 8c).
m. Anno 3 Cane. Xochicalco (Ibid.:180, Fig. 17f).
Anno 5 Cane. Xochicalco (Ibid.:180, Fig. 17e).
n. 3 Jaguar. Unknown (Ibid.:166, Fig. 1).
o. 7 Eagle. Unknown (Ibid.:166, Fig. 1).
p. 5 Owl?. Xochicalco (Ibid.:180, Fig. 17e).
q. 2 Quake. Xochicalco (Ibid.:167, Fig. 2k).
r. Anno 10 Flint. Xochicalco (Ibid.:180, Fig. 17a).
4 Flint. Cerro de los Monos, Xochicalco (Ibid.:175, Fig. 11c).
6 Flint. Xochicalco (Ibid.:171, Fig. 7a; 180, Fig. 17a).
7 Flint. Río Grande, Oax. (Ibid.:175, Fig. 11b).
10 Flint. Unknown (Ibid.:175, Fig. 11a).
s. 4 Rain. Xochicalco (Ibid.:176, Fig. 12a).
5 Rain. Xochicalco (Ibid.:170, Fig. 6b).
7 Rain. Xochicalco (Ibid.:170, Fig. 6a).
t. 2 Lord. Xochicalco (Ibid.:174, Fig. 10a).
10 Lord. Unknown (Ibid.:174, Fig. 10c; 180, Fig. 16)
(See A.D. 426, 569, 768.)

YUCATEC

The Yucatecan Maya are primarily identified with the Tikal calendar, which they shared with the Chol and Kekchi. In 1539, they adopted the Mayapan calendar and in 1752 that of Valladolid. In the Late Classic period they made limited use of the calendar of Palenque (which may have been mostly Chol) and that of Campeche. The calendrical usages of the major Yucatecan dialects of the south (Lacandon, Itza, and Mopan) are not well established. Each of the Yucatecan calendars is described elsewhere in this index.

YUCUÑUDAHUI
(III.Q.0 CR?; V c. to X c. A.D. = Teotihuacan + 1d – 14)*

The Yucuñudahui Mixtec calendar begins the year on 20 Zip Mayapan, uses bar-and-dot numerals, counts the day count from 1 to 13 and the days of the month (probably) 0 to 19, naming the year for its first day. Its senior year bearer and direction colors are unknown. It was the earliest calendar to use Type III year bearers (see A.D. 426) and thus is the apparent ancestor of all the later calendars of central Mexico (see Cholula, Huastec, Tarascan, Toltec, Totonac).

The Yucuñudahui writing system was initially Zapotec, but included some Teotihuacan glyphs (especially the year sign), and apparently some Olmec signs as well. At Xochicalco (q.v.), it seems to have incorporated more Teotihuacan elements. Both Xochicalco and Yucuñudahui invert their bar-and-dot numerals, writing the dots below the bars for 6 to 9 and 11 to 13. In the Mixteca, the day glyphs were probably those of the later Ñuiñe writing system surveyed by Moser (1977):

a. 11 Alligator. Huajuapan (Moser 1977:30, Fig. 7). Possibly Teotihuacan.
b. 3 Wind. Huajuapan (Ibid.:33, Fig. 9a). Mixtec?
c. Anno 11 House. Tequixtepec (Ibid.:49, Fig. 18a). Possibly Teotihuacan.
d. Lizard?
e. 2 Serpent. Tequixtepec (Ibid.:79, Fig. 44a). Zapotec.
f. 1 Death. Chilixtlahuaca (Ibid.:113, Fig. 66). Mixtec?
g. Deer? Cerro Caja, Huajuapan (Ibid.:86, Fig. 48). Zapotec?
h. Anno 10 Rabbit. Miltepec (Ibid.:38, Fig. 11). Olmec?
i. 6 Water. Tequixtepec (Ibid.:65, Fig. 33). Zapotec.
j. 9 Dog. Huajuapan (Ibid.:25, Fig. 2b). Zapotec.

    k. 3 Monkey. Tequixtepec (Ibid.:67, Fig. 34b). Zapotec.
    l. 10 Sun. Huajuapan (Ibid.:30, Fig. 6). Zapotec.
   m. Anno 5 Cane. Tequixtepec (Ibid.:53, Fig. 21). Olmec.
    n. Jaguar. Tequixtepec (Ibid.:76, Fig. 43). Zapotec?
    o. Eagle? Tequixtepec (Ibid.:69, Fig. 36b). Zapotec?
    p. 5 Owl. Tequixtepec (Ibid.:65, Fig. 32). Zapotec.
    q. 10 Quake. Cerro Caja, Huajuapan (Ibid.:86, Fig. 48). Zapotec.
    r. Flint. Tequixtepec (Ibid.:51, Fig. 19b). Zapotec.
    s. Rain. Tequixtepec (Ibid.:27, Fig. 4). Zapotec.
    t. Flower. Tequixtepec (Ibid.:63, Fig. 30). Zapotec.

The ordinary names of the concepts of the days differed in the three main Mixtecan languages (Alta, Baja, and Costa), but there seems to have been an archaic and canonical "day-name" system as well, and that may have been shared by all three (see Mixtec). The glyphs and names for the Yucuñudahui months are totally unknown.

As the Mixtec calendar of its time, the Yucuñudahui calendar seems to have been used throughout the Mixteca and beyond. Its position with respect to the Olmec and Zapotec calendars is established by the correlational date of 545 (q.v.) at Monte Alban and confirmed by the date of 768 (q.v.) at Xochicalco. It also appears at Teotihuacan (q.v.). It was replaced in the Mixteca in 973 by the calendar of Tilantongo, but may be represented in Nahuatl by the later Chalca calendar identified by Jiménez (1961) in Chimalpahin (see A.D. 426, 545, 768, 973).

ZAPOTEC
(III.C.19 CR 1 Piy 1546; VI c. B.C. to XVI c. A.D. = Olmec + 1m + 2d)

The Zapotec calendar begins the year on 5 Mol Mayapan, using Type III New Years and Type II name days. It counts the day count from 1 to 13 and the days of the month from 0 to 19. Its senior year bearer is probably Sun, and its directional colors are unknown. Full calendar-round dating is documented by Stela 12 in Monte Alban I (542 B.C.), using bar-and-dot numerals, and citing dates in the form 'Anno, Year bearer, Coefficient, Day, Coefficient, (Month Name, Coefficient).' Alternatively, the month phrase may be given as '(Month Sign, Ordinal Number),' and it is omitted if redundant (see 528 B.C., which also demonstrates zero counting and terminal naming of the year).

The units of the Zapotec calendar are the sun or day (*chij, chee, copijcha*), moon or month (*peo*), day count (*piye, pije*), year (*yza*), and cycle (*cocijo*). The name 'cycle' was applied to the trecena, the quarter day count, and perhaps to the calendar round as well. The units of the calendar are counted currently but named completively; the current unit is thought of as pending (*nazabi*).

Although there are 33 Zapotec pictorial manuscripts, none of them is pre-Conquest; the *Lienzo de Guevea* is one of the few that is in the native tradition and includes some glyphs of possible calendrical significance (Glass 1975b:75, 131). The Valley Zapotec calendar is not documentable after the sixteenth century.

The day signs of the Zapotecs are attested archaeologically, most of them from Tàniquiecàche, the Hill of Precious Stones, better known as Monte Alban, during Monte Alban I and II. Sources are identified below.

a. Day. Monte Alban (Whittaker 1980:28, Fig. 5).
b. Anno 1 Wind. Monte Alban Stela 15 (Caso 1928:46, Fig. 21IV).
   Anno 4 Wind. Monte Alban I (Caso 1965b:933, Fig. 3).
   Anno 10 Wind. Monte Alban IV (Caso 1932:26).
   Anno 11 Wind. Monte Alban Stela 3 (Caso 1928:46, Fig. 21III).
   1? Wind. Xoxo Burial Plaque (Ibid.:28, Fig. 4VII).
   2 Wind. Monte Alban Stela 9 (Ibid., Fig. 4X).
   3 Wind. Zaachila Lintel 1 (Ibid., Fig. 4IX).
   3 Wind. Monte Alban Lápida 1, Museo Nacional (Ibid.:41, Fig. 18III).
   4 Wind. Monte Alban Stela 12 (Ibid., Fig. 18II).
   5 Wind. Monte Alban Lápida 6, Museo de Oaxaca (Ibid.:28, Fig. 4III).
   5 Wind. Monte Alban Lápida 1, Museo Nacional (Ibid., Fig. 4V).
   6 Wind. Monte Alban III (Whittaker 1980:35, Fig. 8).

6 Wind. Xoxo Lintel (Caso 1928:28, Fig. 4IV).

6 Wind. Monte Alban Lápida 2, Museo Nacional (Ibid., Fig. 4VI).

12 Wind. Monte Alban Stela 1 (Ibid., Fig. 4I).

c. 3 Night. Monte Alban Lápida 12, Museo de Oaxaca (Caso 1928:43, Fig. 20RII).

  4 Night. Monte Alban Lápida 12, Museo de Oaxaca (Ibid., (Fig. 20RI).

  10 Night. Monte Alban II (Caso 1965b:937, Fig. 12).

d. 2 Black. Monte Alban Stela 6 (Caso 1928:43, Fig. 20S).

  5? Black. Monte Alban Lápida 1, Museo Nacional (Ibid.:41, Fig. 19I).

  7 Black. Monte Alban Stela 15 (Ibid., Fig. 19III).

  7 Black. Monte Alban Stela 8 (Ibid., Fig. 19V).

  8 Black. Monte Alban Lápida 1, Museo Nacional (Ibid., Fig. 19II).

  8 Black. Monte Alban Stela 7 (Ibid., Fig. 19V).

  Black. Monte Alban III (Ibid.:935, Fig. 8s).

e. 1 Serpent. Monte Alban Lápida 1, Museo Nacional (Caso 1928:41, Fig. 18IV).

  5 Serpent. (Ibid. Fig. 18IV).

  5 Serpent. Zaachila (Moser 1977:157, Fig. 72b).

  6 Serpent. Monte Alban III Ibid.:157, Fig. 72a).

  10 Serpent. Monte Alban Lápida 1, Museo Nacional (Caso 1928:43, Fig. 20Y).

  Serpent, Monte Alban III (Caso 1965b:935, Fig. 8y).

f. 1? Head. Monte Alban Stela 10 (Caso 1928:35, Fig. 11IV).

  2 Head. Monte Alban Stela 7 (Ibid., Fig. 11I).

  2 Head. Monte Alban Lápida 2, Museo Nacional (Ibid., Fig. 11VIII).

  11 Head. Monte Alban, Lápida 13 (Ibid., Fig. 11III).

  11 Head. Pectoral, Urn, Museo Nacional (Ibid., Fig. 11VI).

  12 Head. Monte Alban Stela 8 (Ibid., Fig. 11II).

  Head. Monte Alban I (Caso 1965b:934, Fig. 5h).

  Head. Monte Alban III (Ibid.:935, Fig. 8h).

  Head. Monte Alban Stela 9 (Caso 1928:35, Fig. 11V).

  Head. Zaachila Lápida 1 (Ibid., Fig. 11VIII).

g. Anno 1 Deer. Monte Alban Stela 6 (Caso 1928:46, Fig. 21XI).

  Anno 6 Deer. Monte Alban Stela 10 (Ibid., Fig. 21XV).

  Anno 6 Deer. Monte Alban Stela 9 (Ibid., Fig. 21XIV).

  Anno 6 Deer. Monte Alban (Moser 1977:150, Fig. 70d).

  Anno 10 Deer. Monte Alban Stela 6 (Caso 1928:46, Fig. 21XII).

  Anno 11 Deer. Monte Alban, Lápida 7, Museo de Oaxaca (Caso 1928:46, Fig. 21XVI).

  1 Deer. Monte Alban Stela 6 (Ibid.:34, Fig. 10V).

  4 Deer. Monte Alban Lápida 1, Museo Nacional (Ibid., Fig. 10II).

  6 Deer. Monte Alban Stela 9 (Ibid., Fig. 10VIII).

  6 Deer. Monte Alban Stela 10 (Ibid., Fig. 10IX).

  6? Deer. Monte Alban Stela 5 (Ibid., Fig. 10III).

  8 Deer. Monte Alban Stela 4 (Ibid., Fig. 10I).

  10 Deer. Monte Alban Stela 6 (Ibid., Fig. 10IV).

  11 Deer. Monte Alban Lápida 7 (Ibid., Fig. 10X).

12 Deer. Monte Alban Jamb 2 (Ibid., Fig. 10VI).

Deer. Monte Alban III (Caso 1965b:935, Fig. 8g).

h. 2 Rabbit. Monte Alban Stela 15 (Caso 1928:35, Fig. 12VI).

4 Rabbit. Monte Alban Stela 10 (Ibid., Fig. 12III).

5 Rabbit. Monte Alban I (Whittaker 1980:50, Fig. 14).

5? Rabbit. Monte Alban Lápida 1, Museo Nacional (Caso 1928:35, Fig. 12VIII).

Rabbit. Monte Alban Stela 9 (Ibid., Fig. 12I).

Rabbit. Monte Alban Stela 15 (Ibid., Fig. 12VII).

Rabbit. Monte Alban Lápida 11, Museo de Oaxaca (Ibid., Fig. 12VI).

i. 1 Water. Monte Alban II (Caso 1965b:937, Fig. 70d).

1 Water? Monte Alban Jamb 2 (Caso 1928:43, Fig. 20ZIV).

2 Water. Monte Alban Lápida 1, Museo Nacional (Ibid., Fig. 20ZIII).

3 Water. Ibidem (Ibid., 41, Fig. 18V).

8 Water. Monte Alban Stela 1 (Ibid., 43, Fig. 20ZII).

8 Water. Monte Alban Stela 12 (Ibid., Fig. 20ZIII).

8 Water. Monte Alban I (Caso 1965b:933, Fig. 3).

13 Water. Cuilapan Idol (Caso 1928:41, Fig. 18VI).

j. 8 Dog. Monte Alban Stela 15 (Caso 1928:43, Fig. 20X).

9 Dog. Monte Alban (Moser 1977:149, Fig. 70b).

Dog. Monte Alban III (Caso 1965b:935, Fig. 8k).

k. 2 Monkey. Monte Alban I (Whittaker 1980:50, Fig. 14).

4 Monkey. Monte Alban, Jamb 5 (Caso 1928:43, Fig. 20U).

10 Monkey. Monte Alban, Lápida 4, Museo Nacional (Ibid., Fig. 20II).

11 Monkey. Monte Alban, Lápida 1, Museo Nacional (Ibid., Fig. 20I).

13 Monkey. Zaachila Lápida 1 (Ibid., Fig. 20III).

l. Anno 6 Sun. Monte Alban II (Caso 1965b:937, Fig. 12).

Anno 6 Sun. Monte Alban II (Ibid., 938, Fig. 13).

Anno 6 Sun. Cuilapan Lintel (Caso 1928:46, Fig. 21X).

Anno 12 Sun. Monte Alban I (Whittaker 1980:50, Fig. 14).

Anno 13 Sun. Monte Alban Stela 2 (Caso 1928:46, Fig. 21IX).

1 Sun. Xoxo Lintel 1 (Ibid., 32, Fig. 8XXIV).

1? Sun. Incomplete Stela, Museo de Oaxaca (Ibid., Fig. 8XXIII).

2 Sun. Monte Alban Jamb 7 (Ibid., Fig. 8IV).

2 Sun. Monte Alban Stela 15 (Ibid., Fig. 8VII).

3 Sun. Monte Alban Stela 9 (Ibid., Fig. 8X).

3 Sun. Monte Alban Lápida 8, Museo Nacional (Ibid., Fig. 8XXI).

3 Sun. Monte Alban Lápida 8, Museo de Oaxaca (Ibid., Fig. 8XIX).

4 Sun. Monte Alban Lápida 1, Museo Nacional (Ibid., Fig. 8XX).

4 Sun. Monte Alban Lápida 4, Museo Nacional (Ibid., Fig. 8XV).

4 Sun. Zaachila Lápida 1 (Ibid., Fig. 8VII).

5 Sun. Monte Alban Lápida 4, Museo Nacional (Ibid., Fig. 8XVI).

5 Sun. Monte Alban Stela 1 (Ibid., Fig. 8XII).

6 Sun. Monte Alban Jamb 1 (Ibid., Fig. 8III).

6 Sun. Monte Alban Jamb 2 (Ibid., Fig. 8V).

6 Sun. Monte Alban Jamb 4 (Ibid., Fig. 8VI).

6 Sun. Monte Alban Stela 15 (Ibid., Fig. 8IX)

6 Sun. Cuilapan Lintel (Ibid., Fig. 8XVII).

6 Sun. Zaachila Lápida 2 (Ibid., Fig. 8XXII).

7 Sun. Monte Alban Stela 9 (Ibid., Fig. 8X).

7 Sun. Cuilapan Lintel (Ibid., Fig. 8XVIII).

8 Sun. Monte Alban Stela 1 (Ibid., Fig. 8I).

12 Sun. Xoxo Lintel II (Ibid., Fig. 8XIII).

13 Sun. Monte Alban Stela 2 (Ibid., Fig. 8II).

m. 2 Cane. Monte Alban Stela 9 (Caso 1928:27, Fig. 3VI).

2 Cane. Monte Alban Stela 15 (Ibid., Fig. 3IV).

3 Cane. Monte Alban Stela 2 (Ibid., Fig. 3II).

7 Cane. Monte Alban Stela 1 (Ibid., Fig. 3I).

13 Cane. Monte Alban Lápida 1, Museo Nacional (Ibid., Fig. 3III).

Cane. Monte Alban Stela 9 (Ibid., Fig. 3V).

Cane. Monte Alban III (Caso 1965b:935, Fig. 8a).

n. 1 Jaguar. Monte Alban I (Whittaker 1980:50, Fig. 14).

o. Eagle. Monte Alban I (Caso 1965b:934, Fig. 6).

p. 1 Crow. Monte Alban Lápida 14 (Caso 1928:33, Fig. 9IV).

1 Crow. Monte Alban Lápida 1, Museo Nacional (Ibid., Fig. 9VII).

2 Crow. Monte Alban II (Caso 1965b:937, Fig. 12).

3 Crow. Monte Alban (Moser 1977:150, Fig. 70d).

3 Crow. Monte Alban Stela 9 (Caso 1928:33, Fig. 9III).

5 Crow. Monte Alban Lápida 4, Museo Nacional (Ibid., Fig. 9VI).

5 Crow. Xoxo Lintel 1 (Ibid., Fig. 9V).

13 Crow. Monte Alban Stela 2 (Ibid., Fig. 9I).

13 Crow. Monte Alban Stela 6 (Ibid., Fig. 9II).

13 Crow. Monte Alban Jamb 5 (Ibid., Fig. 9VIII).

q. 6 Quake. Monte Alban, Lápida 1, Museo Nacional (Caso 1928:39, Fig. 16I).

6 Quake. Ibidem (Fig. 16III).

10 Quake. Ibidem (Fig. 16II).

10 Quake. Monte Alban I (Whittaker 1980:50, Fig. 14).

r. 13 Cold. Huamelulpan (Moser 1977:74, Fig. 41b).

13 Cold. Cuilapan Idol (Caso 1928:43, Fig. 20Q).

s. 7 Cloud. Monte Alban II (Caso 1965b:937, Fig. 12).

t. Anno 4 Flower. Monte Alban Stela 9 (Caso 1928:43, Fig. 20V).

Presumably an Olmec date.

1 Flower. Monte Alban Stela 9 (Ibid., 36, Fig. 13III).

2 Flower. Monte Alban figurine (Ibid., Fig. 13VII).

2 Flower. Monte Alban II (Caso 1965b:935, Fig. 9j).

3 Flower. Monte Alban Stela 8 (Caso 1928:36, Fig. 13IVbis).

7 Flower. Monte Alban Stela 9 (Ibid., Fig. 13III).

9 Flower. Yucuñudahui (Moser 1977:145, Fig. 69).

13 Flower. Monte Alban, Lápida 1, Museo Nacional (Caso 1928:36, Fig. 13IV).

It is obvious that the identification of days a, n, o, and s remains problematic.

The names of the days in Valley Zapotec are reported by Caso (1965b:944) from Córdova (1578):

a. Chi (day)	f. Quíqueni (head)	k. Pilloo (monkey)	p. Quilloo (crow)
b. Pèe (wind)	g. Chìna (deer)	l. Piy (sun)	q. Tixòo (quake)
c. Quèela (night)	h. Pèela (rabbit)	m. Quij (cane)	r. Gopa (cold)
d. Yàce (black)	i. Niça (water)	n. Pèche (jaguar)	s. Gappe (cloud)
e. Pèlla (snake)	j. Tella (dog)	o. Ñaa (eagle)	t. Lào (flower)

Nothing is known of the day names of the other Zapotecan languages (Serrano, Netzichus, Bixanas, Caxonos, Lachiguiri, Logueche, and Loxicha), although Loxicha is still using an alternative day count based on the novena and trecena (see Loxicha). Whittaker (1980) has argued the case for trecena numeration at Monte Alban as well, but that requires moving day glyphs around and invoking alternative calendars for which there is no independent warrant. The assumption of a single continuous Zapotec calendar gives a simpler and more efficient explanation and eliminates the evidence for novena and trecena calculation in ancient Valley Zapotec.

The names of the Zapotec months are unknown. The glyph for Zapotec month P. (the Mayan Zip) is found on Stela 17 in Monte Alban I and is the only month name so far identified:

Month P.   Month in General

Zapotec writing is found only in the Valley of Oaxaca, and principally at Monte Alban, but correlational dates have been found linking its calendar to others: Olmec (563 B.C., A.D. 545 and 798), Yucuñudahui (A.D. 545 and 768), Tenango (A.D. 798), and Tilantongo (A.D. 1465). Monte Alban is at the center of a large area in which Type III calendars were all but universal during the Early Classic period, from Teotihuacan in the north to the borders of Guatemala and Yucatan. The Yucatecan and Guatemalan Maya clung to Type II, whereas Type IV became the rule north of Oaxaca in Postclassic times. The central position of the Zapotec is reflected clearly in their calendar.

Zapotec and Olmec calendrical glyphs and the conceptions underlying them were almost certainly the sources of many later calendrical ideas of the other Middle American peoples (see 542, 531, 528, 229, 209 B.C., A.D. 545, 744, 1445).

ZOQUE
(= Mixe?)

The units of the Zoque calendar were the sun or day (*hama*), moon (*poya*), month (*sepe'*), count or day count (*may*), year (*hame, 'aminte*) and calendar round (?). A number of possible Zoque day names are to be found in González (1672):

a. Vepi (alligator)
b. Sava (wind)
c. Tec, Tzu (house, night)
d. Nato (iguana)
e. Tzan (serpent)

f. Cacuy (death)
g. Mùea (deer)
h. Coya (rabbit)
i. Na (water)
j. Tuy (dog)

k. Tzàui (monkey)
l. Tetz (tooth)
m. Socuy (cane)
n. Tziquin Cang (jaguar)
o. Hon (bird)

p. —
q. Mics (quake)
r. Tza (stone)
s. Hoy (rainstorm)
t. Yùmi (lord)

Mixe, Zoque, Sayula, and Tapachultec were almost certainly a single language in the Preclassic and were very likely responsible for the invention of the Olmec calendar. It is probable that the modern Zoque shared the Mixe calendar, but evidence is altogether lacking.

CONCLUSION

A close and detailed examination of the internal structure, calendrical placement, genealogical interrelationships, and ethnohistorical and archaeological dating of the calendars of Anahuac reveals pluralism and unity. The pluralism is at bottom religious in character; the unity is astronomical.

Underlying the variations is a common preoccupation with the apparent motion of the sun and a common numerology for predicting it. A limited number of calendrical features account for the differences among the calendars: terminal and initial naming, the presence or absence of zero dating, advances in year bearers, changing the month of the New Year, and (very exceptionally) altering the counting of the day count. These mechanisms have been shown to be reflexes of simple mathematical transformations of ± 1 day, ± 5 days, and ± 20 days that function to facilitate the calculation of the solar era. I believe these mechanisms account for all of the transformations that the calendars of Anahuac have undergone.

An important part of the proof of these propositions is the plausibility of the historical reconstruction they imply. Our data for assessing this will forever remain fragmentary, but I have tried to collect it as comprehensively as possible. My guiding hypothesis has been that the native astronomers of Middle America have systematically used the year bearers as markers for the calculation of the solar era. This has led to the identification of ten native eral calendars, presumed to have

Figure 18. EXTRA LEAP YEAR, LEAP YEAR, AND ANTI-LEAP YEAR IN MIDDLE AMERICA
(Eral Calendars in Capitals; Year Bearers in Roman Numerals)

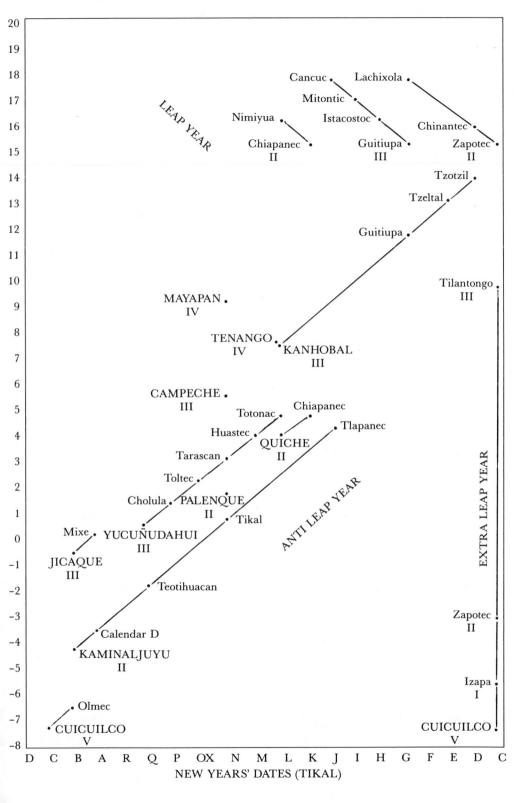

been founded when their New Years' days fell on particular solstices and equinoxes, and the derivation of all remaining calendars from these ten by predictable mathematical transformations. The resulting chronology agrees with the archaeological and ethnohistorical data except in two instances: the dating of the Zapotec and Tilantongo Mixtec calendars. The former dates are seemingly too early by six calendar rounds; the latter by 15 years. At this point, I have more confidence in my dates than in Caso's in these two cases.

Some nineteen calendars, on the other hand, appear to be documented within the first century or so of their existence: Cuicuilco, Olmec, Zapotec, Teotihuacan, Campeche, Tenango, Tilantongo, Tepepulco, Texcoco, Colhua, Cuitlahuac, Colhuacan, Cakchiquel, Iximche, Nimiyua, Mitontic, Cancuc, Lachixola, and Valladolid. Given the vagaries of both archaeological and ethnohistorical dating, this seems to me an impressive confirmation of the hypothesis. There are many ways in which it could be falsified. Perhaps the simplest example would be the clear identification of a Long Count Mayan date in the seventh baktun. A Classic period date in the Tzotzil or Tzeltal calendar would also do very well. These are non-trivial examples: the argument is sufficiently tight that such cases would force rejection of the basic hypothesis. A more direct assault on the argument could also be made by providing an alternative explanation or the calendrical features on which it rests.

Figure 18 presents an overview of the argument. It omits certain calendars that do not bear on the astronomy of the solar year: the Olmec–Maya Long Count and the related calendars of Valladolid and Iximche, the specialized trecena calendars of Loxicha and Mazatlan, the undocumented (and unlikely) Calendar E, and the calendars differentiated only by their frozen correlations with European dating (Mazatec, Pokom, and Kekchi). Omitted also are the Postclassic calendars that are politically rather than astronomically derived by calendrical syncretism: Cakchiquel and the derivatives of Tilantongo. Most of these last were almost certainly originally derived by the processes that generated all the rest of the calendars—calculations of the solar era (see Chapter 3, Figure 8), but we lack the data to assert it for Otomi, Aztec, Texcoco, Tilantongo, Metztitlan, and Colhua.

Four kinds of calendrical derivation are illustrated: (1) the establishment of a new era (usually + 1 day); (2) the "extra" leap-year adjustment (also + 1 day); (3) the anti-leap-year adjustment (– 20 days), and (4) the leap-year adjustment (+ 20 days).

In all of the initial dated calendars, an advance of + 1 day is the marker of a new era. Seven of the ten eral calendars accomplish this by year-bearer advance. Two more (Quiche and Palenque) use zero deletion without advancing the year bearers. The original calendar (Cuicuilco) is of course unmarked.

In the terminally dated calendars, an advance of + 1 day is the marker for the extra leap-year days required (every 219 years) by a 6 day overcalculation of the length of the era. This year-bearer advance is responsible for the genesis of the Izapa, Zapotec, and Tilantongo calendars.

All of the remaining pre-Columbian calendars were generated by the – 20–day anti-leap-year adjustment. Masked in the diagram is the special case of the Teotihuacan and Tikal calendars that also included a – 5-day Uayeb adjustment.

New calendars generated in the Colonial period use the opposite mechanism of a leap year (+ 20-day) adjustment. It seems obvious that this represents a European influence, though the manner of its application (by 20-day months) remains indigenous.

Calendars linked by solid lines in Figure 18 are genetically related. The complete genealogy is represented in Figure 8 (chapter 3). It must be remembered that this is a genealogy of New Years' days. Although it is true that any other day of the year could potentially be used for marking the era, the most plausible alternatives would be the 100th day, the 260th, or the 360th. Either of the first two is ruled out by archaeological evidence: they would shift the chronology of the whole reconstruction by about four centuries. I believe the third is rendered unlikely by the clear association of terminal dating with an inaccurate calculation of the era. The remarkable calendrical synthesis of the Postclassic seems to me to clinch the argument by reuniting the terminal and initial dating traditions, but it is still the New Years' days of the initial calendars (and hence the first name days of the terminal ones) that preserve the astronomically correct calculation. It is difficult to avoid the conclusion that the sages of the Toltec era were aware of this.

I believe the evidence presented documents the fact that the calculation of the solar (tropical) year by the astronomers of Kaminaljuyu of 433 B.C. was identical to the fourth decimal place with the corresponding calculations of modern astronomy. The era of twenty-nine calendar rounds (1,508 years) completes 1,507 tropical years of 365.2422 days each.

But perhaps the most astonishing thing about the calendar of Anahuac is its historicity. By its combination of historical unity and conscious and disciplined diversity, it embodies in the structure of its New Years' dates a remarkably permanent and detailed record of its origin and development.

The day of the New Year stands as a monument to the astronomical date of its birth—a monument on the road of the sun. In the cyclic time of Middle America, it is, by the same token, a nearly perfect prediction. In its elegance, precision, and antiquity, the year count is testimonial to a conquest of time that stands alone in the annals of world civilization.

BIBLIOGRAPHY

Acosta II (cited by Caso 1977).

Acosta, Jorge R.

    1944   La Tercera temporada de exploraciones arqueológicas en Tula, Hdg., 1942. *Revista Mexicana de Estudios Antropológicos* 7:23–64. Mexico.

    1956   Resumen de los informes de las exploraciones arqueológicas en Tula, Hgo. durante las VI, VII y VIII temporadas 1946–1950. *Anales del Instituto Nacional de Antropología e Historia* 8:37–115. Mexico.

    1956–57   Interpretación de algunos de los datos obtenidos en Tula relativos a la época tolteca. *Revista Mexicana de Estudios Antropológicos* 14:75–110.

    1957   Resumen de los informes de las exploraciones arqueológicas en Tula, Hgo. durante las IX y X temporadas 1953–1954. *Anales del Instituto Nacional de Antropología e Historia* 9:119–169. Mexico.

Albornoz, J. de

    1691   *Arte de la lengua chiapaneca.* MS in Bibliothèque Nationale. Paris.

Alcina Franch, José

    1973   Juan de Torquemada, 1564–1624. *Handbook of Middle American Indians* 13:256–75. Austin.

Almanaque Guadalupano.

    1950   *Almanaque del Dr. J. H. McLean, Año 97°.* McLean Medicine Co., St. Louis.

Alvarado, Francisco de

    1962   *Vocabulario en lengua mixteca.* INI and INAH, Mexico.

Alvarado, Pedro de

    1925   *An account of the conquest of Guatemala in 1524 by Pedro de Alvarado.* Sedley J. Mackie, ed. The Cortes Society, New York.

Alvarado Guinchard, Manuel

    1976   *El Códice de Huichapan.* Instituto Nacional de Antropología e Historia, Mexico.

Alvarado Tezozomoc, H.

    1949   *Crónica Mexicayotl.* Universidad Nacional Autónoma de México, Inst. Hist. Pub. 10, Mexico.

Ancona, E.

    1878   *Historia de Yucatán desde la época más remota hasta nuestros días.* Merida.

Anderson, Arthur J. O.

    1985   British and Foreign Bible Society MS 374, Vol. III. Paper read at American Society for Ethnohistory, Nov. 4. New Orleans.

Anderson, Arthur, J. O., and Charles Dibble

    1975–81   *Florentine Codex.* 13 vols. University of Utah Press, Salt Lake City.

Andrews, E. Wyllys IV

    1940   Chronology and astronomy in the Maya era. In *The Maya and Their Neighbors* 150–61. Appleton, New York.

Andrews, James Richard

    c1975   *Introduction to Classical Nahuatl.* University of Texas Press, Austin.

Annals of the Cakchiquels (see Villacorta & Villacorta 1933; Recinos 1950).
Annals of Tlatelolco (see Berlin 1948).
Anónimo de Mendieta (cited by Jimenez 1940).
Anon.
    1685   Calendario de los Indios de Guatemala. Photographic copy of Juan Gavarrete MS copy of 1878, Brigham Young University Library, Provo, Utah. Identified as pages 21–28 from the "Crónica Franciscana," an anonymous 283-page MS in possession of the Sociedad Económica de Guatemala in 1878 (see Rodriguez 1940).
    1931   A Lanquin Kekchi Calendar. *Maya Society Quarterly* 1:29–32.
Antonio de León (Glass 1975b: No. 8).
Aschman, H. P.
    1973   *Diccionario Totonaco de Papantla*. Instituto Lingüístico de Verano, Mexico.
Aubin, J.-M.-A.
    1893   *Histoire de la nation méxicaine depuis le départ d'Aztlan jusqu' à l'arrivée des conquérants espagnols (et au delà 1607)*: Reproduction du codex de 1576 appartenant à la collection de M. E. Eugène Goupil. Leroux. Paris.
Aubin 20 (Glass 1975b: No. 14).
Aubin Tonalamatl (Glass 1975b: No. 15).
Aveni, Anthony F.
    1980   *Skywatchers of Ancient Mexico*. University of Texas Press, Austin.
Aveni, Anthony F., and Horst Hartung
    1984   Archaeoastronomy and the Puuc Sites. In *Simposio sobre Arqueoastronomía y Etnoastronomía de Mesoamérica*, Universidad Nacional Autónoma de México, Mexico.
Azoyú I (Glass 1975b: No. 21).
Azoyú II (Glass 1975b: No. 22).
Barela, Francisco
    ca. 1600   Vocabulario kakchiquel. MS in Museo Nacional de México.
Barlow, Roberto H.
    1951   El manuscrito del calendario matlatzinca. In *Homenaje al Dr. Alfonso Caso* pp. 69–72. Imprenta Nuevo Mundo, Mexico.
Barrera Vásquez, Alfredo
    1948   *El Libro de los libros de Chilam Balam*. Fondo de Cultura Económica, Mexico.
Basseta, Domingo
    1698   Vocabulario de la lengua quiché. MS in Bibliothèque Nationale. Paris.
Baudot, Georges
    1983   *Utopía e Historia en México*. Espasa-Calpe, Madrid.
Becerra, Marcos Ec.
    1933   *El Antiguo Calendario Chiapaneco: estudio comparativo entre éste y los calendarios precoloniales maya, quiché y nahua*. Mexico.
Becker I (Glass 1975b: No. 27).
Becker II (Glass 1975b: No. 28).
Becquelin, Pierre, and Claude F. Baudez
    1982   Tonina, une cité maya du Chiapas (Mexique). *Études Mésoaméricaines* 6(3). Mission Archéologique et Ethnologique Française au Mexique, Paris.
Bell, Betty
    1971   Archaeology of Nayarit, Jalisco, and Colima. *Handbook of Middle American Indians* 11:694–753. Austin.
Belmar, Francisco
    1901   *Lenguas del Estado de Oaxaca. Investigación sobre el idioma amuzgo que se habla en algunos pueblos del distrito de Jamiltepec*. Oaxaca.

Berlin, Heinrich
   1951   The Calendar of the Tzotzil Indians. In *The civilization of ancient America,* ed. Sol
   Tax. Selected papers of the 29th International Congress of Americanists. Chicago.
Berlin, Heinrich, ed.
   1948   *Anales de Tlatelolco.* Robredo, Mexico.
Beyer, Hermann
   1937   Lunar glyphs of the supplementary series. *México Antiguo* 4:75–82.
Boban Calendar (Glass 1975b: No. 30).
Bodley (Glass 1975b: No. 31).
Book of Mimiahuapan (codex).
Borbonicus (Glass 1975b: No. 32).
Boturini (Glass 1975b: No. 34).
Bowditch, C. P.
   1910   *The numeration, calendar systems and astronomical knowledge of the Mayas.* Peabody
   Museum, Harvard University.
Brasseur de Bourbourg, Charles Etienne
   1858   *Histoire des nations civilisées du Méxique et de l'Amérique Centrale durant les siècles antérieurs
   à Christophe Colomb.* Bertrand, Paris. 4 vols.
Breton, Adela C.
   1915   Pocomchi Notes. *International Congress of Americanists* 19:545–48.
Bricker, Victoria R.
   1967   Unpublished field notes, Chamula. MS in possession of author. New Orleans.
   1982   The Origin of the Maya Solar Calendar. *Current Anthropology* 23:101–3.
Brinton, Daniel G.
   1893   The Native Calendars of Central America and Mexico. *Proceedings of the American Philosophical Society* 31 (142):258.
British and Foreign Bible Society Manuscript 374.
   1606   British and Foreign Bible Society, Los Angeles (cited by Anderson 1985).
Broda de Casas, Johanna
   1969   The Mexican Calendar as Compared to Other Mesoamerican Systems. *Acta
   Ethnologica et Linguistica* 15. Vienna.
Brotherston, Gordon
   1983   The Year 3113 B.C. and the Fifth Sun of Mesoamerica: An Orthodox Reading
   of the Tepexic Annals (Codex Vindobonensis obverse). In Aveni, Anthony. F, and
   Gordon Brotherston, *Calendars in Mesoamerica and Peru: Native American Computations of
   Time.* BAR International Series 174. Oxford.
Bunge, O. D. E.
   1940   Contribution à l'astronomie maya. *Journal de la Société des Américanistes* 32:69–92.
Bunting, E. J. W.
   1932   Ixtlavacan Quiche Calendar of 1854. *Maya Society Quarterly* 1:72–75.
Burkitt, Robert
   1930-31   The Calendar of Soloma and Other Indian Towns. *Man* 30–31:103-7, 146–50.
Cabildo 1528 (cited by Caso 1967).
Calderón, Hector M.
   1982   *Correlacion de la rueda de los katunes, la cuenta larga y las fechas cristianas.* Grupo
   Dzibil, Mexico.
Calendario de 1553 (cited by Caso 1967).
Calendario de 1685 (see Anon. 1685).
Calendario del más Antiguo Galván.
   1967   141st ed. Mexico.
Calendario Sna Holobil
   1979   (Tzeltal). Sna Holobil, San Cristobal.

Campbell, Lyle
   1979   Yajalón Libro de Bautismos y Casamientos 1554–1568 (Barbara MacLeod from Lyle Campbell's original data, summer 1979, SCLC).
Carrasco, P.
   1950   *Los Otomíes, cultura e historia prehispánica de los pueblos mesoamericanos de habla otomiana.* UNAM, Inst. de Historia Ser. 1ª., No. 15:168–195, Mexico.
   1951   Una cuenta ritual entre los zapotecos del sur. In *Homenaje al Dr. Alfonso Caso,* pp. 91–100. Imprenta Nuevo Mundo, Mexico.
Carrasco, P., W. Miller, and R. J. Weitlaner
   1961   El Calendario mixe. *México Antiguo* 9:153–72.
Caso, Alfonso
   1928   *Las estelas zapotecas.* Museo Nacional, Mexico.
   1932   Las exploraciones de Monte Albán, temporada 1931–32. *Instituto Panamericano de Geografía e Historia Publicación* 7, Mexico.
   1937   ¿Tenían los Teotihuacanos conocimiento del tonal pohualli? *México Antiguo* 4:131–43.
   1939   La correlación de los años azteca y cristiano. *Revista Mexicana de Estudios Antropológicos* 3:11–45.
   1941   El complejo arqueológico de Tula y las grandes culturas indígenas de México. *Revista Mexicana de Estudios Antropológicos* 5:85–95.
   1943   The Calendar of the Tarascans. *American Antiquity* 9:11–28.
   1946   El calendario matlatzinca. *Revista Mexicana de Estudios Antropológicos* 8:95–109.
   1954   Calendario de los totonacos y huastecos. *Revista Mexicana de Estudios Antropológicos* 13:337–50.
   1956   El Calendario mixteco. *Historia Mexicana* 5:481–97.
   1958   El calendario mexicano. *Memorias de la Academia Mexicana de Historia* 17:41–96.
   1963   El calendario mixe. *Revista Mexicana de Estudios Antropológicos* 19:63–74.
   1965a Mixtec Writing and Calendar. *Handbook of Middle American Indians* 3:948–61.
   1965b Zapotec Writing and Calendar. *Handbook of Middle American Indians* 3:931–47.
   1967   *Los calendarios prehispánicos.* Série Cultura Náhuatl, Monografía 6, UNAM, Mexico.
   1971   Calendrical Systems of Central Mexico. *Handbook of Middle American Indians* 10:333–48.
   1977   *Reyes y reinos de la mixteca.* Fondo de Cultura Económica, Mexico.
Caso, Alfonso, and Ignacio Bernal
   1965   Ceramics of Oaxaca. *Handbook of Middle American Indians* 2:871–95. Austin.
Castillo, Cristobal de
   ca. 1600   *Historia de los mexicanos.* Published by Francisco del Paso y Troncoso, 1908, Florence.
Chacxulubchen (Text quoted identical to *Yaxkukul,* q.v.).
Cervantes de Salazar, Francisco
   1566–75   Crónica de Nueva España. In F. Paso y Troncoso (ed.), *Papeles de Nueva España,* 1914–36, Madrid and Mexico. 3 vols.
Chavero (Glass 1975b: No. 43).
Chavero, Alfredo
   1892   *Homenaje á Cristóbal Colón.* Junta Colombina, Mexico.
Chávez, Gabriel de
   1923   *Relación de la provincia de Meztitlán.* Mexico.
Chimalpahin Quauhtlehuanitzin, Domingo Francisco de San Antón Muñón
   1889   *Annales: sixième et septième relations (1258–1612).* Maisonneuve et Ch. Leclerc, Paris.
   1958   *Das Memorial Breve acerca de la fundación de la ciudad de Culhuacan.* Iberoamerikanischen Bibliothek. Quellenwerke 7. Berlin. Walther Lehmann and Gerd Kutscher, eds.

Chimalpopoca Codex = *Cuauhtitlan* (see Lehmann 1938).
Chol Poval Ahilabal Q'ih (The Count of the Cycle and the Numbers of the Days)
    1722   MS: Photographic copy in Latin American Library, Tulane University, New Orleans.
Chronicle of Acalan-Tixchel (see Scholes & Roys 1968).
Chumayel (see Edmonson 1986).
Clark, Lewis
    1981   *Diccionario popoluca de Oluta*. Instituto Lingüístico de Verano, Mexico.
Clavijero, Francisco Javier
    1945   *Historia antigua de México*. Porrúa, Mexico.
    1964   *Historia antigua de México*. Porrúa. Mexico.
Codex Mexicain 65–71. Bibliothèque Nationale. Paris.
Coe, M. D.
    1957   Cycle 7 Monuments in Middle America: A Reconsideration. *American Anthropologist* 59:597–611.
    1965a   Archaeological Synthesis of Southern Veracruz and Tabasco. *Handbook of Middle American Indians* 2:679–715. Austin.
    1965b   The Olmec Style and Its Distribution. *Handbook of Middle American Indians* 2:739–75. Austin.
Coixtlahuaca I (Glass 1975b: No. 70).
Coixtlahuaca II (Glass 1975b: No. 71).
Colby, Benjamin, and Lore M. Colby
    1981   *The Daykeeper: The Life and Discourse of an Ixil Diviner*. Harvard University Press, Cambridge.
Colombino (Glass 1975b: No. 72).
Colville, Jeffrey K.
    1985   The Structure of Mesoamerican Number Systems: With a Comparison to Non-Mesoamerican Systems. Ph.D. Dissertation, Tulane University. New Orleans.
Compendio de la historia mexicana . . . escrito por Chimalpahin,
    1064–1526 (known primarily from León y Gama 1781–84).
Cook de Leonard, Carmen
    1973   A New Astronomical Interpretation of the Four Ballcourt Panels at Tajin, Mexico. MS (cited by Kelley 1983).
Córdova, J. de
    1578a   *Arte en lengua zapoteca*. Pedro Balli, Mexico (Reprinted 1886, Morelia).
    1578b   *Vocabulario en lengua zapoteca*. Pedro Charle y Antonio Ricardo, Mexico (reprinted 1942).
Cortés, Hernán
    1963   *Cartas y documentos*. Mexico.
Costumbres de Nueva España (cited by Caso 1967).
Cozcatzin (Glass 1975b: No. 83).
Craine, Eugene R., and Reginald C. Reindorp
    1979   *The Codex Pérez and The Book of Chilam Balam of Maní*. University of Oklahoma Press, Norman.
Crónica mexicáyotl (see Alvarado Tezozomoc 1949).
Cruz, Wilfredo C.
    1935   *El Tonalamatl Zapoteco: el mito y la leyenda zapoteca*. Gobierno del Estado de Oaxaca, Oaxaca.
Cuadernillo de San Lucas Camotlan (cited by Weitlaner 1936).
Cuauhtitlan, Annals of (see Lehmann 1938).

Dahlgren de Jordán, Barbro
    1954   *La Mixteca: su cultura e historia prehispánica.* Colección Cultura Mexicana 11.
    Imprenta Universitaria, Mexico.
Davies, Nigel
    1977   *The Toltecs until the Fall of Tula.* University of Oklahoma Press, Norman.
De la Coruña, Martín
    1541   *Relación de Michoacán.* Morelia.
De la Fuente, Beatriz, and Nellie Gutiérrez
    1980   *Escultura huasteca en piedra.* Instituto de Investigaciones Estéticas, Universidad
    Nacional Autónoma de México, Mexico.
De la Torre, Concepción and Antonio Pérez Elias
    1976   *El Tajín: Official Guide.* Instituto Nacional de Antropología e Historia, Mexico.
Dehesa (Glass 1975b: No. 113).
Díaz del Castillo, Bernal
    1904   *Historia verdadera de la conquista de la Nueva España.* Secretaría de Fomento, Mex-
    ico.
Dibble, Charles E.
    1951   *Códice Xolotl.* Inst. Hist. 1st ser. 22, UNAM, Mexico.
Dinsmoor (cited by Satterthwaite 1965:630).
Dittrich, A.
    1936   Die Korrelation der Maya-Chronologie. Reprinted from *Abhandlungen der Preus-
    sischen Akademie der Wissenschaften.* Berlin.
D'Olwer, Luis Nicolau, and Howard F. Cline
    1973   Sahagún and His Works. *Handbook of Middle American Indians* 13:186–206. Aus-
    tin.
Dresden (Glass 1975b: No. 113; see Thompson 1972).
Durán, Diego
    1967   *Historia de las indias de Nueva España.* 2 v. Porrúa, Mexico.
    1971   *Book of the Gods and Rites and the Ancient Calendar.* University of Oklahoma Press,
    Norman.
Dutton, B. P.
    1955   Tula of the Toltecs. *El Palacio 62:195-251.*
Dyk, Anne, and Betty Stroudt
    1965   *Vocabulario mixteco de San Miguel el Grande.* Instituto Lingüístico de Verano, Mex-
    ico.
Edmonson, Barbara W.
    1983   Field notes on Potosino Huastec, 1980–83. In possession of the author.
Edmonson, Munro S.
    1961   Unpublished field notes, Quezaltenango, Guatemala, 1960–61. In possession of
    the author.
    1971   The Book of Counsel: The Popol Vuh of the Quiche Maya of Guatemala.
    Middle American Research Institute Publication 35, New Orleans.
    1976   The Mayan Calendar Reform of 11.16.0.0.0. *Current Anthropology* 17:713-17.
    1982   *The Ancient Future of the Itza: The Book of Chilam Balam of Tizimin.* University of
    Texas Press, Austin.
    1985   The First Chronicle of Yucatan (692–1848). *Quinta Mesa Redonda de Palenque,
    1983* 7:193-210.
    1986   El Calendario de Teotihuacán. *Archaeoastronomy and Ethnoastronomy of Middle
    America* (in press).
    1986   *Heaven Born Merida and Its Destiny: The Book of Chilam Balam of Chumayel.* Univer-
    sity of Texas Press, Austin.
    n.d.   The Baktun Ceremonial of 1618. *Cuarta Mesa Redonda de Palenque, 1980* (in press).

Escalona Ramos, A.
    1943   Cronología y astronomía maya-mexica. Un nuevo sistema de correlación calendárica. *Proceedings of the 27th International Congress of Americanists* 1:623–30. Mexico.
Feldman, Lawrence H.
    1976   Names of Deities in Early Guatemala. *International Congress of Americanists* 51(2), 236–42.
    1983   The Structure of Cholan and Mayan Surnames in Sixteenth and Seventeenth Century Manuscripts. *Mexicon* 5(3):46–53.
Fernández Leal (Glass 1975b: No. 119).
Fernández de Miranda, María Teresa
    1961   *Diccionario del Ixcateco*. Instituto Nacional de Antropología e Historia 7, Mexico.
García, Esteban
    1918   *Crónica de la Provincia Agustiniana del Santísimo Nombre de Jesús, de México*. Madrid (pp. xv, xvii, 301, Otomi correlational date).
García Icazbalceta, J.
    1891   *Nueva colección de documentos para la historia de México*. Mexico.
García Payón, José
    1939   El simbolo del año en el México antiguo. *México Antiguo* 4:241–54.
    1949   Una nueva fecha maya en el territorio veracruzano. *Uni-Ver* 1:403–40. Jalapa.
    1971   Archaeology of Central Veracruz. *Handbook of Middle American Indians* 10:505–42. Austin.
Gemelli Careri, Gio Francesco
    1699–1700   *Giro del Mondo*. Naples. 6 vols.
Girard, Rafael
    1942   *Los Mayas Eternos*. Editorial México Lee, Mexico.
Glass, John B.
    1975a   *Compendio de la historia mexicana. Domingo Chimalpahin. Extracts from a lost manuscript*. Lincoln Center (Conemex Associates) (cited by Prem 1983).
    1975b   A Survey of Native Middle American Pictorial Manuscripts. *Handbook of Middle American Indians* 13:3–80. University of Texas Press, Austin.
Gómara (see López de Gómara).
Gonzáles, Luis
    1672   Arte breve y vocabulario de la lengua zoque. In Grasserie, Raoul de la, *Langue Zoque et lange mixe*. Maisonneuve, Paris, 1898.
González, P. and P. Buenaventura
    1958   Carta a Sahagún (see Caso 1958).
Goodman, J. T.
    1897   The archaic Maya inscriptions. In appendix to Maudslay 1889–1902.
    1905   Maya dates. *American Anthropologist* 7:642–47.
Gossen, Gary H.
    1974   A Chamula Calendar Board. In N. Hammond, ed., *Mesoamerican Archaeology: New Approaches*, pp. 217–53. Duckworth, London.
Goubaud Carrera, Antonio
    1937   *The Guajxaquip Bats: An Indian Ceremony of Guatemala*. Centro Editorial, Guatemala.
Graham, Ian
    1967   Archaeological Explorations in El Peten, Guatemala. *Middle American Research Institute Publication* 33, Tulane University, New Orleans.
    1980   *Corpus of Mayan Hieroglyphic Inscriptions* 2(3). Peabody Museum, Harvard University, Cambridge.

Graham, Ian, and Eric von Euw
    1977   *Corpus of Maya Hieroglyphic Inscriptions* 3(1:Yaxchilan). Peabody Museum, Harvard University, Cambridge.
Graham, John A., R. F. Heizer, and E. M. Shook
    1978   Abaj Takalik 1976: Exploratory Investigations. *University of California Archaeological Research Facility Contribution* 36:85–110. Berkeley.
Graulich, Michel
    1981   The Metaphor of the Day in Ancient Mexican Myth and Ritual. *Current Anthropology* 22:45–60.
Guevea, Lienzo de (Glass 1975b: No. 130).
Guiteras-Holmes, Calixta
    1961   *Perils of the Soul: The World View of a Tzotzil Indian.* Free Press, Glencoe.
Haeserijn V., Esteban
    1979   *Diccionario K'ekchi' Español.* Editorial Piedra Santa, Guatemala.
Hendrichs, P. R.
    1939   Un estudio preliminar sobre la lengua cuitlateca de San Miguel Totolapan, Gro. *México Antiguo* 4:329–62.
Hernández Spina, V.
    1854   Calendario quiché (see Bunting 1932).
Herrera, Antonio de
    1952   *Historia general de los hechos de los castellanos.* Tomo IX. Editorial Maestre, Madrid.
Historia de los mexicanos por sus pinturas (see García Icazbalceta 1891:3).
Historia Tolteca-Chichimeca (see Kirchhoff *et al.* 1976; Preuss and Menghin 1937).
Hochleitner, Franz Joseph
    1970   An attempt at a chronological-astronomical interpretation of the numbers and day-signs of Dresden codex. *Boletín Informativo de Escritura Maya.* Universidad Nacional Autónoma de México. Mexico.
    1972   The Correlation between the Mayan and the Julian Calendar. 40th International Congress of Americanists, Rome.
    1974   MS (cited by Kelley 1983).
    1977   *Cronología e astronomía maia.* Univ. Fed. de Juíz de Fora, Brazil.
Holland, William
    1963   *Medicina maya de los altos de Chiapas.* Instituto Nacional Indigenista, Colección de Antropología Social 2, Mexico.
Huichapan, Annals of (Glass 1975b: No. 142; see Alvarado 1976).
Humboldt Fragment I (Glass 1975b: No. 147).
Ilhuitlan (Glass 1975b: No. 157).
Ixtlilxochitl (Glass 1975b: No. 171).
Ixtlilxochitl, Fernando de Alva
    1952   *Obras históricas.* Editora Nacional, Mexico.
    1958   *Obras históricas.* Editora Nacional, Mexico.
Jamieson, Allen
    1978   *Mazateco de Chiquihuitlán.* Centro de Investigaciones para la Integración Social, Mexico.
Jansen, Maarten E. R. G. N.
    1980   *Tnuhu Niquidza Iya: Temas Principales de la Historiografía Mixteca.* Dirección General de Educación y Bienestar Social, Oaxaca.
    1981   *Huisi Tacu.* 2 v. Centrum voor Studie en Documentatie van Latijn Amerika. Incidentele Publicaties 24, Amsterdam.
    1984   El Códice Ríos y Fray Pedro de los Ríos. MS in possession of the author.
Jiménez Moreno, Wigberto
    1940   *Códice de Yanhuitlán.* Museo Nacional, Mexico.

1961   Diferente principio del año entre diversos pueblos y sus consecuencias para la cronología prehispánica. *México Antiguo* 9:137–52.

Jones, Christopher and Linton Satterthwaite
1982   The Monuments and Inscriptions of Tikal: The Carved Monuments. *Tikal Report* No. 33, Part A. University of Pennsylvania Museum, Philadelphia.

Josserand, Judith Kathryn
1983   Mixtec Dialect History. (Proto-Mixtec and Modern Mixtec Text). Unpublished Ph.D. dissertation. Tulane University. New Orleans.

Kaucher, Carl D.
1980   Maya Chronology and the Conjunction of Mars. MS (cited by Kelley 1983).

Kaufman, Terence
n.d.   Uspantec field notes.

Kelley, David H.
1960   Calendar Animals and Deities. *Southwestern Journal of Anthropology* 3:317–37.
1976   *Deciphering the Maya Script.* University of Texas Press, Austin.
1983   The Maya Calendar Correlation Problem, in Richard M. Leventhal and Alan L. Kolata, eds., *Civilization in the Ancient Americas*:157–208. University of New Mexico Press, Albuquerque.

Kiemele, Mildred
1975   *Vocabulario mazahua-español y español-mazahua.* Biblioteca Enciclopédica del Estado de México, Mexico.

Kirchhoff, Paul
1950   The Mexican Calendar and the Founding of Tenochtitlan-Tlatelolco. *Transactions of the New York Academy of Sciences*, Ser. II, Vol. 12, No. 4:126–32.
1955–56   Calendarios tenochca, tlatelolca y otros. *Revista Mexicana de Estudios Antropológicos* 14(2):257–67.

Kirchhoff, Paul, Lina Odena Güemes, and Luis Reyes Garcia
1976   *Historia Tolteca Chichimeca.* Instituto Nacional de Antropología e Historia, Mexico.

Knowles, Susan
1984   A Descriptive Grammar of Chontal Maya (San Carlos Dialect). Ph.D. Dissertation, Tulane University, New Orleans.

Kreichgauer, Damian
1927   La correspondencia entre la cronología maya y el cómputo europeo. *Investigación y Progreso* 1:7. Madrid.
1932   Maya-Chronologie. (Cited by Andrews 1940:159, note 21).

Kubler, George, and Charles Gibson
1951   The Tovar Calendar: An illustrated Mexican manuscript of ca. 1585. *Memoirs, Connecticut Academy of Arts and Sciences* 11.

LaFarge, Oliver,
1947   *Santa Eulalia: The Religion of a Cuchumatan Indian Town.* University of Chicago Press, Chicago.

LaFarge, Oliver, and Douglas Beyers
1931   The Year Bearer's People. *Middle American Research Institute Publication* 3. New Orleans.

Landa, Diego de
1966   *Relación de las cosas de Yucatán.* Porrúa, Mexico.

Lanquin Calendar (see Anon. 1931).

Larsen, Raymond S.
1955   *Vocabulario huasteco del Estado de San Luis Potosí, México.* Instituto Lingüístico de Verano, Mexico.

Las Navas, Francisco
    1553   El Calendario tlaxcalteca. *Archivos Históricos. Col. Antigua* 210:111–25. Museo Nacional de Antropología e Historia, reprinted by Baudot 1983:467–70.
Laughlin, Robert M.
    1975   The Great Tzotzil Dictionary of San Lorenzo Zinacantán. *Smithsonian Contributions to Anthropology* 19, Washington.
Lehmann, Walter
    1920   *Zentral-Amerika: die Sprache Zentral-Amerikas in ihren Beziehungen zueinander sowie zu Süd-Amerika und Mexiko.* 2 vols. Dietrich Riemer, Berlin.
    1926   Reisebrief aus Puerto Mexic. *Zeitschrift für Ethnologie* 1926:171–77.
    1938   *Die Geschichte der Königreiche von Colhuacan und Mexico.* Ibero-Amerikanischen Institut. Quellenwerke 1. Berlin (=*Códice Chimalpopoca, Anales de Cuauhtitlán*).
Lehmann, Walter, and Gerd Kutscher
    1958   *Chimalpahin: Das Memorial Breve acerca de la Fundación de la Ciudad de Culhuacan.* Quellenwerke 7. Stuttgart.
León, Nicolás
    1903   *Los Tarascos: notas históricas, étnicas y antropológicas.* Imprenta del Museo Nacional, Mexico.
    1933   *Códice Sierra.* Museo Nacional, Mexico.
León y Gama, Antonio de
    1792   *Descripción histórica y cronológica de las dos piedras que con ocasión del nuevo empedrado que se está formando en la plaza principal de México se hallaron en ella el año de 1790.* Mexico.
Lienzo de Guevea (Glass 1975b: No. 130).
Lienzo de Tlapa (Glass 1975b: No. 342).
Lincoln, Jackson Stewart
    1942   The Maya Calendar of the Ixil of Guatemala. *Carnegie Institution of Washington Publication* 528, Contribution 38. Washington.
Lipp, Frank
    1982   The Mije Calendrical System: Concepts of Behavior. Ph.D. dissertation. New School for Social Research, New York.
Lizardi Ramos, César
    1955   ¿Conocían el Xihuitl los Teotihuacanos? *México Antiguo* 8:219–24.
Long, R. C. E.
    1934   The dates in the Annals of the Cakchiquels and a Note on the 260 Period of the Mayas. *Journal of the Royal Anthropological Institute of Great Britain and Ireland* 64:57–68. London.
Longacre, Robert
    1967   Systemic Comparison and Reconstruction. *Handbook of Middle American Indians* 5:117–59.
López de Gómara, Francisco
    1552   La historia de las Indias y conquista de México. Zaragoza.
Lorenzana, Francisco Antonio
    1770   *Historia de Nueva España escrita por su esclarecido conquistador Hernán Cortés, aumentada con otros documentos y notas.* Mexico.
Lothrop, Samuel K.
    1930   A Modern Survival of the Maya Calendar. *XXIII International Congress of Americanists*, New York, 1928:652–55.
Lounsbury, Floyd
    1978   Maya Numeration, Computation and Calendrical Astronomy. *Dictionary of Scientific Biography* 15:759–818.

1980   Some Problems in the Interpretation of the Mythological Portion of the Hiero-glyphic Text of the Temple of the Cross at Palenque. *Third Palenque Round Table, 1978*:99–115.

Lowe, G. W.
1962   Mound 5 and minor excavations, Chiapa de Corzo, Chiapas, Mexico. *Papers New World Archaeological Foundation* 12.

Ludendorff, Hans
1930–37   Untersuchungen zur Astronomie der Maya (cited by Andrews 1940:153, note 10).

McArthur, Harry
1965   Notas sobre el calendario ceremonial de Aguacatan, Huehuetenango. *Folklore de Guatemala* 1:33–38.

McVicker, Donald
1985   The "Mayanized" Mexicans. *American Antiquity* 50:82–99.

Madrid (Glass 1975b: No. 187).

Magliabecchi (Glass 1975b: No. 188).

Makemson, Maude W.
1946   The Maya correlation problem. *Vassar College Observatory* 5.

Malström, Vincent H.
1978   A Reconstruction of Mesoamerican Calendrical Systems. *Journal for Astronomy* 9:105–16.

Mani (see Craine & Reindorp 1979).

Marcus, Joyce
1976   The Origins of Mesoamerican Writing. *Annual Review of Anthropology* 5:35–68.

Martínez Grácida, Manuel
1910   *Civilización Chontal: Historia Antigua de la Chontalpa Oaxaqueña*. Imprenta del Gobierno Federal, Mexico.

Martínez Hernández, Juan
1926   *Crónicas Mayas: Crónica de Yaxkukul*. Tipografía Yucateca, Merida.
1926b   Paralelismo entre los calendarios Maya y Azteca. *Diario de Yucatán*, February 7. Merida.

Mathews, Peter
1982   La date du Tonina fragment 35. *Bulletin de la Mission Archéologique et Ethnologique Française en Méxique*, No. 3.

Matritense (cited by Caso; presumably Glass 1975b: Nos. 271, 272, 273).

Maxwell, Judith M.
1981   Unpublished field notes on Chuh. In possession of the author. New Orleans.

Mazizcatzin (cited by Caso 1967).

Mechling, William H.
1912   The Indian Stocks of Oaxaca, Mexico. *American Anthropologist* 14:643–82.

Meighan, Clement W., and Leonard J. Foote
1968   *Excavations at Tizapan*. Latin American Center, UCLA. Los Angeles.

Melgarejo Vivanco, José Luis
1966   *Los Calendarios de Cempoala*. Cuadernos del Instituto de Antropología 2. Universidad Veracruzana, Jalapa.

Membreño, Alberto
1901   *Nombres Geográficos de la República de Honduras*. Tipografía Nacional, Tegucigalpa.

Memorial Breve (see Lehmann and Kutscher 1958).

Mendoza (Glass 1975b: No. 196; see Ross 1978).

Mexicano (Glass 1975b: No. 205).

Mexicanus (Glass 1975b: No. 207).

Mexicayotl Chronicle (see Alvarado Tezozomoc 1949).

Miles, S. W.
  1952   An Analysis of Modern Middle American Calendars: a Study in Conservation. In S. Tax, ed. *Acculturation in the Americas*. University of Chicago Press, Chicago.

Miller, Walter S.
  1952   Algunos manuscritos y libros Mixes en el Museo Nacional. *Tlatoani* 1(2):34–35.
  1956   Folklore Mixe. *Biblioteca de Folklore Indígena* 2. Instituto Nacional Indigenista, Mexico.

Mock, Carol
  1977   *Chocho de Santa Catarina Ocotlán*. Centro de Investigaciones para la Integración Social, Mexico.

Morán, Francisco
  1720   Arte breve y compendiosa de la lengua pocomchi. MS in Bibliothèque Nationale, Paris.
  1725? Diccionario pocomchi castellano y castellano pocomchi de San Cristóbal Cahcoh. Photographic copy of MS, Latin American Library, Tulane University. New Orleans.

Moran, Hugh A., and David H. Kelley
  1969   *The Alphabet and the Ancient Calendar Signs*. Daily Press, Palo Alto. (1st ed. 1953.)

Morley, Sylvanus Griswold
  1910   The Correlation of Maya and Christian Chronology. *American Journal of Archaeology*, Archaeological Institute of America, 2d ser., 14:193–203.
  1920   The Inscriptions at Copan. *Carnegie Institution of Washington Publication* 219. Washington.

Moser, Christopher L.
  1977   Ñuiñe Writing and Iconography of the Mixteca Baja. *Vanderbilt University Publications in Anthropology* 19. Nashville.

Motolinía, Toribio
  1549   Calendario de toda la índica gente . . . MS. C-IV-5 of the Monastery Library of the Escorial, fols. 141 r.–143 v. In Baudot 1983:428–30, identified as an eighteenth-century copy of a lost original.

Mukerji, Dhirendra Nath
  1936   A Correlation of the Mayan and Hindu calendars. *Indian Culture* 2:685–92.

Muñoz Camargo, D.
  1892   *Historia de Tlaxcala*. Mexico.

Narciso, Vicente A.
  1906   El calendario poconchí (cited by Termer 1930:394–95).
  1932   Pokonchi Calendar. *Maya Society Quarterly* 1:75–77.

Nash, Manning A.
  1957   Cultural Persistence and Social Structure: The Meso-American Calendar Survivals. *Southwestern Journal of Anthropology* 13:149–55.

Nativitas (Glass 1975b: No. 232).

Navarrete, Carlos
  1986   The Sculptural Complex at Cerro Bernal on the Coast of Chiapas. *Notes of the New World Archaeological Foundation* 1. Provo.

Navarrete, Carlos, and Ana María Crespo
  1971   Un Atlante mexica y algunas consideraciones sobre los relieves del Cerro de la Malinche, Hidalgo. *Estudios de Cultura Náhuatl* 9:11–15. Mexico.

Nicholson, H. B.
  1971   Major Sculpture in Pre-Hispanic Central Mexico. *Handbook of Middle American Indians* 10:92–134. Austin.

Nicholson, H.B. *et al.*
   1971   *Ancient Art of Veracruz.* Exhibit catalogue, Los Angeles County Museum of Natural History. Los Angeles.
Nóguez, Xavier
   1978   *Tira de Tepechpan.* Estado de México, Mexico.
Norman, Garth
   1976   *Izapa Sculpture. Part 2: Text.* New World Archaeological Foundation, Provo.
Nowotny, Karl Anton
   1958   Der Bilderfolge des Codex Vindobonensis und verwandter Handschriften. *Archiv für Völkerkunde* 13:210–21. Vienna.
Ñunaha 1 (codex cited by Jansen 1984).
Núñez de la Vega, Francisco
   1702   *Constituciones diocesanas del Obispado de Chiappa.* Caietano Zenobi, Roma.
Oakes, Maud
   1951   *The Two Crosses of Todos Santos: Survivals of Mayan Religious Ritual.* Pantheon, New York.
Ochoa, Lorenzo, and Thomas A. Lee, Jr., eds.
   1983   *Antropología e Historia de los Mixe-Zoques y Mayas (Homenaje a Frans Blom).* Centro de Estudios Mayas e Instituto de Investigaciones Filológicas, Universidad Nacional Autónoma de México, Mexico, and Brigham Young University, Provo.
Olmos, Andrés de
   1912   Proceso seguido por Fray Andrés de Olmos en contra del cacique de Matlatlán. *Publicaciones del Archivo General de la Nación* 3:205–15. Mexico.
Oviedo y Valdés, Gonzalo Fernández
   1535   *Historia general y natural de las Indias.* Seville.
Owen, Nancy K.
   1975   The Use of Eclipse Data to Determine the Maya Correlation Number. In A. Aveni, ed. *Archaeoastronomy in Pre-Columbian America*: 237–46. University of Texas Press, Austin.
Oxkutzcab (see Morley 1920:470–71, 507–9).
Paddock, John
   1970   A Beginning in the Ñuiñe, salvage excavations at Ñuyoo, Huajuapan. *Boletín de Estudios Oaxaqueños* 26. Museo Frissell de Arte Zapoteca, Mitla.
Paris (Glass 1975b: No. 247).
Parsons, Lee Allen
   1986   The Origins of Maya Art: Monumental Stone Sculpture of Kaminaljuyu, Guatemala, and the Southern Pacific Coast. *Studies in Pre-Columbian Art and Archaeology* 28. Dumbarton Oaks, Washington.
Paso y Troncoso, Francisco del (ed.)
   1904–7   *Papeles de Nueva España.* Madrid. 6 v.
Pereyra, Dionycio
   1723   Copy of Rodaz 1688 (Cited by Berlin 1951). Biblioteca Na Bolom, San Cristobal Las Casas.
Pineda, E.
   1845   *Descripción geográfica de Chiapas y Soconusco.* Mexico.
Pineda, Vicente
   1888   *Historia de la sublevaciones indígenas habidas en el Estado de Chiapas.* Tipografía del Gobierno, Chiapas, San Cristobal de las Casas.
Pogo, Alexander
   1937   Maya Astronomy. *Carnegie Institution of Washington, Year Book* 36:24–25. Washington.

Popol Vuh (See Edmonson 1971).

Porfirio Díaz (Glass 1975b: No. 255).

Pozas, Ricardo
    1959   Chamula: un pueblo indio de los altos de Chiapas. *Memorias del Instituto Nacional Indigenista* 8. Mexico.

Prem, Hanns J.
    1983   Las fechas calendáricas completas en los textos de Ixtlilxóchitl. *Estudios de Cultura Náhuatl* 16:225–31.

Prescott, William
    1863   *The Conquest of Mexico*. Lippincott, Philadelpha.

Preuss, K. Th., and E. Menghin, eds.
    1937   *Historia Tolteca-Chichimeca*. Baessler Archiv 9. Berlin.

Pride, Leslie, and Kitty Pride
    1970   *Vocabulario chatino de Tataltepec*. Instituto Lingüístico de Verano, Mexico.

Primeros Memoriales (Glass 1975b: No. 271).

Probanza de Miltepec 1622 (cited by Jansen 1980:2).

Probanza de Sahagún (cited by Caso 1967, Table X).

Proskouriakoff, Tatiana, and J. E. S. Thompson
    1947   Maya Calendar Round Dates such as 9 Ahau 3 Mol. *Carnegie Institution of Washington Notes on Middle American Archaeology and Ethnology* 79. Washington.

Quecholac, Anales de. Museo Nacional, Mexico.

Quintana, Agustín de
    1729   *Arte de la lengua mixe*. Imprenta del Comercio, Oaxaca. Repr. 1891.

Ramírez (Glass 1975b: No. 365).

Ramírez, Francisco
    1600   *Historia del Colegio de la Compañía de Jesús de Pátzcuaro*.

Recinos, Adrian
    1950   *Memorial de Sololá: Anales de los Cakchiqueles seguido del Título de los Señores de Totonicapán*. Fondo de Cultura Económica, Mexico.

Relación Genealógica
    1534   (cited by Jiménez 1961).

Relación de Metztitlán (see Chávez 1923).

Relación de Michoacan (see de la Coruña 1541).

Relación de Tecciztlán (see Paso y Troncoso 1905:4).

Relación de Teotitlán (see Paso y Troncoso 1905:4:213).

Rios 106 (Glass 1975b: No. 270).

Robertson, Donald, and Martha Barton Robertson
    1975   Catalogue of Techialoyan Manuscripts and Paintings. *Handbook of Middle American Indians* 14:265–280. University of Texas Press, Austin.

Robicsek, Francis
    1981   *The Mayan Book of the Dead: The Ceramic Codex: The Corpus of Codex Style Ceramics of the Late Classic Period*. University of Oklahoma Press, Norman.

Rodaz, Juan de
    1688   *Arte de la lengua Tzotzlem ó Tzinacanteca con explicación de Año solar y un Tratado de las quentas de los Indios en Lengua Tzotzlem*. Bibliothèque Nationale Manuscrit Mexicain 411. (Microfilm in Museo Nacional de Antropología, Serie Chiapas, Spool 94, Mexico.)

Rodríguez, Leopoldo Alejandro
    1948   Estudio geográfico, histórico, etnográfico, filológico y arqueológico de la República de El Salvador en Centro América. *Sociedad de Geografía e Historia de Guatemala* 23:146–229. Guatemala.

Rodríguez, Raquel
    1957   Calendario cakchiquel de los indios de Guatemala, 1685. Copiado en la ciudad

de Guatemala por C. H. Berendt, Marzo de 1878. *Antropología e Historia de Guatemala* 9(2):17–29 (see Anon. 1685).

Romero Castillo, Moisés
    1980   Correlation of the Christian and Mayan Calendars with Observations on Mayan Predictions. Chilam Balam of Tizimin. *Latin American Indian Literature* 4:12–28.

Ronan, Colin Alistair
    1966 Calendar, *Encyclopedia Britannica* 4:611–19. Encyclopedia Britannica, Chicago.

Rosales, Juan de Dios
    1939   Notes on San Pedro la Laguna. University of Chicago Library Microfilm No. 25. Chicago.

Ross, Kurt
    1978   *Codex Mendoza: Aztec Manuscript Productions*. Liber, Fribourg.

Roys, Ralph L.
    n.d.   Libro de Chilam Balam de Ixil. Copies of Maya Documents, Folio 43–67. Latin American Library, Tulane University. New Orleans.

Ruz Lhuillier, Alberto
    1945   *Guía arqueológica de Tula*. Instituto Nacional de Antropología e Historia, Mexico.

Sahagún, Bernardino de
    1975–81   *Relación de las cosas de la Nueva España* (see Anderson and Dibble 1975–81).

Santiago, Felipe de
    1632   *Códice de Huichapan* (original C. C. James, Mexico City; copy A. Caso; in Caso 1967:211–25).

Sapper, Karl Theodor
    1897   *Das nördliche Mittel-Amerika nebst einem Ausflug nach dem Hochland von Anahuac*. Braunschweig.
    1912   Ueber einige Sprachen von Südchiapas. *International Congress of Americanists Publication* 17(2):295–320.

Satterthwaite, Linton
    1951   Moon Ages of the Mayan Inscriptions: The Problem of Their Seven-Day Range of Variation from Calculated Mean Ages. In Tax, Sol, ed., *The Civilizations of Ancient America* 142–54.
    1961   Maya Long Count. *México Antiguo* 9:125–33.
    1965   Calendrics of the Maya Lowlands. *Handbook of Middle American Indians* 3:603–31.

Schele, Linda, and Mary Ellen Miller
    1986   *The Blood of Kings: Dynasty and Ritual in Maya Art*. Kimbell Art Museum, Fort Worth.

Scholes, France V., and Eleanor B. Adams
    1960   *Relaciones histórico-descriptivas de las Verapaz, el Manche y Lacandón, en Guatemala*. Editorial Universitaria, Guatemala.

Scholes, France V., and Ralph L. Roys
    1968   *The Maya Chontal Indians of Acalan-Tixchel*. University of Oklahoma Press, Norman.

Schove, Derek J.
    1976   Maya Chronology and the Spectrum of Time. *Nature* 261:471–73.
    1977   Maya Dates, A.D. 352–1296. *Nature* 268:670.
    1980   Maya Correlations Quantitatively Evaluated. MS (cited by Kelley 1983).

Schram, Robert
    1908   *Kalendariographische und Chronologische Tafeln*. J. C. Hinrichs, Leipzig.

Schultze-Jena, Leonhard Sigismund
    1933   *Leben, Glaube und Sprache der Quiche von Guatemala*. Indiana 1. Jena.
    1938   *Bei den Azteken, Mixteken und Tlapaneken der Sierra Madre del Sur von Mexiko*. Indiana 3. Jena.

Schulz, R. P. C.
  1936 Beiträge zur Chronologie und Astronomie des alten Zentralamerika. *Anthropos* 31:758–88.
  1942 Apuntes sobre cálculos relativos al calendario de los indígenas de Chiapas. *México Antiguo* 6:6–14.
  1953 Nuevos datos sobre el calendario tzeltal y tzotzil de Chiapas. *Yan* 2:114–16.
  1955 Dos variantes nuevas del calendario chinanteco. *México Antiguo* 8:233–46.
Schulz Friedemann, Ramon P. C.
  1972 El punto cero de la cuenta larga maya y las inscripciones astronómicas de Palenque. *Estudios de Cultura Maya* 8:167–74.
  1972b Nine lords of the night (Zapotec). *Archiv für Völkerkunde* 26:197–204.
Sedat, Guillermo
  1955 *Nuevo diccionario de las lenguas k'ekchi' y española*. Guatemala.
Selden (Glass 1975b: No. 283).
Selden Roll (Glass 1975b: No. 284).
Seler, Eduard
  1904 The Mexican Chronology, with special reference to the Zapotec calendar. *Bureau of American Ethnology Bulletin* 28:11–56.
Serna, J. de la
  1892 Manual de ministros de indios para el conocimiento de sus idolatrías y extirpación de ellas. *Anales del Museo Nacional de México*. Epoca 1, vol. 6:263–475.
Shook, E. M.
  1960 Tikal Stela 29. *Expedition Bulletin, University of Pennsylvania Museum* 2(2):29–35.
Sierra (Glass 1975b: No. 289; see León 1933).
Sigüenza y Góngora, Carlos de
  1960 *Obras Históricas*. Porrúa, Mexico (1st ed. 1944).
Smiley, Charles H.
  1960a A New Correlation of the Mayan and Christian Calendars. *Nature* 188:215–16.
  1960b The antiquity and precision of Maya astronomy. *Journal of the Royal Astronomical Society of Canada* 54:222–26.
  1961 Bases astronómicas para una nueva correlación entre los calendarios maya y cristiano. *Estudios de Cultura Maya* 1:237–42.
Smith, Mary Elizabeth
  1973 *Picture Writing from Ancient Southern Mexico*. University of Oklahoma Press, Norman.
Smith-Stark, Thomas Cedric
  1982 Jilotepequeño Pocomam Phonology and Morphology. Unpublished Ph.D. dissertation. University of Chicago.
Solís Alcalá, Ermilo
  1949 *Códice Pérez*. Oriente, Merida.
Soustelle, Jacques
  1937 La Famille Otomi-Pame du Mexique Central. *Mémoires de l'Institut d'Ethnologie* 26. Paris.
Spinden, Herbert J.
  1924 The reduction of Maya dates. *Papers Peabody Museum, Harvard University* 6(4).
  1930 Maya dates and what they reveal. *Brooklyn Institute of Arts and Sciences* 4(1).
  1957 *Maya art and civilization*. Indian Hills, Colorado.
Stairs Kreger, Glenn Albert, and Emily Florence Scharfe de Stairs
  1981 *Diccionario huave de San Mateo del Mar*. Instituto Lingüístico de Verano, Mexico.
Starr, Frederick
  1900 Notes upon the Ethnography of Southern Mexico. *Proceedings of the Davenport Academy of Sciences* 8:102–98.

Stirling, Matthew W.
    1943   Stone Monuments of Southern Mexico. *BAE Bulletin* 138.
    1965   Monumental Sculpture of Southern Veracruz and Tabasco. *Handbook of Middle American Indians* 3:716–38. Austin.

Stoll, Otto
    1884   *Zur Ethnographie der Republik Guatemala.* Bruck von Orell Füssli, Zürich.
    1889   *Die Ethnologie der Indianerstämme von Guatemala.* P. W. M. Trap. Leiden.

Suchtelen, Berthe C. C. M. M. van
    1957   *Maya zero 583.919.* Leyden.

Swadesh, Maurice
    1969   *Elementos del tarasco antiguo.* Universidad Nacional Autónoma de México, Mexico.

Tax, Sol
    1947   Notes on Santo Tomás Chichicastenango. University of Chicago Library Microfilm No. 16.

Tecamachalco (Glass 1975b: No. 300).

Tedlock, Barbara
    1982   *Time and the Highland Maya.* University of New Mexico Press, Albuquerque.

Teeple, John D.
    1926   Maya Inscription: The Venus calendar and another correlation. *American Anthropologist* 28:108–15.

Teletor, Celso Narciso
    1959   *Diccionario Castellano-Quiche y voces Castellano-Pokomam.* Re-Rajau, Guatemala.

Telleriano (Glass 1975b: No. 308).

Tepechpam, Tira de (Glass 1975b: No. 317; see Nóguez 1978).

Termer, Franz
    1930   Zur Ethnologie und Ethnographie der nördlichen Mittel-Amerika. *Ibero-Amerikanisches Archiv* 4:303–492.
    1930   Über die Mayasprache von Chicomucelo. *International Congress of Americanists Proceedings* 23:926–36.

Tezozomoc (see Alvarado Tezozomoc).

Thompson, John Eric Simpson
    1932a   A Maya Calendar from the Alta Vera Paz, Guatemala. *American Anthropologist* 34:449–54.
    1932b   The Solar Year of the Mayas at Quirigua, Guatemala. *Field Museum of Natural History, Anthropological Series* 17(4).
    1934   A correlation of the Mayan and European calendars. *Field Museum of Natural History Publication* 241, Anthropological Series 1. Chicago.
    1935   Maya chronology: the correlation question. *Carnegie Institution of Washington Publication* 456, Contribution 14.
    1937   A New Method of Deciphering Yucatecan Dates with Special Reference to Chichen Itza. *Carnegie Institution of Washington Publication* 483, *Contributions to American Archaeology* 22. Washington.
    1950   Maya hieroglyphic writing: introduction. *Carnegie Institution of Washington Publication* 589. Washington, D.C.
    1952   The Introduction of Puuc Style of Dating at Yaxchilan. *Carnegie Institution of Washington Notes on Middle American Archaeology and Ethnology* 110.
    1962   *A Catalogue of Maya Hieroglyphs.* University of Oklahoma Press, Norman.
    1965   Archaeological Synthesis of the Southern Mayan Lowlands. In R. Wauchope, *Handbook of Middle American Indians* 2:331–59.
    1971   *Maya Hieroglyphic Writing.* University of Oklahoma Press, Norman. Second edition.

1972   A Commentary on the Dresden Codex: a Maya Hieroglyphic Book. *American Philosophical Society Memoirs* 93. Philadelphia.

Tizayuca (Glass 1975b: No. 342).

Tizimin (see Edmonson 1982).

Tlatelolco, Annals of (see Berlin 1948).

Torquemada, Juan de
1976   *Monarquía Indiana*. Universidad Nacional Autónoma de México, Mexico.

Toscano, Salvador
1943   Los códices tlapanecas de Azoyú. *Cuadernos Americanos*, año 2, 10(4):127–36.

Tovar (Glass 1975b: No. 366).

Tovar Calendar (Glass 1975b: No. 364).

Tovilla, M. A. de
1631   Relación histórica-dyscreptiva de las provincias de la Verapaz y de la Manché de el Reyno de Guatemala (see Scholes & Adams 1960).

Tozzer, Alfred M.
1941   Landa's Relación de las Cosas de Yucatán: a translation. *Papers of the Peabody Museum of American Archaeology and Ethnology, Harvard University* 18. Cambridge.

Tschohl, Peter
1972   *Catálogo Arqueológico y Etnohistórico de Puebla-Tlaxcala, México*. 2 v. Peter Tschohl and Herbert J. Nickel, Köln.

Tola, Títulos de Santa Isabel (Glass 1975b: No. 358).

Tula, Anales de (Glass 1975b: No. 369).

Tulane (Glass 1975b: No. 370)

Tulane Calendar
n.d. Useful Calendar from the Year 1776 to 2000 Inclusive. Tulane University Press, New Orleans.

Turner, Paul and Shirley
1971   *Chontal to Spanish, Spanish to Chontal Dictionary*. University of Arizona Press, Tucson.

Vaillant, George C.
1935   Chronology and stratigraphy in the Maya area. *Maya Research* 2:129–43.

Valadés, Didacus
1579   *Rhetorica christiana*. Perugia.

Vaticanus A (Glass 1975b: No. 270).

Vaticanus 3738 (Glass 1975b: No. 270).

Vetancurt, Agustín de
1679   *Chrónica de la provincia del Santo Evangelio de México*. Mexico.

Veytia (Glass 1975b: No. 394).

Veytia, Mariano Fernández de Echeverría y
1836   *Historia antigua de Mejico*. Juan Ojeda, Mexico.
1973   *Los Calendarios mexicanos*. San Angel Ediciones, Mexico.

Vienna (Glass 1975b: No. 395).

Villacorta, José Antonio and C. A. Villacorta R.
1933   *Memorial de Tecpán-Atitlán (Anales de los Cakchiqueles)*. Tipografía Nacional, Guatemala.

Vollemaere, Antoon
1972   Problème des calandriers mayas et la corrélation. *Proceedings of the 40th International Congress of Americanists*, Rome 1:419–26.
1984   De leiden-plaat en de maya-korrelatie. *Oud-Amerika* 17:219–46. Vlaams Instituut voor Amerikaanse Kulturen. Mechelen.

Wasson, R. Gordon, George M. and Florence Cowan, and Willard Rhodes
1974   *Maria Sabina and Her Mazatec Mushroom Velada*. Harcourt Brace Jovanovich, New York.

Waterman, Thomas Talbot
    1924   On Certain Antiquities in Western Guatemala. *Bulletin Pan American Union* 58(4):341–61. Washington.
Wauchope, Robert
    1947   An approach to the Maya correlation problem through Guatemala highland archaeology and native annals. *American Antiquity* 13:59–66.
Weitlaner, Irmengard
    1936   A Chinantec Calendar. *American Anthropologist* 38.
Weitlaner, Robert J.
    1939a   Notes on the Cuitlatec language. *México Antiguo* 4:363–73.
    1939b   Beitrag zur Sprache der Ocuilteca von San Juan Azingo. *México Antiguo* 4:297–328.
    1956   Un calendario ritual entre los zapotecos del sur. *Proceedings of the 32nd International Congress of Americanists* 296–99.
Weitlaner, Robert J. and Irmengard Weitlaner Johnson
    1946   The Mazatec Calendar. *American Antiquity* 11:194–97.
    1963   Nuevas versiones sobre calendarios mixes. *Revista Mexicana de Estudios Antropológicos* 16:183–209.
Weitzel, R. B.
    1947   Yucatecan chronological systems. *American Antiquity* 13:53–58.
Whelan, Frederick G. III
    1967   The Passing of the Years. Unpublished MS, Harvard Chiapas Project.
Whittaker, Gordon
    1980   The Hieroglyphics of Monte Alban. Ph.D. dissertation. Yale University, New Haven.
Willson, Robert W.
    1924   Astronomical notes on the Maya codices. *Peabody Museum Papers in American Archaeology and Ethnology* 6(3).
Ximénez, Francisco
    1720   *Historia de la provincia de San Vicente de Chiapa y Guatemala de la Orden de Predicadores.* Biblioteca Goathemala, 3 vols., 1929–31, Guatemala.
Xolotl (Glass 1975b: No. 412; see Dibble 1951).
Yanhuitlan (Glass 1975b: No. 415; see Jiménez 1940).
Yaxkukul (See Martínez 1926).
Zacatepec 25 (Glass 1975b: No. 422).
Zapata (cited by Caso 1967)
Zimmermann, Günter
    1960   El Cotoque, la lengua mayense de Chicomucelo. *Traducciones Mesoamericanistas* 1:27–71. Sociedad Mexicana de Antropología. (Original in *Zeitschrift für Ethnologie* 80:59–87, 1955, Braunschweig.)
Zouche-Nuttall (Glass 1975b: No. 240).

INDEX